Computational Fluid Dynamics Modelling of Solid Fuel Co-Combustion

The growing need for sustainable energy solutions has driven research into co-combustion processes, where coal is blended with biomass to reduce emissions and enhance combustion efficiency. This book overviews best practices to optimise the combustion processes through the use of computational fluid dynamics (CFD) simulations, focusing particularly on solid fuel combustion modelling. It equips the reader with an understanding of combustion parameter optimisation and modelling from a thermodynamic-fluid dynamic-mass transfer perspective.

Key features

1. Offers a specialised approach to solid fuel co-combustion, using CFD to model and optimise combustion processes for sustainable energy solutions
2. Combines theoretical and practical applications, with a focus on energy optimisation and emissions reduction
3. Includes modelling of complex phenomena such as multiphase and reaction modelling
4. Gives insights into important sub-models required during combustion modelling, forming the basis of reliable models
5. Contains detailed case studies and real-world applications

This book is aimed at graduate students and researchers in thermodynamics, combustion, energy engineering, fluid dynamics, and clean energy processes.

Africa Circular Economy Series

CRC Press (Taylor and Francis Group) introduces **Africa Circular Economy Series** (with a Science and Technology outlook but preferably more interdisciplinary). Under this initiative, we invite scholars, academicians, researchers, and professionals to contribute to this series. We are interested in discussing monographs, references, textbooks, shortform books and handbooks to add to our book programs, at undergraduate, postgraduate and doctoral level. All the books published under the series will be part of our Global Publishing program.

Circular and Transformative Economy
Advances towards Sustainable Socio-economic Transformation
Edited by Luxon Nhamo, Sylvester Mpandeli, Stanley Liphadzi and Tafadzwanashe Mabhaudhi

Municipal Solid, Agricultural, and Mining Waste in Sub-Saharan Africa
Sustainable and Circular Valorization Opportunities
Freeman Ntuli, Tebogo Mashifana, Thandiwe Sithole and Thabo Falayi

Computational Fluid Dynamics Modelling of Solid Fuel Co-Combustion
Garikai Tawanda Marangwanda, Daniel Makundwaneyi Madyira, and Innocent Musonda

For more information about this series, please visit: www.routledge.com/Africa-Circular-Economy-Series/book-series/CRCACES

Computational Fluid Dynamics Modelling of Solid Fuel Co-Combustion

Garikai Tawanda Marangwanda,
Daniel Makundwaneyi Madyira,
and Innocent Musonda

CRC Press
Taylor & Francis Group
Boca Raton London New York

CRC Press is an imprint of the
Taylor & Francis Group, an **informa** business

First edition published 2027
by CRC Press
2385 NW Executive Center Drive, Suite 320, Boca Raton FL 33431

and by CRC Press
4 Park Square, Milton Park, Abingdon, Oxon, OX14 4RN

CRC Press is an imprint of Taylor & Francis Group, LLC

ISBN: 9781041088127 (hbk)
ISBN: 9781041090014 (pbk)
ISBN: 9781003647942 (ebk)

DOI: 10.1201/9781003647942

Typeset in Times
by Deanta Global Publishing Services, Chennai, India

"This book is dedicated to my wife, Modesta, and to our children, Mali and Gabriella. I thank Modesta for making a home for our children and me. I thank her for her support, her companionship, and her love. She is my soulmate." Dr Garikai Tawanda Marangwanda

Contents

About the Authors

Garikai Tawanda Marangwanda is a registered mechanical engineer with more than eight years of combined academic and industry experience. He holds a PhD in mechanical engineering from the University of Johannesburg, South Africa, and MSc and BSc degrees in mechanical engineering from the University of Science and Technology of Oran-Mohammed Boudiaf, Algeria. He is currently a postdoctoral researcher in the Centre for Applied Research and Innovation in the Built Environment (CARINBE) at the University of Johannesburg. He has taught core mechanical engineering courses related to thermo-fluid-energetic principles, and his speciality lies in combustion modelling, computational fluid dynamics, thermal systems analysis, and the integration of machine learning in energy engineering applications. He has contributed to several funded research and industry-linked projects such as the development of a digital twin for a utility boiler, a feasibility study for a biofuel processing plant, and experimental research related to glass fibre reinforced polymers. He has also led and participated in applied engineering projects such as designing solar off-grid and irrigation systems, commissioning bio digesters, developing pyrolysis units, and conducting industrial energy audits. As an active researcher, he has published journal papers and refereed conference papers in the areas of combustion modelling and experimental thermal analysis, supervised undergraduate and postgraduate research projects, and actively engages in industry-oriented engineering activities to ensure that his academic work remains practically relevant.

Daniel Makundwaneyi Madyira is a registered professional engineer who is currently an associate professor of mechanical engineering in the Department of Mechanical Engineering Science, Faculty of Engineering, and the Built Environment at the University of Johannesburg, South Africa. He has taught a wide range of core mechanical engineering courses, including machine design, fluid dynamics, thermodynamics, and strength of materials. He is a highly experienced mechanical engineer and academic with more than 26 years of academic and industrial experience. He is passionate about engineering design with special expertise in machine component stress analysis and thermal systems analysis. He has developed several mechanical designs addressing a range of contemporary societal problems, such as biomass briquetting and solar crop drying, to advanced mechanical systems such as rotating bending fatigue testing machines. His research interests range from high-speed machining of titanium, fatigue of titanium and composites, natural composites, biomass briquetting and biomass combustion modelling, fracture behaviour of materials produced using modern manufacturing techniques such as additive manufacturing, and post-weld heat treatment of steels using weld beads. He is also involved in several industry-based activities, solving industry-based problems. This makes his academic work relevant to industry and fruitful to his students. He is a very hard-working engineer, as evidenced by the range of his research publications. He has edited multiple conference proceedings, published book chapters, journal papers, and refereed conference papers, registered one patent, graduated multiple PhD students as supervisors and co-supervisors, and graduated multiple MSc, MPhil, and MTech students, and supervised multiple international MSc interns. As a registered professional engineer, he is active in ECSA activities, including assessment and professional review interviews of professional engineer applicants.

Innocent Musonda is a registered civil engineer and professional construction manager, currently serving as a full professor in Construction Management and Quantity Surveying in the Department of Construction & Management–Quantity Surveying at the University of Johannesburg (UJ). He is also the director of the Centre for Applied Research and Innovation in the Built Environment (CARINBE). He has taught a broad spectrum of courses in construction project management, engineering management, and digital built-environment technologies. As an experienced academic and practitioner, he combines more than two decades of industrial experience across Botswana, Zambia, and South Africa with deep scholarly engagement. Prof Musonda is particularly passionate about advancing engineering design for sustainable, safe, and digitally enabled infrastructure. His expertise spans building information modelling (BIM), digital twins, digital transformation in the built environment, sustainable infrastructure, and construction of health and safety. He has made substantial contributions to both academia and practice, serving as a thought leader in digitalisation and transformation in infrastructure. His applied work addresses societal and industry challenges, including infrastructure safety, sustainability, and the integration of 4IR technologies in construction. On the research side, Prof Musonda has published extensively and supervised numerous postgraduate students. At his inauguration, it was noted that he has supervised 8 PhD students and over 20 master's students, and has authored more than 60 journal articles, 4 book chapters, and over 50 conference papers. He is rated by the National Research Foundation (NRF) in South Africa (C2). His work is strongly aligned with the UN's Sustainable Development Goals, particularly around sustainable infrastructure and digital innovation. Beyond research and teaching, Prof Musonda is deeply engaged with professional bodies. He is a full member of the Chartered Institute of Building (CIOB-UK) and the International Council for Research and Innovation in Building and Construction (CIB). He also contributes to national and international peer review processes, acting as an assessor for bodies such as the Australian Research Council (ARC) and the NRF. He has chaired and co-edited influential conferences, most notably the Development and Investments in Infrastructure (DII) series, helping drive discourse on infrastructure development, sustainability, and investment across Africa. Prof Musonda is a very hard-working scholar-practitioner, committed to bridging the gap between academic research and real-world infrastructure transformation. His leadership at CARINBE helps make his academic work industry-relevant, and his mentorship of students drives capacity-building in the built environment in Africa.

Preface

This book is written for researchers, mechanical, chemical, thermal power, and environmental engineers who want to understand the real physics of solid fuel combustion and apply that knowledge directly to practical systems. Rather than offering abstract theory, the chapters connect thermodynamic principles, chemical reaction theory, and fluid dynamic theory towards co-combustion of solid fuels. Of interest to engineers and researchers are furnace parameters such as ignition, particle motion, flame structure, and emissions. The mathematical relations that underpin computational fluid dynamics are paired with clear physical explanations to make the modelling process intuitive. Complex modelling ideas are condensed into practical steps supported by case studies, validation examples, and real combustion scenarios. From fuel characterisation to turbulence and optimisation, the focus throughout is on giving readers the tools to build, evaluate, and improve models for coal biomass co-combustion. The early chapters introduce the foundations, the fuel characteristics that govern reactivity, the chemistry that defines conversion pathways, and the governing equations from mass, momentum, energy, and species transport. These principles are then connected to the numerical modelling principles, particularly computational fluid dynamics. From there, the book extends into the modelling of physical interactions that drive co-combustion behaviour. Turbulence modelling is treated as a living discipline, not a catalogue of equations. Particle dynamics are explored with attention to the real-world irregularity of biomass, where sphericity, size distribution, and drag all influence the flame structure. Radiation modelling is unpacked beyond the basics, revealing the spectral characteristics that different gaseous species contribute to heat transfer within a furnace. Each concept is paired with industrial case studies that show how theory intersects with operational reality. The following chapters take this foundation further by examining modern trends, alternative firing strategies such as oxy-fuel and chemical looping combustion, and the steadily growing influence of artificial intelligence in assisting, accelerating, or even reshaping traditional physics-based modelling. Machine learning is introduced not as a replacement for computational fluid dynamics but as a complementary tool, one capable of predicting fuel properties, optimising designs, or identifying hidden correlations that conventional models may overlook. Case studies from international and regional contexts, especially those relevant to sub-Saharan Africa, are integrated to demonstrate the scale of opportunity. The potential for municipal solid waste to generate electricity equivalent to powering millions of households illustrates both the magnitude of the challenge and the feasibility of addressing it using scientifically grounded approaches. These examples give the book practical relevance for industries seeking cleaner operation without discarding capital-intensive systems.

Acknowledgements

The authors would like to acknowledge those who have supported them during the process of writing this book. Special mention goes to the staff at the Mechanical Engineering Science Department and the Center for Applied Research and Innovation in the Built Environment at the University of Johannesburg, South Africa.

1 Introduction to Solid Fuels

1.1 INTRODUCTION

The laws of thermodynamics are centred around the study of energy and how it is transformed from one form to another. With heat transfer being a form of energy due to temperature difference, combustion is a very important process that is used to induce these temperature variations. Research into combustion systems will always be relevant since it is the most common method of inducing heat energy. Unfortunately, the exploitation of fuels through combustion has led to a rise in global temperatures, hence the need to sustainably exploit these fuels. Sustainable development goals (SDGs) number 7 and 13 outline the need to have affordable, reliable, sustainable, and modern energy for all whilst mitigating the effects of climate change [1]. Gradual movement from hard fossil fuels has been proposed by the United Nations for all its members. Inasmuch as the Paris climate conference (COP21) further highlighted the plight of climate change as influenced by high carbon dioxide (CO_2) emitting fuels, only a few ratifications have been implemented. As such, most developing countries, which in turn contribute a small fraction towards CO_2 emissions on a global scale, continue to rely on high CO_2 emitting fuels and conversion technologies that have poor efficiency. One solution towards mitigating this problem for developing countries is to use biomass waste as a supplement in the generation of energy, thus achieving SDGs 7 and 13 goals simultaneously, since biomass by nature is carbon neutral with regard to combustion [2]. Conventional coal combustion on its own is known to produce undesired products in large quantities, such as CO_2, nitrogen oxides (NOx), sulphur oxides (SOx), and particulate matter [3]. Co-combustion of coal and biomass waste offers a combined approach to emission reduction and combustion efficiency improvement for boilers.

By applying a well-to-wheel analysis, co-combustion of biomass waste with coal results in the reduction of overall CO_2 because biomass is carbon neutral. Applications of co-combustion are generally skewed towards developed countries. A combined heat and power (CHP) plant, Amager, situated in Denmark, which was traditionally considered a coal-fired plant, was successfully retrofitted to accommodate more than 50% of biomass on a mass basis [4]. Power-generating stations within the United States have also been retrofitted to accommodate biomass, mainly wood waste, at different blending ratios, either through direct or indirect co-firing. Of particular importance is a 469-MW station fitted with cyclone boilers located within Michigan City capable of handling co-fired coal and biomass, justifying the applicability of this technology on a large scale [5]. China also demonstrated the

DOI: 10.1201/9781003647942-1

applicability of co-firing between coal and various biomass wastes, such as municipal solid wastes and even agricultural residues, with different kinds of boilers with varied success [6]. Finland's FORTUM's Naantali 3 315-MW CHP plant experiences with co-firing emphasised the need to have an efficient fuel handling and feeding system that ensures maximum homogeneity between coal and biomass for maximum benefits [7]. Some industrial tests have even gone further to employ co-combustion of biomass and coal under oxyfuel (CO_2 and O_2 as oxidant) conditions, achieving a reduction of CO_2 emissions [8].

Combustion is the most prominent process used to harness energy from coal and biomass waste. Combustion is the oxidation, rapid or slow, of a fuel leading to the generation of heat accompanied by light in most cases [9]. In the quest to improve combustion efficiency, routine maintenance and optimisation are performed so as to assure optimum combustion characteristics are exhibited. The optimisation process utilises an experimental approach, which tends to be expensive for large industrial processes such as thermal power plants.

Inasmuch as coal and biomass co-firing results in the reduction of emissions, important combustion parameters such as ignition, burnout, intensity, and stability tend to be affected as well. As reported by Savolainen [7], the variation in ignition and burnout parameters during the co-firing of coal and pine sawdust requires modification of the burner design so as to avoid flashback flames. This is mainly due to the tendency of biomass having the ability to ignite more easily than coal. Rüdiger et al. [10] also concurred with these findings through their investigations with hard coal, wood, and straw. Ignition is an important combustion parameter that has been reported from a relative point of view with respect to boiler furnaces by just comparing ignition temperatures. When Yin et al. [11] reported changes in ignition properties with regard to blended fuel samples, they focused on comparing the ignition temperatures. Comparison of ignition temperatures is deemed to be technical, thus its wide use by various researchers. Xing et al. [12] employed a more scientific approach of comparing the ignition index, a derived parameter from ignition temperature and rate of mass loss during combustion. This approach is deemed to be more robust when used to compare blended and unblended cases, thus its recent use by Lü et al. [13].

Burnout, a measure used to monitor the level of fuel oxidation, is also one of the most common parameters that is investigated during combustion. Conclusions on change in burnout due to blending are mainly based on measuring the carbon in ash content at the exit of the furnace [7, 10]. Kruczek et al. [14] reported increased burnout due to blending of hard coal with sawdust biomass. Guo et al. [15] also reported increased burnout whilst investigating blending of bituminous coal with biomass pellets at a blending ratio of 50% by mass. Pérez-Jeldres et al. [16] did a modelling study on the change in burnout when coal is blended with pine sawdust. Their findings were quite elaborate through the use of appropriate computational fluid dynamics (CFD) techniques such as visualising the contours of burnout rates at different furnace positions. Bhuiyan et al. [17] explained further how other researchers have employed CFD in studying the variation of burnout as blending increased. As with ignition, a more appropriate parameter called burnout index has also been propounded to investigate the extent of combustion [18, 19]. The main advantage associated with using ignition and burnout indexes as comparative parameters is that they both incorporate appropriate reaction kinetic parameters. As such, Jiang et al. [20] reported a gain in burnout index after coal was blended with carbonaceous waste, which meant the blended fuel was able to attain complete combustion at lower temperatures.

Reaction kinetic parameters are quite useful in determining the expected mechanisms during the ignition and burnout of fuels. Few researchers have deemed it necessary to determine the overall kinetic parameters that are exhibited by blended fuels. The interaction between the individual coal and biomass reaction kinetic parameters has been proven to either enhance or suppress certain combustion parameters. Wnorowska et al. [19] reported on the synergistic effects of blending hard coal with biomass towards combustion indices. Synergistic coefficients make it easier to incorporate all relevant combustion indices as a tool to advise on the optimum blending ratios. Most combustion indices can have their synergistic coefficients analysed if enough reaction kinetic data is available [21, 22]. Wu et al. [23] actually managed to evaluate the overall reaction kinetic parameters exhibited by blending cellulose with low rank coal. Certain researchers tend to use the slagging propensity of biomass as a measure of optimum blending ratio. Unfortunately, there is no universal slagging or fouling coefficient that has been employed in industry. Consequently, CFD studies have focused on analysing the heat flux through the walls which is directly related to slagging through heat transfer principles [3, 17, 24].

Another combustion parameter that has been employed to advise on the optimum blending ratio is combustion stability as studied by Baukal Jr [25]. Stability is commonly linked to the flame behaviour exhibited during the combustion process. Blending is expected to alter the flame behaviour, as rightly noted by Mohanna [26]. Unstable flames can result in incomplete combustion that causes poor burnout, flashback flames that cause explosions due to flames tracking back into the burner tubes, as well as uncontrolled high temperatures that can reduce burner integrity [27]. Savolainen [7] reported on the increased tendency to experience flashback when coal is blended with pine sawdust at a blending ratio of 50% by volume. Investigations on countering combustion instability are more directed towards burner modification [28]. An extension of combustion stability, intensity has also been studied as a parameter useful in determining optimum combustion parameters. Ma et al. [29] demonstrated that blending South African coal with either olive waste or wood creates zones of high combustion intensity by measuring the temperatures within those areas. It is useful to note that, since heat is released during the combustion of both volatiles and char, it makes sense that combustion intensity varies with blending ratio [30].

What has also affected the determination of optimum blending ratios with respect to co-combustion is the technology used to fire the fuels. Tillman [5] studied the different co-firing methods employed in industry, highlighting how most applications prefer direct co-firing of coal and biomass. No pre-treatment is extended towards the biomass other than drying before introduction into the furnace. Pre-treated biomass or processed biomass tends to be easily blended with coal, making it possible to employ large blending ratios. The ability to capture almost all of these combustion parameters by CFD makes it an indispensable and affordable tool with regard to co-combustion optimisation.

1.2 OVERVIEW OF SOLID FUELS USED IN CO-FIRING

In existing coal-fired boilers, co-firing using biomass waste provides an attractive approach to pollution control and efficient use of both the fossil fuel and biomass waste [31]. In the context of combustion, co-firing can be defined in various ways depending on the fuels being employed, the method of combustion, as well as the industrial application being considered. Tilliman [5] defined co-firing as "the practice of supplementing a base fuel with a dissimilar fuel", whilst Leckner [31] defined co-firing when applied to industrial boilers as "combustion of two or more fuels together in one boiler". Co-firing is an extension of fuel blending practices common to the fuel community. However, the use of biomass for co-firing in existing coal-fired power plants is not an easy task since issues to do with fouling, efficiency, and fuel handling problems become more apparent.

The proportion range of co-firing with biomass that is acceptable is introduced on a mass basis depending on the type of boiler. This is mainly because biomass is fibrous in nature and contains different elements compared to coal. As such, the balance between cost incurred due to an increase in slagging and fouling of boiler material and the reduction of SOx and NOx emissions is not constant [32]. Tillman [5] suggested that for every tonne of biomass co-fired with coal, an equivalent amount of CO_2 is removed from the atmosphere [33].

Biomass waste has a high waste-to-energy (WtE) potential, especially when co-fired with coal. WtE as an industry had a projected value of US$35.1 billion globally in 2019. Figure 1.1 shows the projected growth of WtE as an industry from 2019 up to 2027, as adapted from the projections of the World Bank [34].

As the global population increases, biomass waste will also grow simultaneously. Unfortunately, sub-Saharan African economies are yet to harness the potential of WtE. This can be attributed to high capital costs coupled with a lack of long-term policies towards investment in WtE technologies [34]. Investigation into alternative energy sources and cleaner energy production techniques is considered an active research area; hence, the various funding agencies and research institutions are dotted across the globe [35]. Biomass waste as an alternative clean energy source has always been a promising energy source, though it is limited by its seasonal nature and how it is difficult to find in concentrated locations. In general, biomass waste as a resource can be classified as municipal waste, industrial waste, forest residues, and agricultural residues, as shown in Figure 1.2. The exploitation of biomass waste is necessary for obvious economic and environmental reasons [36]. To illustrate the regional opportunities and constraints of co-firing, Zimbabwe is presented as a case study. The country's diverse agricultural, municipal, and industrial waste streams provide a representative picture of the challenges facing sub-Saharan Africa. A summary of various biomass waste that can be combusted to produce energy with respect to the Zimbabwean context is summarised in Table 1.1.

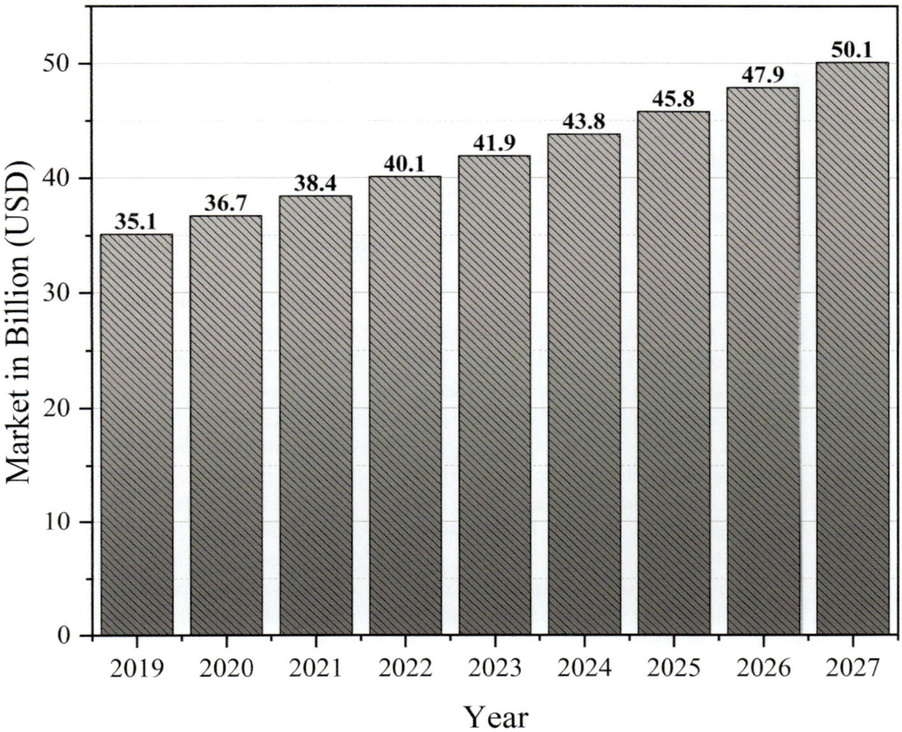

FIGURE 1.1 Global forecast of waste to energy.

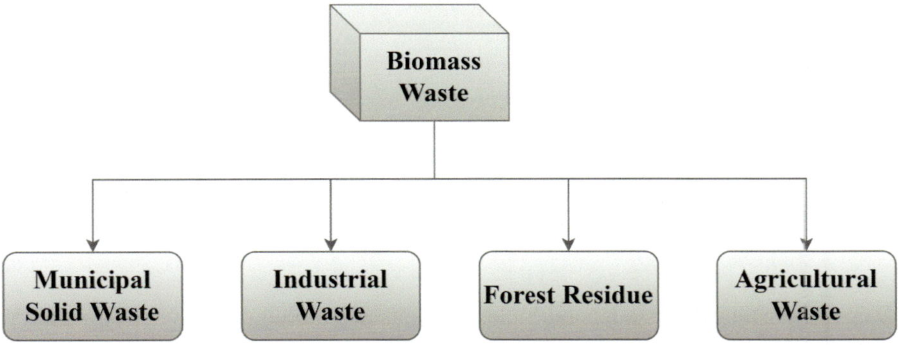

FIGURE 1.2 Biomass waste classification.

1.2.1 Sawmill Waste

During timber processing, various waste products are produced such as bark, shavings, wood chips, and sawdust, all of which can be referred to as sawmill waste.

TABLE 1.1
Zimbabwean Biomass Waste Characteristics

Name	Characteristics	Potential
Sawmill waste	Timber processing produces sawmill waste (bark, shavings, wood chips, and sawdust) [37].	In Manicaland province, Zimbabwe, sawmill waste accumulates at a rate of 70,000 tonnes per year, equivalent to 1,330 GJ per year of energy which is enough to cater for the needs of approximately 150 households [38, 39].
Agricultural waste	Poultry, swine, and cattle production contribute significantly towards waste generation due to animals (manure, feathers, spilled food, chip woods, and straw) Crop harvesting produces mainly maize stover (cobs, husks, leaves, and stalks) and sugarcane bagasse [40, 41]. Sugarcane bagasse is centralised since sugar juice extraction is centralised as well [42, 43].	Maize stover amounts around 5 tonnes per hectare for high rainfall conditions whilst that for low rainfall is around 2.5 tonnes per hectare [44]. Poultry contributes about 0.819 million tonnes of manure for Zimbabwe compared to 35 million tonnes produced in the United States [45]. Crop residues have potential to power roughly 840,000 households in Zimbabwe whilst poultry waste has potential to power 670000 households [36].
Municipal solid waste	Refers to the materials thrown away in municipal urban (paper, plastics, food waste, metals, textiles, yard waste, and all those miscellaneous objects thrown away at household level) [46]. Accumulation of MSW within a landfill can cause CO_2 threefold more compared to fossil fuel [47].	Zimbabwean energy potential of MSW is averaged at 8,432 GJ per year though it is highly dependent on community lifestyle [48].
Industrial waste	Constitutes food processing (bakery, citrus, dairy, and meat waste) and refinery waste (acid tar waste) Developed countries prefer to produce biogas through anaerobic digestion [49].	The energy content of acid tar waste can go as high as 33.5 MJ/kg with an ability to produce 99% combustion efficiency when used in fluidised beds infused with limestone [50].

Generally, sawmill waste has multiple functions depending on the type of industry. The sub-Saharan region only utilises about 10% of the sawmill waste for their process boilers and lumber drying kilns, leaving the remainder to be stockpiled [51, 52]. Most of these sawmill waste stockpiles are generally left to decompose, which poses an environmental risk through pollution or, in extreme cases, spontaneous combustion that produces hazardous greenhouse emissions. Woody biomass possesses a calorific value of around 19.5 MJ/kg depending on species [52, 53]. The uptake of the sawmill waste is not similar in the sub-Saharan region as it is in developed countries, where almost all the waste is utilised by downstream industries. Taking Australia as an example, after the uptake of sawmill waste by downstream industries, only 1% is left unutilised [54].

The timber industry in Zimbabwe, for example, is reported to be generating over 750 kilotons of sawmill waste each year, which possesses an energy potential of around 232 kilotons of oil equivalent [37, 53]. This translates to about 15.84 kg of oil equivalent per capita against a requirement of 845 kg of oil equivalent per capita according to World Bank data [55]. Eventually the developing countries must contend with the menace caused by sawmill waste, mainly sawdust, which accumulates yearly. Preliminary surveys put the accumulation rate at 70,000 tonnes per year of the sawmill waste in Manicaland Province, Zimbabwe. This is equivalent to 1,330 GJ per year of energy, which is enough to cater for the needs of approximately 150 households based on the World Bank data of power consumption per capita [38, 39].

Since the use of sawmill waste requires information on the quantities and location of the waste product within the Manicaland Province, geographical information systems (GIS) can be employed as a mapping tool. Generally, biomass waste is not found in a centralised place unless it has been transported to that location. Though biomass waste is relatively cheap, the major constraints around biomass waste are shared amongst researchers, which are feedstock supply chain dynamics and biomass waste conversion technologies. As a result, GIS becomes an indispensable tool in terms of optimisation of location, size of plant, and supply chain dynamics [52].

1.2.2 Agricultural Waste

Crop harvesting produces a lot of agricultural process waste. The major contributors of crop waste in Southern Africa are maize stover, sugarcane bagasse, as well as waste from other crop species. On average, maize stover consisting of stalks, leaves, sheaths, husks, shanks, cobs, tassels, lower ears, and silks accounts for 70% of the agricultural residue from a maize plant [43, 44]. Sugarcane produces lignocellulosic residues after the extraction of cane juice. Fortunately, most ethanol-processing plants utilise this residue, bagasse, as a fuel for process heating [42]. On average, plant waste has a calorific value of 17.65 MJ/kg [56].

Another major contributor to agricultural waste is animal husbandry waste such as chicken litter, which is a potential biomass fuel of animal origin produced from poultry farming. It consists of manure, feathers, spilled food, chip woods, and straw. Developing countries are heavily dependent on agriculture for their gross domestic product (GDP), which in turn contributes significantly to the amount of agricultural waste. Poultry, swine, and cattle production contribute significantly towards this animal waste. The calorific value of poultry waste is between 9.0 and 13.5 MJ/kg, depending on the moisture content of the litter. On a yearly basis, the United States of America (USA) produces 35 million tonnes of poultry waste. In contrast, Zimbabwe produces about 0.819 million tonnes of poultry waste, which translates roughly to an equivalent energy for 840,000 households in Zimbabwe [36, 39, 45, 57].

1.2.3 Municipal Solid Waste

Municipal Solid Waste (MSW) generally refers to the materials thrown away in municipal urban areas, usually collected, transported, and dumped in landfills [47]. This material usually contains food waste, paper, cardboard, glass, ceramics, metals, plastics, leather, rubber, textiles, yard waste, wood, putrescible, and miscellaneous objects [46]. The calorific value of MSW can be calculated based on the average calorific value of the different constituents, giving a typical range between 9.3 and 15.1 MJ/kg [58]. A summary of the data compiled by Kaza et al. [59] for the World Bank is summarised in Figure 1.3 for the MSW quantities produced from different regions of the world.

Maqhuzu et al. [48] suggested using Equation (1.1) to estimate the annual energy that can be produced from MSW.

$$MSW_{AE} = 365.Pop_U.\eta_{eff}.MSW_{gpc}.MSW_{CV} \tag{1.1}$$

where MSW_{AE} is the MSW Annual Energy, Pop is the population, η_{eff} is the collection efficiency, MSW_{gpc} is the MSW generation per capita, and MSW_{CV} is the MSW calorific value.

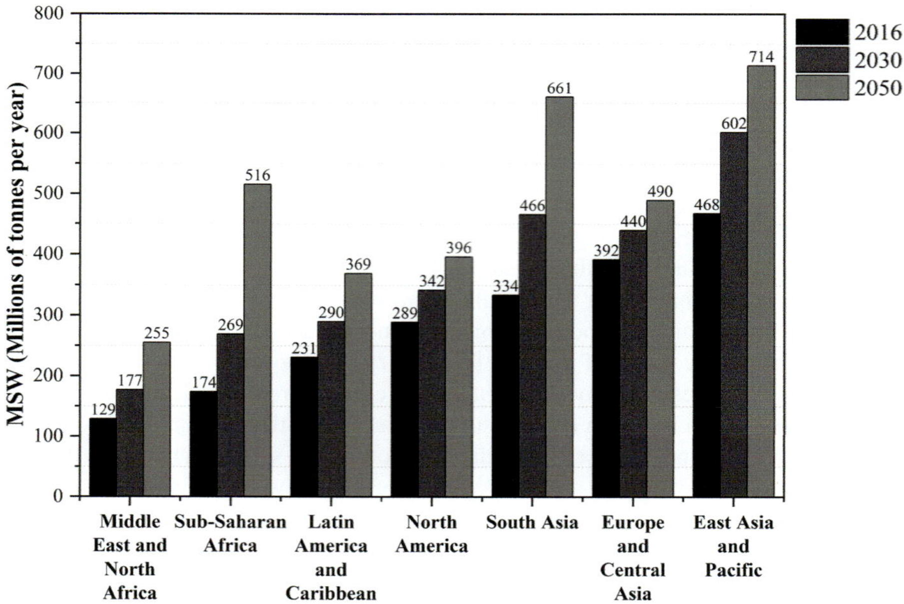

FIGURE 1.3 Municipal solid waste projections by region.

1.2.4 Industrial Waste

Food processing waste contributes a great deal to the biomass waste mix due to inefficiencies within the citrus, dairy, meat, and bakery industries. This waste comes with a considerable quantity of energy that can be harnessed for combustion purposes, provided the correct technology is applied. Generally, developed countries prefer to anaerobically digest the food processing waste to further supplement the energy needs of the food processing plant in the form of biogas [49].

1.2.5 Coal

Conventional coal combustion on its own is known to produce undesired products in large quantities, such as CO_2, NOx, SOx, and particulate matter [3]. Unfortunately, within the sub-Saharan region, technologies related to clean coal exploitation require further years of development so they can reduce the carbon footprint. Table 1.2 summarises the classification of coal with regard to rank according to ASTM standards.

A case study on Zimbabwean coal reserves shows that an estimated 30 billion tons of coal are still available for exploitation, which translates to around 3 million tons a year for all the collieries combined. Hwange Colliery, Makomo Resources, and Zambezi Gas are the main producers of coal, and the Mid Zambezi basin located northwest of Zimbabwe, which is part of the Karoo system, plays host to the most exploited coal reserves within the country [60]. On average, all the thermal-grade coal that is produced can be classified as low-grade bituminous coal with a low calorific value of 25.43 MJ/kg [61–64]. Since it is quite difficult from a policy perspective to convert thermal power plants from pure coal to co-fired plants, small to medium processing industries offer the largest spectrum through which co-firing technology can be implemented in Zimbabwe.

The composition characteristics of biomass and coal are not similar; hence, they portray different combustion behaviours [2]. A summative comparison between biomass and coal is given in Table 1.3. As noted by various authors, biomass contains a higher volatile content and lower ash percentage as compared to coal. Part of the fuels investigated by Murredu et al. [18] contained Pinus wood chips and

TABLE 1.2
Coal Classification

Name	Peat	Lignite	Sub-bituminous	Bituminous	Anthracitic
Rank		Low rank coal		Medium rank coal	High rank coal
Description	Not a coal	Brown coal		Soft coal	Hard coal
Heat		Low	⟶		High
Carbon		Low	⟶		High
Sulphur		High	⟵		Low

TABLE 1.3
Biomass and Coal Specifications [18, 65]

Specification	Coal	Biomass (wood)
Moisture content [% wt.]	6–10	5–60
Net calorific value [J/kg]	26–28.3	18.5–20
Ash content [% wt.]	8.5–10.9	0.4–0.5
Carbon [% wt.] *	76–87	48–52
Hydrogen [% wt.]	3.5–5	6.2–6.4
Nitrogen [% wt.]	0.8–1.5	0.1–0.5
Oxygen [% wt.]	2.8–11.3	38–42
Sulphur [% wt.]	0.5–3.1	<0.05
Clorine, Potasium, Calcium	Trace	Trace

* On a dry basis.

South African bituminous coal, and those researched by Zhou et al. [65] contained sub-bituminous coal and sawdust blends, with both agreeing on the chemical characterisation of the fuels.

1.3 SIGNIFICANCE OF ENERGY TOWARDS SUSTAINABLE DEVELOPMENT GOALS

Achieving SDGs 7 and 13 requires technologies that reduce emissions while ensuring energy security. Co-firing of coal with biomass provides one such pathway, particularly for developing economies. Conventional coal combustion on its own is known to produce undesired products in large quantities. Co-combustion of coal and biomass waste offers a combined approach to emission reduction and combustion efficiency improvement for boilers.

1.4 INTRODUCTION TO COMPUTATIONAL FLUID DYNAMICS MODELLING

Experimental optimisation of large combustion systems is costly and limited; computational fluid dynamics (CFD) is one of the tools that has become indispensable. It allows engineers to simulate combustion processes at scale, reducing trial-and-error in industry. CFD is an approach used by engineers to model thermo-fluid phenomena mathematically so it can be analysed on a minute scale before it gets integrated back to represent the overall system behaviour. Generally, the process of CFD modelling involves a couple of steps, as illustrated in Figure 1.4. The steps can be outlined as follows:

1. Identifying the physical boundaries of a system that needs to be modelled.
2. Modelling the geometry of the physical system using various techniques such as computer-aided design.

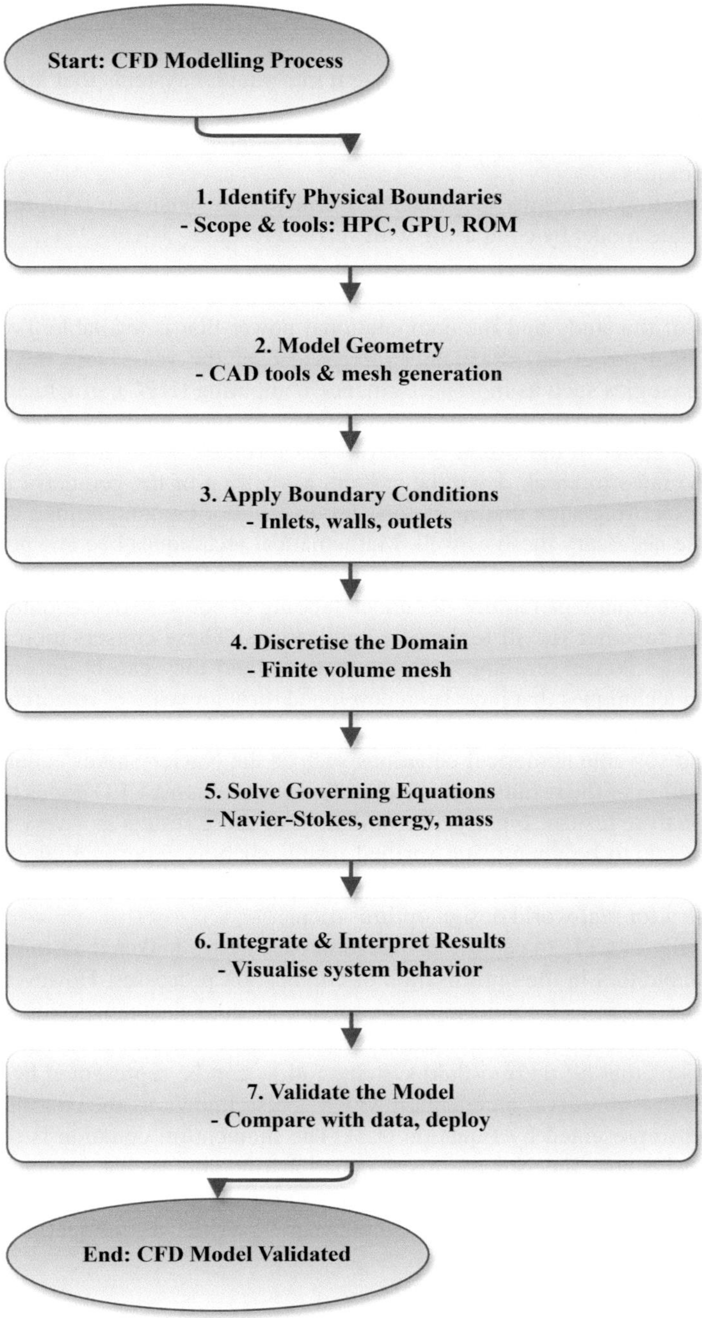

FIGURE 1.4 Simplified flowchart of setting up a CFD model.

3. Applying boundary conditions to the modelled system using prior physics-informed knowledge.
4. Discretisation of the modelled system into smaller systems that are linked to one another.
5. Solving the discretisation equations that are formed using various techniques.
6. Integrating the results such that the overall system behaviour is represented.
7. Validate model by comparing with verified results.

The process of identifying the physical boundaries is generally governed by the objectives of the study and the computational power that is available. Fortunately, computational power has drastically increased over the years with the advent of other technologies such as high-performance computing (HPC), graphical processing unit (GPU) and reduced order model (ROM). As such, analysis of large systems has become easier to the extent that a full gas turbine can be modelled in one go without the need to break down the system. Modelling of the geometry is streamlined with meshing such that most software capable of computer-aided design can produce the necessary mesh as well. Mathematical techniques become pronounced during the discretisation process as differential methods are applied to the system. Solving these equations requires an understanding of conservation equations which are coupled together for all real-world applications. These conservation equations include mass, momentum, and energy equations, and they can be solved through specialised techniques that have been developed through years of research.

Discretisation with regard to fluid analysis uses finite volume elements to form simpler and solvable numerical equations such as the Navier-Stokes equations [66]. The integration of the results is also streamlined within most CFD codes to produce ready-to-analyse results. Finally, the validation of these results is a very important step that needs to be done using verified results such as experimental readings or previously verified modelled results. Having validated the model, it is then deployed into industry for real-world design optimisation.

The ability of CFD to capture combustion parameters makes it an indispensable and affordable tool in the optimisation of combustion processes. However, a mathematical description of combustion is complex as fluid mechanics, mass transfer, chemical kinetics, and thermodynamics issues all come into play [67]. The conservation equations that all thermo-fluid systems follow can be represented by Equation (1.2). The mass conservation equation which is also known as the continuity equation is then represented by Equation (1.3). The momentum equation is sometimes reduced to become the Navier-Stokes equation (incompressible, irrotational, and Newtonian fluid flow) as represented by Equations (1.4) to (1.6). The internal energy equation which is part of the energy equation can be written as in Equation (1.7) [68].

Overall:

$$\frac{\partial}{\partial t}(\rho\phi)+\frac{\partial}{\partial x_i}(\rho c_i\phi)=\frac{\partial}{\partial x_i}\left(\Gamma\frac{\partial\phi}{\partial x_i}\right)+S_\phi \tag{1.2}$$

Mass:

$$\frac{\partial}{\partial t}(\rho)+\frac{\partial}{\partial x_i}(c_i)=0 \tag{1.3}$$

Momentum (e.g. Navier-Stokes):

$$\frac{\partial}{\partial t}(\rho u)+\frac{\partial}{\partial x_i}(\rho c_i u)=-\frac{\partial P}{\partial x}+\frac{\partial}{\partial x_i}\left(\mu\frac{\partial u}{\partial x_i}\right)+S_{\phi,u} \tag{1.4}$$

$$\frac{\partial}{\partial t}(\rho v)+\frac{\partial}{\partial x_i}(\rho c_i v)=-\frac{\partial P}{\partial y}+\frac{\partial}{\partial x_i}\left(\mu\frac{\partial v}{\partial x_i}\right)+S_{\phi,v} \tag{1.5}$$

$$\frac{\partial}{\partial t}(\rho w)+\frac{\partial}{\partial x_i}(\rho c_i w)=-\frac{\partial P}{\partial z}+\frac{\partial}{\partial x_i}\left(\mu\frac{\partial w}{\partial x_i}\right)+S_{\phi,w} \tag{1.6}$$

Energy (e.g. Internal Energy):

$$\frac{\partial}{\partial t}(\rho c_v T)+\frac{\partial}{\partial x_i}(\rho c_i c_v T)=-P\frac{\partial c_i}{\partial x_i}+\frac{\partial}{\partial x_i}\left(k\frac{\partial T}{\partial x_i}\right)+\Psi+S_{c_vT} \tag{1.7}$$

where:

ϕ: variable (e.g. temperature (K), pressure (Pa), species mass fraction),
c_i: velocity vector (u: x component, v: y component, w: z component) (m/s),
x_i: direction (x_1: x direction, x_2: y direction, x_3: z direction for Cartesian coordinates),
ρ: density (kg/m^3),
Γ: diffusion coefficient,
S_ϕ: difference between variable source and sink,
P: pressure (Pa),
μ: dynamic viscosity (Pa.s),
c_v: specific heat capacity at constant volume (J/kg),
T: temperature (K), and
Ψ: dissipation function.

The adaptation of these equations for CFD application requires an understanding of various physical phenomena such as turbulence and boundary layer interaction. The equations that are formed are generally complicated (multiple unknowns) and extremely coupled. The strength of a CFD application is generally measured by its ability to solve these equations. Certain closure techniques have been used by various researchers to help solve these equations. One such method is the use of turbulent models such as Large Eddy Simulations, k-omega (k-ω), k-epsilon (k-ε), SST, and other equally capable models specially adapted for each application. Detailed theory and application of these techniques are discussed in Chapter 5 of this book.

1.5 TRENDS IN MODELLING OF SOLID FUEL CO-COMBUSTION

Co-combustion is the process of combustion that takes place when two or more fuels are present in the combustion chamber at the same time. From a theoretical perspective, a solid fuel particle goes through drying, devolatilisation, char combustion, volatile combustion, and ash formation during the combustion process [9, 69]. Good combustion models must capture the details of each stage to give better agreement between experimental and modelled results (validation). Combustion modelling requires a good grasp of the theory behind fluid mechanics, thermodynamics, chemical kinetics, heat transfer, and mass transfer. Figure 1.5 describes these stages explicitly with corresponding models that need to be developed.

Drying entails the removal of moisture (surface and inherent) when the particle is heated up to around 105°C in the absence of oxygen, whilst devolatilisation involves the removal of light volatiles and heavy volatiles when the heating continues further to around 800°C in the absence of oxygen. The quantitative (mass of volatiles released and the rate of release) and qualitative (chemical composition of volatiles) analysis of volatiles is rather complicated, making this stage one of the most important ones in the combustion modelling process [70]. Table 1.4 summarises the characteristics of each stage with relation to combustion modelling.

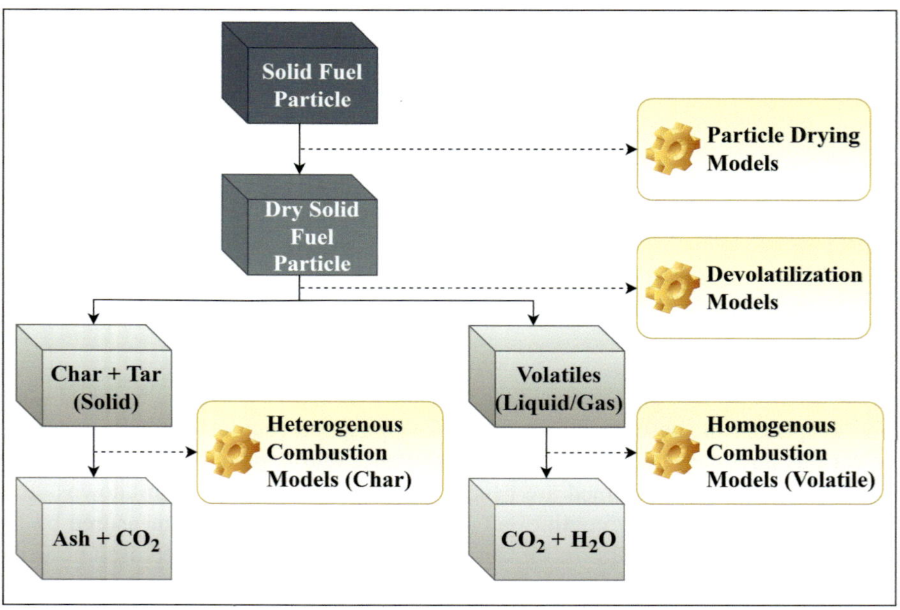

FIGURE 1.5 Solid fuel combustion stages.

TABLE 1.4
Characteristics of Solid Fuel Combustion Stages with Relation to CFD [71]

Stage	Characteristics	Numerical Equations Used in CFD
Drying	Removal of water molecules within the solid fuel. Particle drying is based mainly on heating and transfer of the H_2O molecules from the particle surface. Little thought has been placed on this stage in terms of variations in models.	$T_p < T_{vap}$ (1.8) $m_p c_{p,p} \frac{dT_p}{dt} = hA_p(T_\infty - T_p) + \varepsilon_p A_p \sigma\left(T_w^4 - T_p^4\right) - Q_v$ (1.9)
Devolatilisation	Removal of volatile substances which takes place when the fuel particle is heated in the absence of oxygen. The simplest models are empirical and employ global kinetics, where Arrhenius expressions are used to correlate rates of weight loss with temperature. The simple models can accurately predict the volatile yield rates depending on the input parameters whilst the more complex network models are good at predicting the composition of volatiles.	$T_p = T_{vap}$ (1.10) $m_p c_{p,p} \frac{dT_p}{dt} = h_{conv} A_p(T_\infty - T_p) + \varepsilon_p A_p \sigma\left(T_w^4 - T_p^4\right) + \frac{dm_p}{dt} h_{fg}$ (1.11)
Char combustion	Combustion of the products of devolatilisation. Also known as heterogeneous char combustion.	$T_p > T_{vap}$ (1.12) $\text{Char}(s) + \text{oxidant}(g) \rightarrow \text{products}(g)$ (1.13)
Volatile combustion	Combustion of the products of devolatilisation, homogeneous volatile combustion.	$CH_yO_x + (1 + \frac{y}{2} - \frac{x}{4})O_2 \rightarrow CO_2 + \frac{y}{2} H_2O$ (1.14)

with:

T_p: particle temperature (K),
T_{vap}: local vapourisation temperature of the volatiles (K),
T_∞: bulk fluid temperature (K),
T_w: furnace wall temperature (K),

m_p: particle mass (kg),
$c_{p,p}$: specific heat capacity of particle (J/kg.K),
A_p: surface area of particle (m²),
ε_p: emissivity of particle,
h_{fg}: specific latent heat of vaporisation of the particle (J/kg),
h_{conv}: convective heat transfer coefficient (W/m².K), and
σ: Stefan Boltzmann constant.

CFD software capable of handling combustion uses a discrete phase model (DPM) or particle phase model (PPM) to capture the multiphase nature of the combustion of solid fuels. A DPM model makes it possible to analyse particle–particle and continuum–particle interaction between the gaseous substances that constitute the continuous phase and the solid particles which constitute the discrete phase [72]. Validation of CFD combustion models by use of industrial boilers is rather difficult as it requires access to closely controlled and expensive boilers usually run by large institutions. Fortunately, experimental test facilities such as drop tube furnaces (DTF) and entrained flow reactors (EFR) have been developed to simulate the operation of industrial boilers [73].

1.5.1 Applications in Thermal Powerplants

A combustion model was applied to the 210-MW thermal power plant in Indonesia to ascertain the potential benefits of co-firing waste biomass in the form of olive waste with coal on combustion parameters. Various substitution ratios were employed with the most promising deduction showing that substituting by 5% biomass had the potential of reducing the CO_2 emissions by 30% [74]. Huang et al. [75] by use of a combustion model showed that substituting biomass by 40% on a thermal basis to an 840-MW pulverised fluidised bed boiler leads to contradicting results that need optimisation. It is important to note that the overall CO_2 emitted from the system increases as biomass is increased; however, the proportion of CO_2 that is due to coal decreases. Inasmuch as biomass emits a higher amount of CO_2 per unit energy when compared to coal, the CO_2 from biomass is considered green or carbon neutral. Generally, depending on the properties of the biomass, a reduction in SOx and NOx is experienced for the various biomass fuels that were used in the study. Karampinis et al. [76] used the 300-MW suspension-fired furnace located at Kardia power plant in Greece as a case study to investigate the effect of co-firing lignite and cardoon biomass. Since CFD allows users to study intricate parameters generally difficult to access experimentally, the findings of this study showed how the fuel particles progress from each burner level to the other [77].

The effect of direct and indirect co-firing was successfully investigated by Amirabedin and McIlveen-Wright [78] during the studies of co-firing poor quality coal with agricultural residues in the form of olive waste. Of interest was the empirical rather than the numerical modelling approach that was used for the circulating fluidised thermal power plant. As expected, both methods of co-firing resulted in a decrease in fuel consumption, CO_2, SO_2, and particulate matter. Numerical models require sufficient boundary conditions in the form of intricate chemical fuel properties such as Arrhenius equation properties during the combustion process to have successful models [79]. This makes it very difficult to validate most combustion models currently in existence. Gubba et al.[80] demonstrated that modifying certain boundary conditions (e.g. wall temperature) such that they are different from experimental conditions improves the overall accuracy of the model. Datta et al. [81] employed artificial neural networks to optimise the co-firing potential of coal with various biomass residues. Machine learning, when used in conjunction with experimental and numerical modelling, provides a physics-informed approach that can be rapidly deployed for most applications [82].

As an extension to co-firing coal with biomass, in-situ emission reduction techniques such as wet scrubbing by use of ammonia have also been successfully studied by some researchers. The chemical interaction between ammonia and combustion products (CO_2, NOx, and SOx) was modelled by Adi Saputra et al. [83] with the results showing the best concentration to achieve optimum reductions without affecting combustion parameters.

1.5.2 Applications in Industrial Thermal Systems and Pilot Furnaces

Due to the small scale of industrial thermal systems as compared to thermal power plants, it is easier to apply the technology of biomass substitution on process heaters. Some even prefer to utilise 100% biomass rather than coal for their process heaters. Upgrading of biomass waste before combustion is usually carried out through physical or chemical densification processes. Li et al. [84] first upgraded forest willow residue into briquettes before co-firing them with coal. Inasmuch as this approach might increase the overall cost of the whole process, it helps improve the grindability of biomass, which is a major constraint for most applications. Wu et al. [85] chemically upgraded paper mill rejects into syngas, a biofuel, before co-firing with coal. Unfortunately, their study showed that CO increases at the furnace exit with no significant changes in NOx and SOx emissions.

Pilot furnace studies occupy a very large space with regard to combustion studies, as parameters can be easily manipulated when compared to thermal power plants. These furnaces include drop tube furnaces, down-fired furnaces, entrained flow furnaces, and pressurised furnaces. By use of a 350-kW pilot furnace, Gurel et al. [86] monitored the combustion of chicken manure with lignite in a circulating fluidised bed combustor. The study concluded that secondary air temperature and inlet position play a significant role in maximising the benefits of co-firing coal with biomass. Burner design is in turn critical during the retrofitting of biomass co-firing systems. Dunye et al.'s [87] interest in detailed burner design of co-fired systems demonstrated that biomass results in high localised temperatures which need to be catered for during the design process. Table 1.5 gives a summary of co-firing applications found in industry.

TABLE 1.5
Co-firing Applications in Industry and Corresponding Modelling Code

Fuels	Furnace Type	Max Biomass Substitution Level	Code	Outcomes	Reference
Coal and agricultural residue	315-MW suspension-fired powerplant	50% on a mass basis	Numerical (Fluent)	An optimum ratio of 5% was deduced as CO_2 emissions were reduced by 30% and increased combustion intensity.	Ihsan et al. [74]
Coal and rice husk	Fluidised bed powerplant	100% on a mass basis	Empirical	Use of rice husks required increase of the fluidisation velocity.	Gaba et al. [88]
Coal and biomass	360-MW fluidised-bed powerplant	40% on thermal basis	Numerical (Fluent)	The CO_2 emitted due to coal reduced by an average of 38%. All cases showed a reduction in the SOx by around 30% whilst NOx emissions either increased with blending for high nitrogen-enriched biomass or decreased for the other biomass samples.	Huang et al. [75]
Lignite and biomass	300-MW suspension-fired furnace powerplant	10% thermal basis	Numerical (Fluent)	A reduction in NOx of 10% was demonstrated. This particular case showed an increase in unburnt carbon.	Karampinis et al. [76]
Lignite and forest residue	150-MW suspension-fired furnace	43% on mass basis	Numerical (Fluent)	Variations in the velocity, temperature, H_2O, CO, and CO_2 were successfully evaluated for each burner level.	Deniz et al. [77]
Lignite and agricultural residues	Circulating fluidised bed boiler	10% on mass basis	Empirical (THERMOFLEX)	Both direct and indirect co-firing resulted in the reduction of coal fuel consumption, CO_2 and SO_2 and particulate matter.	Amirabedin and McIlveen-Wright [78]

(*Continued*)

TABLE 1.5 (CONTINUED)
Co-firing Applications in Industry and Corresponding Modelling Code

Fuels	Furnace Type	Max Biomass Substitution Level	Code	Outcomes	Reference
Coal and wheat straw	300-MW suspension-fired powerplant	12% thermal basis	Numerical (Fluent)	The study focused on good modelling practices with regard to combustion so as to improve the performance of the models.	Gubba et al. [80]
Coal and biomass waste	70-MW suspension-vortex burner	10% mass basis	Numerical (Fluent)	This study focused on the biomass burner design of the co-firing process.	Dunayevska et al. [87]
Coal and torrefied biomass	220-MW suspension-fired boiler	100% mass basis	Numerical (Fluent)	Torrefied biomass improves the physical properties of raw biomass which in turn can be co- fired with coal using the same burners.	Li et al. [84]
Lignite and chicken manure	350-kW fluidised bed pilot furnace	25% on mass basis	Numerical (Fluent)	The study resolved that secondary air temperature has a huge impact on the NOx emissions that are produced.	Gürel et al. [86]
Coal and paper mill syngas	70 tonnes/h suspension fired furnace	5% on thermal basis	Numerical (GLACIER)	Through chemical upgrading of paper mill waste into syngas, various combustion scenarios such as reburning were tested for optimum conditions.	Wu et al. [85]

1.6 SUMMARY

This chapter introduced solid fuels as an energy source that can be combusted to produce heat. Focus was placed on introducing solid fuels that can be co-fired, which are coal and biomass waste in the form of forest residue, agricultural waste, and municipal solid waste. Co-firing as a technology can help address sustainable development goals related to emission reduction (SDG 7) and clean energy production (SDG 13). CFD can be applied to model combustion in an approach that combines mathematical principles with fluid mechanics, reaction kinetics, and thermodynamic theory. CFD is currently being used by engineers to optimise systems that are complex and difficult to analyse; hence, its wide use in industry. Trends on how CFD has been applied to combustion modelling show that there is a need to understand intricate combustion processes such as drying, devolatilisation, homogeneous volatile combustion, and heterogeneous char combustion sub-models in detail.

REFERENCES

1. United Nations. *The Sustainable Development Goals Report 2023: Special Edition*; New York, 2023. Available online: https://sdgs.un.org/sites/default/files/2023-07/The-Sustainable-Development-Goals-Report-2023_0.pdf
2. Veijonen, K.; Vainikka, P.; Järvinen, T.; Alakangas, E. *Biomass Co-Firing an Efficient Way to Reduce Greenhouse Gas Emissions*; European Bioenergy Networks: Finland, 2003. Available online: https://publications.vtt.fi/julkaisut/muut/2003/cofiring180603.pdf.
3. Ma, W.; Zhou, H.; Zhang, J.; Zhang, K.; Liu, D.; Zhou, C.; Cen, K. Behavior of Slagging Deposits during Coal and Biomass Co-Combustion in a 300 KW Down-Fired Furnace. *Energy & Fuels* **2018**, *32*, 4399–4409, doi:10.1021/acs.energyfuels.7b03050.
4. Mandø, M.; Rosendahl, L.A.; Yin, C.; Sørensen, H. Pulverized Straw Combustion in a Low-NOx Multifuel Burner: Modeling the Transition from Coal to Straw. *Fuel* **2010**, *89*, 3051–3062, doi:10.1016/j.fuel.2010.05.016.
5. Tillman, D.A. Biomass Cofiring: The Technology, the Experience, the Combustion Consequences. *Biomass Bioenergy* **2000**, *19*, 365–384, doi:10.1016/S0961-9534(00)00049-0.
6. Cong, K.; Han, F.; Zhang, Y.; Li, Q. The Investigation of Co-Combustion Characteristics of Tobacco Stalk and Low Rank Coal Using a Macro-TGA. *Fuel* **2019**, *237*, 126–132, doi:10.1016/j.fuel.2018.09.149.
7. Savolainen, K. Co-Firing of Biomass in Coal-Fired Utility Boilers. *Applied Energy* **2003**, *74*, 369–381, doi:10.1016/S0306-2619(02)00193-9.
8. Liu, Q.; Zhong, W.; Tang, R.; Yu, H.; Gu, J.; Zhou, G.; Yu, A. Experimental Tests on Co-Firing Coal and Biomass Waste Fuels in a Fluidised Bed under Oxy-Fuel Combustion. *Fuel* **2021**, *286*, 119312, doi:10.1016/j.fuel.2020.119312.
9. Turns, S. *An Introduction to Combustion: Concepts and Applications*; McGraw-Hill series in mechanical engineering; 2nd ed.; McGraw-Hill: New York, 1996.
10. Rüdiger, H.; Kicherer, A.; Greul, U.; Spliethoff, H.; Hein, K.R.G. Investigations in Combined Combustion of Biomass and Coal in Power Plant Technology. *Energy & Fuels* **1996**, *10*, 789–796, doi:10.1021/ef950222w.

11. Yin, C.; Rosendahl, L.A.; Kær, S.K. Towards a Better Understanding of Biomass Suspension Co-Firing Impacts via Investigating a Coal Flame and a Biomass Flame in a Swirl-Stabilized Burner Flow Reactor under Same Conditions. *Fuel Processing Technology* **2012**, *98*, 65–73, doi:10.1016/j.fuproc.2012.01.024.
12. Xing, X.; Wang, S.; Zhang, Q. Thermogravimetric Analysis and Kinetics of Mixed Combustion of Waste Plastics and Semicoke. *Journal of Chemistry* **2019**, *2019*, 10, doi:10.1155/2019/8675986.
13. Lü, H.F.; Deng, J.; Li, D.J.; Xu, F.; Xiao, Y.; Shu, C.M. Effect of Oxidation Temperature and Oxygen Concentration on Macro Characteristics of Pre-Oxidised Coal Spontaneous Combustion Process. *Energy* **2021**, *227*, 120431, doi:10.1016/j.energy.2021.120431.
14. Kruczek, H.; Rczka, P.; Tatarek, A. The Effect of Biomass on Pollutant Emission and Burnout in Co-Combustion with Coal. *Combustion Science and Technology* **2006**, *178*, 1511–1539, doi:10.1080/00102200600721297.
15. Guo, F.; He, Y.; Hassanpour, A.; Gardy, J.; Zhong, Z. Thermogravimetric Analysis on the Co-Combustion of Biomass Pellets with Lignite and Bituminous Coal. *Energy* **2020**, *197*, 1–9, doi:10.1016/j.energy.2020.117147.
16. Pérez-Jeldres, R.; Cornejo, P.; Flores, M.; Gordon, A.; García, X. A Modeling Approach to Co-Firing Biomass/Coal Blends in Pulverized Coal Utility Boilers: Synergistic Effects and Emissions Profiles. *Energy* **2017**, *120*, 663–674, doi:10.1016/j.energy.2016.11.116.
17. Bhuiyan, A.A.; Blicblau, A.S.; Naser, J. Co-Firing of Biomass and Slagging in Industrial Furnace: A Review on Modelling Approach. *Journal of the Energy Institute* **2017**, *90*, 838–854, doi:10.1016/j.joei.2016.08.010.
18. Mureddu, M.; Dessì, F.; Orsini, A.; Ferrara, F.; Pettinau, A. Air-and Oxygen-Blown Characterization of Coal and Biomass by Thermogravimetric Analysis. *Fuel* **2018**, *212*, 626–637, doi:10.1016/j.fuel.2017.10.005.
19. Wnorowska, J.; Ciukaj, S.; Kalisz, S. Thermogravimetric Analysis of Solid Biofuels with Additive under Air Atmosphere. *Energies (Basel)* **2021**, *14*, 2257, doi:10.3390/en14082257.
20. Jiang, X.; Wang, C.; Wang, X.; Liu, Y.; Song, Z.; Lin, Q. Co-Combustion of Coal and Carbonaceous Wastes: Thermogravimetric Analysis. *Energy Sources, Part A: Recovery, Utilization, and Environmental Effects* **2020**, 1–15, doi:10.1080/15567036.2019.1704314.
21. Florentino-Madiedo, L.; Vega, M.F.; Barriocanal, C. Evaluation of Synergy during Co-Pyrolysis of Torrefied Sawdust, Coal and Paraffin. A Kinetic and Thermodynamic Study. *Fuel* **2021**, *292*, 120305, doi:10.1016/j.fuel.2021.120305.
22. Wang, H.; Tian, Y.; Chen, X.; Li, J. Kinetic and Synergetic Effect Analysis of the Co-Combustion of Coal Blended with Polyurethane Materials. *ACS Omega* **2020**, *5*, 26005–26014, doi:10.1021/acsomega.0c03449.
23. Wu, Z.; Zhang, J.; Zhang, B.; Guo, W.; Yang, G.; Yang, B. Synergistic Effects from Co-Pyrolysis of Lignocellulosic Biomass Main Component with Low-Rank Coal: Online and Offline Analysis on Products Distribution and Kinetic Characteristics. *Applied Energy* **2020**, *276*, 115461, doi:10.1016/j.apenergy.2020.115461.
24. Zhang, X.; Ghamari, M.; Ratner, A. Numerical Modeling of Co-Firing a Light Density Biomass, Oat (Avena Sativa) Hulls, and Chunk Coal in Fluidized Bed Boiler. *Biomass Bioenergy* **2013**, *56*, 239–246, doi:10.1016/j.biombioe.2013.05.006.
25. Baukal Jr, C. *Industrial Combustion Pollution and Control*; Baukal Jr, C., Ed.; Environmental Science & Pollution; CRC Press: Boca Raton, 2003; Vol. 27.
26. Mohanna, H. Combustion of Pulverized Biomass: Impact of Fuel Preparation and Flow Conditions, Normandie Université, 2021. Available online: https://theses.hal.science/tel-03272429/file/MOHANNA.pdf.

27. Colannino, J. *Modeling of Combustion Systems*; 1st ed.; CRC Press: Boca Raton, 2006; ISBN 978-0-8493-3365-1.
28. Londerville, S.; Baukal Jr., C. *The Coen & Hamworthy Combustion Handbook*; Londerville, S., Baukal Jr., C.E., Eds.; 1st ed.; CRC Press: London, 2013.
29. Ma, L.; Gharebaghi, M.; Porter, R.; Pourkashanian, M.; Jones, J.M.; Williams, A. Modelling Methods for Co-Fired Pulverised Fuel Furnaces. *Fuel* **2009**, *88*, 2448–2454, doi:10.1016/j.fuel.2009.02.030.
30. Niu, S.L.; Lu, C.M.; Han, K.H.; Zhao, J.L. Thermogravimetric Analysis of Combustion Characteristics and Kinetic Parameters of Pulverized Coals in Oxy-Fuel Atmosphere. *Journal of Thermal Analysis and Calorimetry* **2009**, *98*, 267–274, doi:10.1007/s10973-009-0133-1.
31. Leckner, B. Possibilities and Limitations of Co-Firing of Biomass. In Proceedings of the 1st Project Conference AGS; 1st Project Conference AGS: Stockholm, 2006. Available online: https://www.energy-pathways.org/pdf/Leckner.pdf.
32. Fernando, R. *Fuels for Biomass Cofiring*; London, 2005. Available online: https://www.sustainable-carbon.org/wp-content/uploads/dlm_uploads/reports/powergeneration/fuels-for-biomass-cofiring-ccc-102.pdf.
33. Tillman, D.A. Cofiring Benefits for Coal and Biomass. *Biomass Bioenergy* **2000**, *19*, 363–364, doi:10.1016/S0961-9534(00)00048-9.
34. Grand View Research. *Waste To Energy (WTE) Market Share, Size & Trend Analysis Report By Technology (Biological, Thermal), by Region (North America, Europe, Asia Pacific, CSA, MEA), and Segment Forecasts, 2025–2030*; 2024. Available online: https://www.grandviewresearch.com/industry-analysis/waste-to-energy-technology-industry.
35. Khwamman, S.; Chaiyat, N. Energy, Economic, and Environmental Analysis of a Waste-to-Energy-to-Zero System. *Cleaner Chemical Engineering* **2025**, *11*, 100142, doi:10.1016/J.CLCE.2024.100142.
36. Weldekidan, H.; Strezov, V.; Kan, T.; Town, G. Waste to Energy Conversion of Chicken Litter through a Solar-Driven Pyrolysis Process. *Energy & Fuels* **2018**, *32*, 4341–4349, doi:10.1021/acs.energyfuels.7b02977.
37. Jingura, R.M.; Musademba, D.; Kamusoko, R. A Review of the State of Biomass Energy Technologies in Zimbabwe. *Renewable and Sustainable Energy Reviews* **2013**, *26*, 652–659, doi:10.1016/j.rser.2013.05.036.
38. Meincken, M.; Tyhoda, L. Biomass Quality. In *Bioenergy from Wood*; Seifert, T., Ed.; Managing Forest Ecosystems; Springer Netherlands: Dordrecht, 2014; Vol. 26, pp. 169–187, ISBN 978-94-007-7447-6.
39. The World Bank. Electric Power Consumption (KWh per Capita) – Zimbabwe Available online: https://data.worldbank.org/indicator/EG.USE.ELEC.KH.PC?locations=ZW (accessed on 7 January 2020).
40. Moyo, A. Command Agric Financing Succeeds. *The Herald* 2020. Available online: https://www.herald.co.zw/command-agric-financing-succeeds/.
41. The Herald. Zimbabwe Expects 2,7m Tonnes of Maize. *The Herald* 2017. Available online: https://www.herald.co.zw/zimbabwe-expects-27m-tonnes-of-maize/.
42. Parsons, S.; McManus, M.C.; Taylor, C.M. Second-Generation Ethanol from Lignocellulose. In *Greenhouse Gases Balances of Bioenergy Systems*; Thornley, P.; Adams, P., Eds.; Elsevier, 2018; pp. 193–206, doi: 10.1016/B978-0-08-101036-5.00013-6.
43. Pennington, D. Bioenergy Crops. In *Bioenergy*; Dahiya, A., Ed.; Elsevier, 2015; pp. 111–134.
44. Mtambanengwe, F.; Chikowo, R.; Mapfumo, P. Availability and Use of Maize Stover Residues on Smallholder Farms in Zimbabwe and Implications for Conservation Agriculture; Regional Conservation Agriculture (CA) Symposium, 2014. Available

online: https://www.researchgate.net/publication/237100373_Availability_and_Use_of_Maize_Stover_Residues_on_Smallholder_Farms_in_Zimbabwe_and_Implications_for_Conservation_Agriculture.

45. Zimbabwe Poultry Association. *Zimbabwe Poultry Association*; Harare, 2018. Available online: https://www.livestockzimbabwe.com/Updates/ZPA%20newsletter%20%20February%202018.pdf.
46. Sango, I. Analysis of the Solid Waste Management Practices in Chinhoyi: Bridging the Missing Link. *Economia: Seria Management* **2010**, *13*, 332–347. Available online: https://www.management.ase.ro/reveconomia/2010-2/9.pdf.
47. Cheng, H.; Zhang, Y.; Meng, A.; Li, Q. Municipal Solid Waste Fueled Power Generation in China: A Case Study of Waste-to-Energy in Changchun City. *Environmental Science & Technology* **2007**, *41*, 7509–7515, doi:10.1021/es071416g.
48. Maqhuzu, A.B.; Yoshikawa, K.; Takahashi, F. The Effect of Coal Alternative Fuel from Municipal Solid Wastes Employing Hydrothermal Carbonization on Atmospheric Pollutant Emissions in Zimbabwe. *Science of The Total Environment* **2019**, *668*, 743–759, doi:10.1016/j.scitotenv.2019.03.050.
49. Borup, M.B.; Muchmore, D.R. Food-Processing Waste. *Water Environment Research* **1992**, *64*, 413–417, doi:10.1002/j.1554-7531.1992.tb00021.x.
50. Danha, C.; Chihobo, C.H.; Musademba, D.; Simbi, D.J.; Kuipa, P.K.; Jonathan, E. Characterization and Utilization of Acid Tar Waste from Crude Benzol Processing for Environmental Sustainability. *IOSR Journal of Environmental Science, Toxicology and Food Technology* **2014**, *8*, 16–21, doi:10.9790/2402-08131621.
51. Mtshali, L.; Dlamini, M.; Kohler, N.A. The Opportunities and Challenges of Wood Waste Biomass and a Case Study on Biofuels in South Africa. In *Proceedings of the Institute of Waste Management of Southern Africa*; Johannesburg, 2016; pp. 350–355.
52. Charis, G.; Danha, G.; Muzenda, E. A Review of the Application of GIS in Biomass and Solid Waste Supply Chain Optimization: Gaps and Opportunities for Developing Nations. *Detritus* **2019**, *6*, 1, doi:10.31025/2611-4135/2019.13819.
53. Charis, G.; Danha, G.; Muzenda, E. A Review of Timber Waste Utilization: Challenges and Opportunities in Zimbabwe. *Procedia Manufacturing* **2019**, *35*, 419–429, doi:10.1016/j.promfg.2019.07.005.
54. Dean, G.; Peck, M. *Opportunities for Using Sawmill Residues in Australia*; FWPA: Melbourne, 2012, Retrieved from: www.fwpa.com.au.
55. World Bank Group. Energy Use (Kg of Oil Equivalent per Capita) – Zimbabwe Available online: https://0-data-worldbank-org.ujlink.uj.ac.za/indicator/EG.USE.PCAP.KG.OE?locations=ZW (accessed on 1 September 2021).
56. Pordesimo, L.O.; Hames, B.R.; Sokhansanj, S.; Edens, W.C. Variation in Corn Stover Composition and Energy Content with Crop Maturity. *Biomass Bioenergy* **2005**, *28*, 366–374, doi:10.1016/j.biombioe.2004.09.003.
57. William, C.M. *Poultry Waste Management in Developing Countries*; FAO; 2010, Retrieved from: https://www.fao.org/4/i3531e/i3531e05.pdf.
58. Tchobanoglous, G.; Kreith, F. *Handbook of Solid Waste Management, Second Edition*; 2nd ed.; McGraw-Hill Education: New York, 2002.
59. Kaza, S.; Yao, L.; Bhada-Tata, P.; Van Woerden, F. *What a Waste 2.0: A Global Snapshot of Solid Waste Management to 2050*; World Bank Publications, 2018.
60. Hancox, P.J. The Coalfields of South-Central Africa: A Current Perspective. *Episodes* **2016**, *39*, 407–428, doi:10.18814/epiiugs/2016/v39i2/95785.
61. Makomo Resource. Coal Product. Available online: https://makomoresources.com/coal-products/ (accessed on 7 March 2026).

62. Moyana, F. ZPC Wants to Mine Own Coal . . . Says Existing Firms Failing to Meet Demand. *Chronicle*, 2019. Available online: https://www.heraldonline.co.zw/zpc-wants-to-mine-own-coal-says-existing-firms-failing-to-meet-demand/ (accessed on 7 March 2026).
63. Zimbabwe Power Company. *Annual Report 2017*; Harare, 2019.
64. Zimbabwe Power Company. Zimbabwe Power Company. Available online: http://www.zpc.co.zw/ (accessed on 7 March 2026).
65. Zhou, C.; Liu, G.; Wang, X.; Qi, C. Co-Combustion of Bituminous Coal and Biomass Fuel Blends: Thermochemical Characterization, Potential Utilization and Environmental Advantage. *Bioresource Technology* **2016**, *218*, 418–427, doi:10.1016/j.biortech.2016.06.134.
66. Zhou, P.; Zhou, S.; Clelland, D. A Modified Quasi-Dimensional Multi-Zone Combustion Model for Direct Injection Diesels. *International Journal of Engine Research* **2006**, *7*, 335–346, doi:10.1243/14680874JER02604.
67. Yin, C.; Rosendahl, L.; Kær, S.K.; Condra, T.J. Use of Numerical Modeling in Design for Co-Firing Biomass in Wall-Fired Burners. *Chemical Engineering Science* **2004**, *59*, 3281–3292, doi:10.1016/j.ces.2004.04.036.
68. Versteeg, H.K.; Malalasekera, W. *An Introduction to Computational Fluid Dynamics: The Finite Volume Method*; 2nd ed.; Pearson Education Ltd: Harlow, England, 2007.
69. Marangwanda, G.T.; Madyira, D.M.; Chihobo, H.C.; Babarinde, T.O. Modelling Co-Combustion of Bituminous Coal and Pine Sawdust: Thermal Behaviour. *Fuel Communications* **2021**, *9*, 100035, doi:10.1016/j.jfueco.2021.100035.
70. Marangwanda, G.T.; Madyira, D.M.; Babarinde, T.O. Combustion Models for Biomass: A Review. *Energy Reports* **2020**, *6*, 664–672, doi:10.1016/j.egyr.2019.11.135.
71. Sharma, A.; Pareek, V.; Zhang, D. Biomass Pyrolysis–A Review of Modelling, Process Parameters and Catalytic Studies. *Renewable and Sustainable Energy Reviews* **2015**, *50*, 1081–1096, doi:10.1016/j.rser.2015.04.193.
72. ANSYS. *ANSYS FLUENT User's Guide*; ANSYS, 2021, 1–170.
73. Riaza, J.; Gil, M.V.V.; Álvarez, L.; Pevida, C.; Pis, J.J.J.; Rubiera, F. Oxy-Fuel Combustion of Coal and Biomass Blends. *Energy* **2012**, *41*, 429–435, doi:10.1016/j.energy.2012.02.057.
74. Ihsan, S.; Prabowo; Widodo, W.A.; Saputra, I.N.A.A.; Hariana. Utilization of Palm Frond Waste as Fuel for Co-Firing Coal and Biomass in a Tangentially Pulverized Coal Boiler Using Computational Fluid Dynamic Analysis. *Biomass* **2024**, *4*, 1142–1163, doi:10.3390/biomass4040063.
75. Huang, Y.; McIlveen-Wright, D.; Rezvani, S.; Wang, Y.D.; Hewitt, N.; Williams, B.C. Biomass Co-Firing in a Pressurized Fluidized Bed Combustion (PFBC) Combined Cycle Power Plant: A Techno-Environmental Assessment Based on Computational Simulations. *Fuel Processing Technology* **2006**, *87*, 927–934, doi:10.1016/j.fuproc.2006.07.003.
76. Karampinis, E.; Nikolopoulos, N.; Nikolopoulos, A.; Grammelis, P.; Kakaras, E. Numerical Investigation Greek Lignite/Cardoon Co-Firing in a Tangentially Fired Furnace. *Applied Energy* **2012**, *97*, 514–524, doi:10.1016/j.apenergy.2011.12.032.
77. Deniz, C.; Boke, Y.; Aydin, O.; Benim, A. Computational Analysis of Pulverised Coal Co-Firing with Biomass in 150MWe Unit of Tuncbilek Thermal Power Plant. *Isı Bilimi ve Tekniği Dergisi* **2021**, *41*, 37–50, doi:10.47480/isibted.979314.
78. Amirabedin, E.; McIlveen-Wright, D. A Feasibility Study of Co-Firing Biomass in the Thermal Power Plant at Soma in Order to Reduce Emissions: An Exergy Approach. *International Journal of Environmental Research* **2013**, *7*.

79. Kumar, H.; Mohapatra, S.K.; Singh, R.I. Review on CFD Modelling of Fluidized Bed Combustion Systems Based on Biomass and Co-Firing. *Journal of The Institution of Engineers (India): Series C* **2018**, *99*, 449–474, doi:10.1007/s40032-017-0361-2.
80. Gubba, S.R.; Ingham, D.B.; Larsen, K.J.; Ma, L.; Pourkashanian, M.; Tan, H.Z.; Williams, A.; Zhou, H. Numerical Modelling of the Co-Firing of Pulverised Coal and Straw in a 300MWe Tangentially Fired Boiler. *Fuel Processing Technology* **2012**, *104*, 181–188, doi:10.1016/j.fuproc.2012.05.011.
81. Datta, A.; Gajera, B.; Saikia, M.; Patra, T.; Sarma, A.K. A Comprehensive Analysis and Optimization of Coal–Biomass Mixed Fuel for Sustainable Power Generation Using the Design of Experiments and Artificial Neural Network. *Clean Technologies and Environmental Policy* **2024**, doi:10.1007/s10098-024-03048-x.
82. Pant, R.; Bisht, D. The Smart Computation of Biomass Co-Firing in Small Scale Power Plants by Using Artificial Intelligence Based Analysis Model. In Proceedings of the 2023 IEEE International Conference on Integrated Circuits and Communication Systems (ICICACS); IEEE, February 24 2023; pp. 1–6.
83. Adi Saputra, I.N.A.; Manurung, T.D.; Yuliadi, A.E.; Prabowo; Nugroho, G.; Kusumadewi, T.V.; Hariana, H.; Chan, S.H. Numerical Simulation of Co-Firing LRC and Ammonia in Pangkalan Susu 3 & 4 Coal-Fired Steam Power Plant (CFSPP) Capacity 210 Megawatts. *Case Studies in Thermal Engineering* **2024**, *63*, 105230, doi:10.1016/j.csite.2024.105230.
84. Li, J.; Brzdekiewicz, A.; Yang, W.; Blasiak, W. Co-Firing Based on Biomass Torrefaction in a Pulverized Coal Boiler with Aim of 100% Fuel Switching. *Applied Energy* **2012**, *99*, 344–354, doi:10.1016/j.apenergy.2012.05.046.
85. Wu, K.T.; Lee, H.T.; Juch, C.I.; Wan, H.P.; Shim, H.S.; Adams, B.R.; Chen, S.L. Study of Syngas Co-Firing and Reburning in a Coal Fired Boiler. *Fuel* **2004**, *83*, 1991–2000, doi:10.1016/J.FUEL.2004.03.015.
86. Gürel, B.; Kurtuluş, K.; Yurdakul, S.; Varol, M.; Keçebaş, A.; Gürbüz, H. Numerical and Experimental Investigation of Co-Combustion of Chicken Manure and Lignite Blends in a CFBB with Novel Compact Combustion Chamber. *Energy* **2023**, *285*, 129482, doi:10.1016/j.energy.2023.129482.
87. Dunayevska, N.I.; Bondzyk, D.L.; Nehamin, M.M.; Miroshnichenko, Y.S.; Beztsennyi, I.V.; Evtukhov, V.Y.; Shudlo, T.S. Technology of Anthracite and SoliD BiofUelS Co-Firing in PUlVerizeD Coal BoilerS of Tpp and Chp. *Science and Innovation* **2020**, *16*, 79–89, doi:10.15407/scine16.05.079.
88. Gaba, M.; Kumar, H.; Mohapatra, S.K. Modeling of Bubbling Fluidized Bed Combustor Based on Biomass & Co-Firing. *Materials Today: Proceedings* **2017**, *4*, 1515–1625, doi:10.1016/j.matpr.2017.01.185.

2 Fundamentals of Solid Fuel Combustion

2.1 INTRODUCTION

Understanding combustion and its relevant control techniques is essential, especially for heat producing systems. Converting potential energy (chemical energy) into heat energy is generally done through combustion, with 80% of all fuels being converted through combustion [1]. A lot of industrial processes such as metal refining furnaces, boilers, oil refineries, glass melters, dryers, as well as rotary kilns are dependent on combustion systems. Inasmuch as combustion control techniques are heavily focused on safety, at present most of the research is on emission reduction as industries try to align with global warming reduction goals.

The fuel industry is well known to be a multi-billion industry. It comprises synthetic and non-synthetic fuels in the form of solid, liquid, and gas fuels. Natural gas is probably the cleanest non-synthetic fuel; hence, its extensive use in residential and industrial heating systems. With a global push to adopt renewable energy, biogas adoption has increased drastically, mainly influenced by dedicated policies that promote low- and medium-scale deployment of biogas systems [2]. The automobile industry accounts for the bulk consumption of liquid fuels in the form of petrol and diesel. Inasmuch as electric vehicles are being adopted on a large scale, internal combustion engines still occupy a healthy 86% with regard to the market share as of 2025 [1].

Combustion is a broad field; hence, basic classifications can be drawn up based on flame structure, fuel type, as well as firing methods. The flame structure can be regarded as either laminar or turbulent, depending on the quasi-fluid nature of the flame. The flame structure can also be classified as a lift-off flame or a flashback flame, depending on where the flame is located with reference to the burner. The firing method can be premixed or diffusion, depending on how the reactants are introduced into the combustion chamber. All these different classifications highlight how combustion is a multi-disciplinary domain that requires knowledge of thermodynamics, reaction kinetics, and fluid mechanics to understand what happens pre, during, and post combustion [3].

The focus of this book is on solid fuel combustion, mainly propelled by coal combustion. A lot of research has been carried out with regard to coal combustion. Figure 2.1 illustrates the trends in the SCOPUS database for articles with a keyword string ["Combustion" AND ("Coal" OR "Biomass") AND ("Computational Fluid Dynamics" OR "CFD" OR "Modelling")]. Publications related to coal or biomass combustion and computational fluid dynamics (CFD) modelling increased

DOI: 10.1201/9781003647942-2

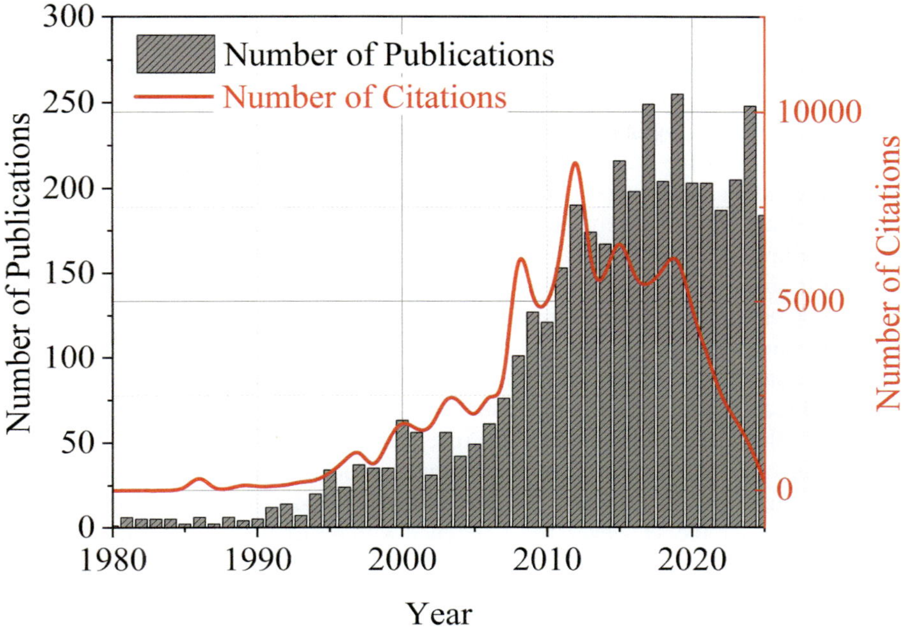

FIGURE 2.1 Indexed articles with combustion, biomass, and coal appearing as keywords.

drastically from the year 1990 to about 2015. The articles then plateaued after the year 2015 at an average of 200 publications per year. The increase in the number of publications can be attributed to the advancement in computational power and ease of access to CFD tools. The plateauing of publications after 2015 can be attributed to the increased investment in renewable energy, which has pulled resources from combustion research. Most research institutions have recalibrated to focus more on renewable energy, such as solar energy and energy storage systems. The main research CFD themes that have been explored over these years can be clustered into biomass combustion towards emission reduction, coal combustion towards emission reduction, as well as alternate combustion systems such as fluidised bed combustors.

It is quite difficult to encounter engineers who are purely focused on combustion but rather as a hybrid between classic engineering fields such as mechanical and chemical engineering or electrical and chemical engineering. Having given a brief background on solid fuels and combustion, this chapter explains in detail the essential principles that are needed to understand the modelling aspects of solid fuel combustion. Typical physical properties that are needed during CFD modelling are particle size, shape, porosity, and density, whilst typical chemical properties that are needed during CFD modelling are fuel composition, heating values, and chemical reaction constants. The properties that are of interest to a combustion modeller are generally dependent on the model complexity and results needed from the combustion process.

2.2 CHEMICAL AND PHYSICAL PROPERTIES OF SOLID FUELS

To set up a successful combustion model, accurate inputs of chemical and physical compositions are probably the most basic step in establishing boundary conditions. The chemical and physical characterisation of solid fuels is highly dependent on the capabilities of the measuring instruments. Online monitoring of chemical and physical solid fuel properties is not yet standardised; hence, there is reliance on tried-and-tested standards that require sampling solid fuels from a common storage point [4]. To sample coal stored in a bunker or on a conveyor belt, appropriate sampling standards such as ASTM D7430/D7430M-22 need to be used. One critical theme that becomes apparent during sampling is the need to have consistency. This implies that if the engineer takes five samples from each coordinate, future sampling should follow the same approach by taking five samples from each coordinate.

The characterisation is generally labelled as received, air dried, dry basis, and dry ash free. When a sample is labelled "as received", all the constituents and variables are considered when determining the total weight of the sample. When a sample is labelled "air dried", the surface moisture is disregarded when computing the total weight of the sample. However, the difference with a sample labelled "dry basis" is that all the moisture, which includes surface and inherent moisture, is disregarded when calculating the total weight of the sample. Some characterisations prefer to use the "dry ash-free" label, which typically neglects all moisture and ash constituents when computing the total weight of the sample.

2.2.1 Proximate Analysis

Proximate analysis is probably one of the most widely applied characterisation methods for solid fuels, especially coal and biomass. The goal of proximate analysis is to know the moisture, volatile matter, fixed carbon, and ash content. The typical standard used to perform a proximate analysis for coal is the ASTM D3172-13(2021)e1 or ISO 17246:2024 when using a muffle furnace. When a thermogravimetric analyser is used, the ASTM D7582-15 standard is better positioned for solid fuel proximate analysis [5–7].

The first step involves heating the sample to a temperature of 105°C in an oxygen-free environment preferably, holding the temperature for at least 7 minutes, then comparing initial and final sample masses. The difference in mass is correlated to the moisture content of the sample. From a more technical point, moisture can exist as inherent, surface, decomposition, or mineral moisture. Surface moisture represents the water held on the surface of coal particles or rather solid fuel particles. Inherent moisture is attributed to the water that is held by capillary action within the pores of the coal. Decomposition moisture is attributed to the moisture evolved during the thermal decomposition (combustion) of organic constituents. Finally, mineral moisture is attributed to the hydrous water, which is part of the crystal structure for hydrous silicates such as clays and inorganic minerals in coal. The proximate analysis determines the total or residual moisture content which is a combination of surface and inherent moisture. The other forms of moisture are chemically bound to the structure of the solid fuel particle; hence, they can only be released effectively beyond a temperature of 105°C.

High moisture content reduces the effective heating value of a fuel, since additional energy is consumed in evaporating the total moisture before combustion reactions can proceed. Inasmuch as various studies have been carried out on this topic, Li et al. [8] managed to monitor the effect of moisture content on lignite by use of a drop tube furnace in detail. Their results quantified the effect of moisture as a function of particle size, moisture content, and combustion parameters such as ignition delay and burnout time. Unfortunately, coal dust suppression methods typically used in industry and mining practices involve water spraying of bunkered coal [9]. Since most biomass fuels have a high content of moisture, innovative ways of drying biomass by use of renewable energy sources tend to offer an alternative route for sustainability.

Volatile matter from coal can be classified under heavy, medium, and light volatiles. This second step involves heating the sample to a temperature of 700–900°C in an oxygen-free environment, holding the temperature for at least 15 minutes, then comparing the initial and final sample masses. The main constituents of these volatiles are hydrocarbons such as methane, tars, hydrogen, and other gases. High-ranking coals like anthracite have lower volatile content, which results in a slower ignition. Volatile matter participates in the homogeneous volatile combustion reactions, which are basically reactions that occur within the same phase. Fuels with higher volatile content tend to exhibit high combustion intensity, a parameter that is related to the instantaneous release of heat over a short period of time. Simplified CFD models usually prefer to treat coal volatiles as a single entity that can be represented as methane. However, to have an accurate CFD model, using the actual chemical formula of the volatile is very important, though overlooked by some. It is imperative to acknowledge the actual structure of the volatile molecule as it affects the chemical reaction mechanism that needs to be employed. For advanced CFD modelling purposes, each volatile is lumped into a form that can be represented as $C_aH_bO_cN_dS_e$ regardless of the various individual constituents of volatiles such as CH_4, CO, etc. Using proximate, ultimate, and calorific value data obtained from the experiments, the lumped volatile form is determined using analytical methods [10].

Ash is the incombustible mineral residue remaining after complete combustion. This third step involves heating the sample to a temperature of 700–900°C in an oxygen environment, holding the temperature for at least 15 minutes, then comparing the initial and final sample masses. Ash primarily consists of oxides of silicon, aluminium, iron, calcium, magnesium, and trace elements. Although ash does not contribute to the heating value, it plays an important role in combustion system performance. High ash content can lead to slagging (formation of molten deposits on furnace walls) and fouling (deposits on heat exchanger surfaces), reducing efficiency and increasing maintenance costs.

Fixed carbon is the solid combustible residue other than the volatile matter within the fuel sample. Fixed carbon is a critical determinant of burnout time in furnaces and boilers. Coals generally have higher fixed carbon, while biomass tends to have much lower levels due to its fibrous structure and higher volatile yield. Fuels with higher fixed carbon tend to exhibit higher combustion stability. During proximate analysis, the fixed carbon content is determined using the difference method, which implies it is indirectly obtained by applying Equation (2.1).

$$MC_{ar} + VM_{ar} + Ash_{ar} + FC_{ar} = 100 \tag{2.1}$$

where,

MC_{ar}: moisture content as received,
VM_{ar}: volatile matter as received,
Ash_{ar}: ash content as received, and
FC_{ar}: fixed carbon as received.

2.2.2 Ultimate Analysis

Ultimate or elementary analysis is useful when determining the elemental composition of a fuel. It quantifies the percentage by weight of carbon, hydrogen, nitrogen, sulphur, oxygen, ash, and other minor components as shown in Equation (2.2).

$$C_{daf} + H_{daf} + N_{daf} + S_{daf} + O_{daf} = 100 \tag{2.2}$$

where,

C_{daf}: carbon content dry ash free,
H_{daf}: hydrogen content dry ash free,
N_{daf}: nitrogen content dry ash free,
S_{daf}: sulphur content dry ash free, and
O_{daf}: sulphur content dry ash free.

The common name given to the equipment used to obtain the elementary composition is CHNS-O analyser. Each manufacturer tends to propose their own set of experimental procedures, though they need to adhere to certain standards such as the ASTM D3176. Coal tends to have a high carbon-to-hydrogen ratio, which tends to ensure high energy density. Biomass generally has negligible sulphur, making it environmentally advantageous compared to coal. Oxygen content influences reactivity, as high oxygen content increases volatile release, promotes lower ignition temperatures, and reduces the net heating value since part of the fuel is already oxidised. Ultimate analysis is critical for stoichiometric calculations, as knowing the carbon content allows engineers to calculate CO_2 emissions, hydrogen informs H_2O, and sulphur informs SO_2. CFD models often combine ultimate and proximate analysis data as modelling inputs.

The composition characteristics of biomass and coal are not similar; hence, they portray different combustion behaviours. Marangwanda et al. [11] performed proximate and ultimate analysis on coal and sawdust fuel samples, as well as their blends. Table 2.1, Figure 2.2, and Figure 2.3 summarise the respective properties of 100%

TABLE 2.1
Chemical Properties of Coal, Sawdust, and Fuel Blends

	Proximate Analysis* (weight %)				Ultimate Analysis* (weight%)				
	**Fixed Carbon	Volatile Matter	Ash	HHV (MJ/kg)	C	H	**O	N	S
100HC	53.97	23.10	22.93	22.51	58.671	2.946	13.245	1.613	0.593
90HC 10PS	48.21	29.91	21.88	21.54	58.489	3.367	14.246	1.448	0.570
80HC 20PS	46.35	31.82	21.83	21.18	56.954	3.734	15.632	1.296	0.555
70HC 30PS	46.02	33.74	20.24	21.37	56.885	4.093	17.100	1.143	0.535
100PS	15.62	80.68	3.70	18.07	49.504	6.035	40.404	0.358	0

*On a dry basis, **By difference.

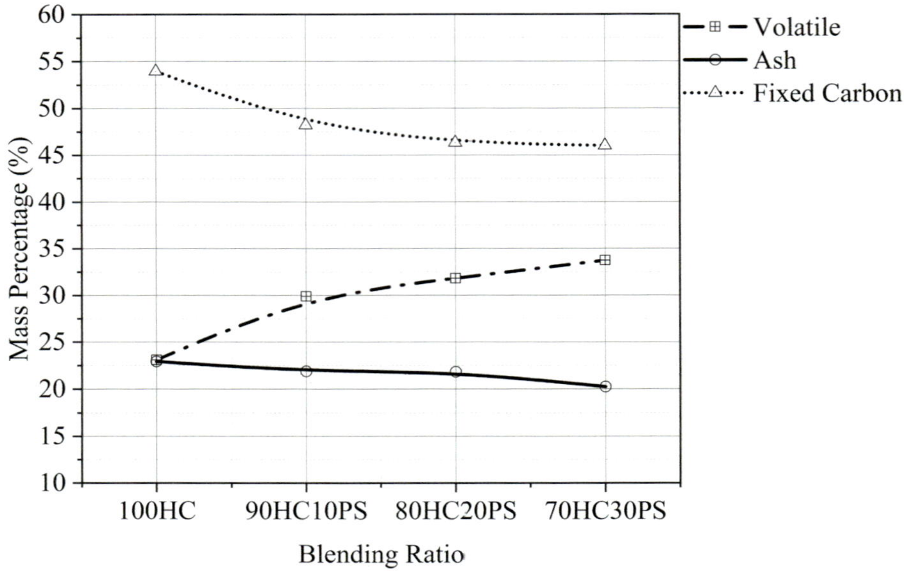

FIGURE 2.2 Effect of blending coal with biomass on proximate analysis.

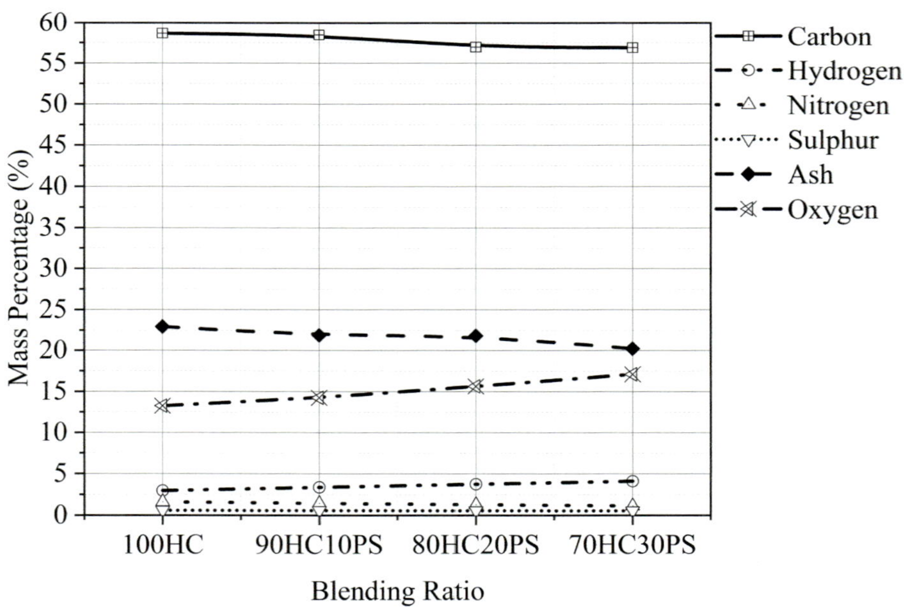

FIGURE 2.3 Effect of blending coal with biomass on ultimate analysis.

coal (100HC), 90% coal and 10% biomass (90HC10PS), 80% coal and 20% biomass (80HC20PS), 70% coal and 30% biomass (70HC30PS), and 100% biomass (100PS).

The fixed carbon was higher for coal than biomass, and it decreased as blending increased. There was no linear variation between the volatile content, ash, and fixed carbon with the blending ratio. Substances with high volatile content have a higher ignition tendency, whilst those with high fixed carbon have better combustion stability [12]. Mishra and Mohanty [13] also highlighted how reduced ash content in the fuel blends results in better combustion properties and reduced fuel preparation costs. Wnorowska et al. [12] further supported the notion by demonstrating how combustion stability decreases as the biomass blending ratio increases due to the increased volatile matter.

The composition of carbon showed an increase with blending from 0.292 up to 0.579 for a 30% blending ratio, as summarised in Table 2.2. Similar trends were also experienced with hydrogen and oxygen, though to a lesser magnitude. Nitrogen decreased with blending from 0.086 to 0.052 as blending increased, whilst sulphur content decreased marginally with blending. This is supported by other authors indirectly through the assertion that biomass contains higher volatile matter than coal. This translates to a gradual decrease in energy content since the H/C ratio of blended volatiles is lower than that of the unblended sample [14]. Maisyarah et al. [15] also suggested that a lower H/C ratio tended to result in a higher mass loss for the overall sample at the end of the combustion process. As such, a higher mass loss was expected for the blended fuel samples. Using Equations (2.1), (2.2), and (2.3), the lumped volatile composition of each fuel blend is determined as shown in Table 2.2 to give a representative $C_aH_bO_cN_dS_e$. The volatile compositions of the blends were determined and presented in Table 2.3.

$$\left(\frac{VM_{ar}}{VM_{ar}+FC_{ar}}\right)+\left(\frac{FC_{ar}}{VM_{ar}+FC_{ar}}\right)=VM_{daf}+FC_{daf}=100 \tag{2.3}$$

TABLE 2.2
Volatile Compositions of Fuel Blends

	Molar Composition	Volatile Molar Mass
100HC	$C_{0.292}H_{2.200}O_{0.618}N_{0.086}S_{0.014}$	17.24
90HC10PS	$C_{0.535}H_{2.102}O_{0.556}N_{0.0646}S_{0.0111}$	18.68
80HC20PS	$C_{0.559}H_{2.36}O_{0.618}N_{0.0585}S_{0.011}$	20.11
70HC30PS	$C_{0.579}H_{2.615}O_{0.683}N_{0.052}S_{0.011}$	21.55
100PS	$C_{1.107}H_{2.37}O_{0.99}N_{0.01}$	31.63

TABLE 2.3
Volatile Composition Determination

	%Weight	%Mol
C	$\frac{C_{daf} - FC_{daf}}{VM_{daf}}$	$\frac{1}{12} * \left[\frac{C_{daf} - FC_{daf}}{VM_{daf}}\right] * Mr_{volatile} = a$
H	$\frac{H_{daf}}{VM_{daf}}$	$\frac{1}{1} * \left[\frac{H_{daf}}{VM_{daf}}\right] * Mr_{volatile} = b$
O	$\frac{O_{daf}}{VM_{daf}}$	$\frac{1}{16} * \left[\frac{O_{daf}}{VM_{daf}}\right] * Mr_{volatile} = c$
N	$\frac{N_{daf}}{VM_{daf}}$	$\frac{1}{14} * \left[\frac{N_{daf}}{VM_{daf}}\right] * Mr_{volatile} = d$
S	$\frac{S_{daf}}{VM_{daf}}$	$\frac{1}{32} * \left[\frac{S_{daf}}{VM_{daf}}\right] * Mr_{volatile} = e$

2.2.3 Thermal Characterisation

Thermal characterisation of fuels is indispensable for every combustion study. Combustion is defined as the rapid oxidation of a fuel producing heat in the process. Each fuel has its own unique characteristics with regard to thermal characteristics. The calorific value, also called heating value, represents the total energy released when a fuel undergoes complete combustion. The bomb calorimetry instrument uses the ASTM D5865 or ISO 1928:2009 standard to determine solid fuel heating values, whilst the ASTM E1269 standard can be used when a differential scanning calorimeter is used [16, 17]. When using a bomb calorimeter, a known mass of fuel is combusted in a sealed vessel surrounded by water, and the resulting temperature rise is measured.

Since combustion of hydrocarbons results in the formation of water vapour, some of the heat produced during the combustion process is absorbed as latent heat of vapourisation. If the latent heat of vapourisation is included in the final evaluation of heating value, the parameter is reported as higher heating value (HHV) or higher calorific value (HCV). If the latent heat of vapourisation is neglected, which generally represents the practically usable energy, the parameter is reported as lower heating value (LHV) or lower calorific value (LCV). The higher heating values of coal, biomass, and their blends are reported in Table 2.1. The HHV decreased with blending as expected, showing the role of moisture in decreasing the practical heating value as well.

FIGURE 2.4 Pulverised Hwange coal and Pinus sawdust samples.

2.2.4 Particle Size Distribution and Shape

Solid fuel combustion is fundamentally a surface reaction process, which means the more the available surface area, the more a fuel particle can rapidly interact with oxygen. To perform particle size analysis, sieves based on the ISO 3310-1 standard are used to perform sieve analysis, which in turn can be used to classify fuel particles based on diameter. Usually, pulverised coal used in power plants has a mean diameter of 100 μm [18]. It however requires more effort to achieve such a diameter with regard to biomass. Coal particles can be simply taken as spherical, whilst biomass particles are better regarded as cylindrical. As such, biomass particles can use a shape factor of 0.83. Marangwanda et al. [11] analysed the particle sizes of coal and biomass shown in Figure 2.4 using ASTM sieves with sizes ranging from 25 μm to 250 μm.

After carrying out a sieve analysis, the most common distribution used to describe the diameter of solid fuel particles is the Rosin Rammler distribution. Equation (2.4) is used to trace the particle diameter distribution as summarised in Table 2.4 and plotted in Figure 2.5 for Hwange coal and Pinus sawdust. Y_d represents the mass

TABLE 2.4
Rosin Rammler Particle Distribution Parameters

	Hwange Coal	Pinus Sawdust
Mean diameter (μm)	189.97	202.89
Spread parameter	3.714	5.754

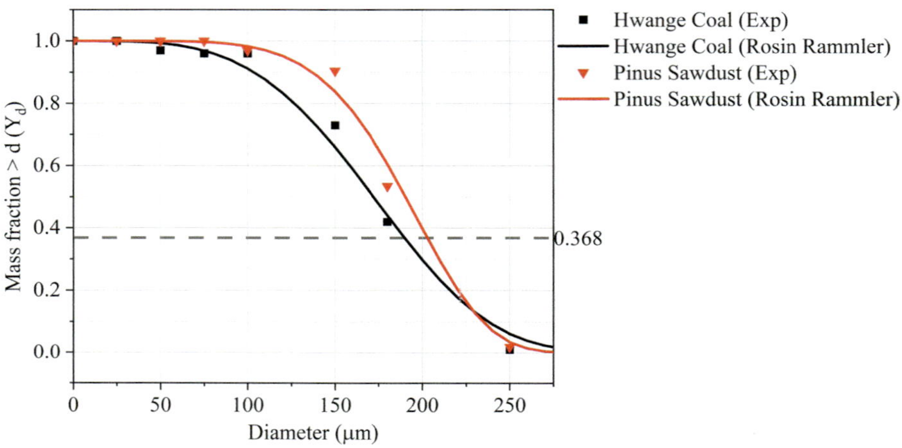

FIGURE 2.5 Particle distribution of the Hwange coal and Pinus sawdust.

fraction of particles with a diameter greater than d_p. The mean particle diameter is represented as $d_{p,mean}$. The spread parameter is represented with an s.

$$Y_d = \exp\left[-\left(\frac{d_p}{d_{p,mean}}\right)^s\right] \tag{2.4}$$

2.2.5 Particle Porosity

The surface area of a solid fuel particle available for combustion is dependent on particle porosity. Porosity is a measure of the fuel particle's volume that is occupied by voids or pores. The higher the porosity, the higher the diffusion of oxygen into the particle and the escape of volatiles during combustion. The higher the ranking of coal, the lesser the void fraction (lower porosity), which makes gas diffusion slower. This leads to longer burnout times and higher activation energies for char heterogeneous combustion [19]. Biomass tends to have a higher porosity, which in turn allows oxygen and volatiles to rapidly diffuse, hence a better ignition index [20, 21]. Of particular interest to a combustion engineer is the change in structure of a fuel particle as it heats up. Coal softens during the devolatilisation stage that increases the pores, which may enhance combustion but also cause irregular burning patterns [22].

Considering the difficulty in measuring porosity during the combustion process, CFD allows engineers to monitor the progression of particle porosity during the combustion process. Wardach-Święcicka and Kardaś [23] showed that modelling solid fuel particles without considering the porosity reduces the model's accuracy significantly. Hosseini Rahdar [24] explained why it becomes important to consider shrinkage factors as a way of accounting for the porosity change during fixed-bed combustion. Shrinkage factors, when modelled correctly, can account for the change in particle diameter as the solid fuel particles deplete during the combustion process. Typical instruments used for porosity measurement of coal are small angle x-ray scattering (SAXS), nuclear magnetic resonance (NMR), scanning electron microscopy (SEM), and mercury intrusion porosimetry (MIP) machines. SAXS and NMR are preferred when a non-destructive analysis needs to be carried out. SEM is good at obtaining high-resolution images with precise measurements of up to 15 nm. Unfortunately, SEM only works with small samples; hence, a lot of statistical work is required to determine the bulk porosity of large sample sizes. It is imperative that appropriate experimental measurements are conducted to determine the void fraction of solid fuel particles [25].

2.3 CONSERVATION LAWS WITH RESPECT TO COMBUSTION

Since combustion entails thermodynamic, fluid mechanics, mass transfer, and chemical kinetics principles, conservation laws are typically applicable when studying combustion. Of importance are conservation of mass or species during combustion reactions and conservation of energy. Conservation of mass or species during combustion reactions allows the engineer to accurately monitor the air-to-fuel ratio and the species involved in the chemical reactions. This means that the total mass of each species is conserved before and after the reaction. The conservation of energy allows the engineer to accurately monitor the heat evolved during the combustion process. This means the total energy stored within the reactants must equate to the total energy liberated during combustion plus the energy stored within the products. Combustion is highly exothermic and self-accelerating; hence, it is widely used in the heating industry, automobiles, power plants, and aviation. The study of chemical reactions forms an important pillar on which combustion depends. During combustion, fuel chemical bonds break and rearrange, resulting in the production of heat energy as represented by a simple Equation (2.5) [14, 29].

$$\text{Fuel} + \text{Oxidant} \rightarrow \text{Products} \tag{2.5}$$

2.3.1 Mass and Species/Stoichiometry

When just the right amount of oxidant (usually air) is used to react with the fuel, it results in a stoichiometric air to fuel ratio $(A/F)_S$ which usually does not guarantee complete combustion in practice. As a result, an actual air-to-fuel ratio $(A/F)_A$ is employed by most thermal applications which assures good intimacy between the oxidant and fuel molecules. A comparison between the stoichiometric and actual fuel ratios is referred to as the equivalence ratio (ϕ) as represented in Equation (2.6).

$$\text{Equivalence Ratio}(\phi) = \frac{(A/F)_S}{(A/F)_A} \tag{2.6}$$

$$\phi > 1\text{:rich mixture}; \phi < 1\text{:lean mixture}$$

An example of the stoichiometric equation for the combustion of 100HC volatile species with air is represented in Equation (2.7).

$$C_{0.292}H_{2.2}O_{0.618}N_{0.086}S_{0.014} + 1.389(O_2 + 3.76N_2) \rightarrow 0.292CO_2 + 1.1H_2O + 5.149N_2 + 0.014SO_2 \tag{2.7}$$

2.3.2 Energy

The conservation of energy is based on the first law of thermodynamics, which stipulates that for a system, the overall heat transfer and work done by the system are equivalent to the change in enthalpy, potential energy, and kinetic energy, as given by Equation (2.8) [29].

$$\Delta\left(h + g.z + \frac{1}{2}c^2\right) = q + w \tag{2.8}$$

where:

- h is specific enthalpy (J/kg),
- z is altitude (m),
- g is gravitational acceleration (m/s^2),
- c is velocity (m/s),
- q is specific heat transfer (J/kg), and
- w is specific work done (J/kg).

To analyse combustion chemical reactions, specific absolute enthalpy is defined in full as given by Equation (2.9) [29].

$$\underbrace{h}_{\text{absolute enthalpy}} = \underbrace{h^0_{f,Tr}}_{\text{formation enthalpy}} + \underbrace{\int_{T_r}^{T} c_p dT}_{\text{sensible enthalpy}} \tag{2.9}$$

where:

- h_f is specific enthalpy formation (J/kg),
- c_p is specific heat capacity at constant pressure for the species at a temperature T, and
- T_r is reference temperature (K).

At any given temperature, each species has an absolute enthalpy, which is useful in determining the overall heat transfer and thus the temperature profiles, using the heat of reaction as given by Equation (2.10) [29].

$$H_{reaction} = \sum_{products} n_i h_i - \sum_{reactants} n_i h_i \tag{2.10}$$

where:

h_i is absolute specific enthalpy of species i and
n_i is the number of moles of species i.

When the products absorb all the heat released during a chemical reaction, the adiabatic flame temperature is attained. To determine the enthalpy of formation for volatile ($H_{f,vol}$), Equation (2.11) was employed after rearranging Equations (2.12) and (2.13).

$$H_{f,fuel} = FC_{daf} * H_{f,char} + VM_{daf} * H_{f,vol} \tag{2.11}$$

$$H_{f,char} : C + O_2 = CO_2 \left(-32.8 MJ / kg\right); H_{f,char} = LCV_{coal} \tag{2.12}$$

$$H_{f,char} = LCV_{coal} = \frac{HCV_{coal} - h_{latent,H20} * MC_{ar}}{1 - MC_{ar} - Ash_{ar}} - \frac{H_{ar} * Mr\left(H_2O\right)}{2 * Mr\left(H\right)} * h_{latent,H20} \tag{2.13}$$

This approach provided a more complete method capable of capturing the fuel properties rather than assuming the fuel sample is composed of methane or butane, as assumed by several authors [26, 27]. The volatile compositions of the blends were determined and presented in Table 2.5.

The molar ratio of carbon showed an increase with blending from 0.292 up to 0.579 for a 30% blending ratio as summarised in Table 2.5. Similar trends were also experienced with Hydrogen and Oxygen, though to a lesser magnitude. Nitrogen in volatile decreased with blending from 0.086 to 0.052 as blending increased, whilst

TABLE 2.5
Volatile Compositions of Fuel Blends

	Molar Composition	Volatile Molar Mass	Enthalpy of Formation for Volatile ($H_{f,vol}$)
100HC	$C_{0.292}H_{2.200}O_{0.618}N_{0.086}S_{0.014}$	17.24	–5.904E+07
90HC 10PS	$C_{0.535}H_{2.102}O_{0.556}N_{0.0646}S_{0.0111}$	18.68	–7.862E+07
80HC 20PS	$C_{0.559}H_{2.36}O_{0.618}N_{0.0585}S_{0.011}$	20.11	–9.819E+07
70HC 30PS	$C_{0.579}H_{2.615}O_{0.683}N_{0.052}S_{0.011}$	21.55	–1.178E+08
100PS	$C_{1.107}H_{2.37}O_{0.99}N_{0.01}$	31.63	–2.548E+08

sulphur content decreased marginally with blending. This is supported by other authors indirectly through the assertion that biomass contains higher volatile matter than coal. This translates to a gradual decrease in energy content since the H/C ratio of blended volatiles is lower than that of the unblended sample [14]. Maisyarah et al. [15] also suggested that a lower H/C ratio tended to result in a higher mass loss for the overall sample at the end of the combustion process. As such, a higher mass loss was expected for the blended fuel samples.

2.4 CHEMICAL KINETICS OF COMBUSTION

Since combustion is mainly dependent on chemical reactions, it is important to note that a series of elementary reactions lead to the formation of final combustion products. These elementary reactions or reaction mechanisms are coupled and thus depend on each other to the extent that the slowest of these reactions is the limiting elementary reaction. Equations (2.14) and (2.15) represent the chemical kinetics equations as applied to combustion [29].

$$A + B \xrightarrow{k} C + D \tag{2.14}$$

$$\left(\frac{d[A]}{dt}\right) = -k[A][B] \tag{2.15}$$

The rate of consumption of species A is proportional to the product of concentrations of A and B. A constant of proportionality k is introduced, which is the kinetic rate. Determination of the rate constant of different elementary reactions and overall global reactions forms the basis of reaction kinetics. The Arrhenius law, as given in Equation (2.16), relates the absolute temperature to the kinetic rate [28]. The pre-exponential factor or frequency factor is related to the probability of collision between the reacting molecules.

$$k = A.\exp\left(-\frac{E}{\Re T}\right) \tag{2.16}$$

where k is the kinetic rate, A is the pre-exponential factor, E is the activation energy (J), $\Re$ is the universal gas constant (J/mol·K), and T is the absolute temperature (K).

2.5 MECHANISMS OF SOLID FUEL COMBUSTION

When a solid fuel particle is introduced into a furnace, it first goes through drying, which is followed by devolatilisation/pyrolysis/gasification, then homogeneous combustion which occurs concurrently with heterogeneous char combustion, and finally ash formation. These steps tend to mimic the proximate analysis steps as shown in Figure 2.6.

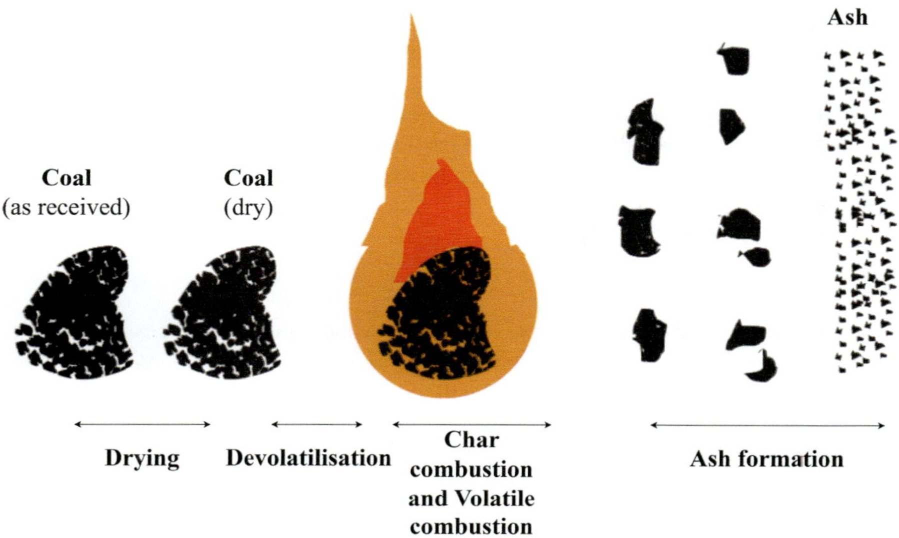

FIGURE 2.6 Schematic of solid fuel combustion.

2.5.1 Drying

During drying, moisture is released as the particle heats up to around 100°C. The process is highly endothermic, hence it can influence the heating value of fuel samples. By inference, wet samples tend to have lower heating values. The typical energy evolved during drying can be measured using a bomb calorimeter. This energy is indirectly quantified by measuring the mass of moisture loss for the sample as represented by Equation (2.17) and using that mass to evaluate the heat used to remove the moisture from the sample as represented by Equation (2.18) [29].

$$m_{evap} = m_{init} - m_{dry} \tag{2.17}$$

$$Q_{dry} = m_{evap} * h_{fg} \tag{2.18}$$

where:

m_{evap} represents the amount of moisture in the sample,
m_{init} is initial sample mass,
m_{dry} is final sample mass after drying, and
h_{fg} represents the latent heat of vapourisation.

On average, anthracite has a moisture content ranging between 2.0% and 4.5%, whilst bituminous and sub-bituminous coals have ranges between 1.0 to 12.2% and 14.1 to 31.0%, respectively, depending on a plethora of factors. Lignite and brown coal due to their porous nature can have moisture content between 25 and 35% or even greater [30]. During drying, as moisture evaporates, physical changes are also experienced by the particle, such as shrinking, reduction in volume, mainly a function of the initial moisture within the particle. During drying, if the moisture within the pores of the particle is not rapidly allowed to migrate from the particle interior to the surface, pressure builds up within the particle which can lead to particle breakup. This particle breakup is mainly influenced by heating rate with pore characteristics also playing a part [31, 32].

2.5.2 Devolatilisation

Continued heating of the fuel sample results in devolatilisation, a process that results in the release of volatiles from the sample leaving behind solid char. Volatiles comprise a mixture of many gases such as methane, light hydrocarbon volatiles, medium hydrocarbon volatiles, heavy hydrocarbon volatiles, and inorganic volatiles [14]. The term pyrolysis is also used in place of devolatilisation in certain applications, whilst others loosely use the term gasification that should be technically reserved for processes where partial oxidation takes place. Technically, devolatilisation progresses from outside of the particle towards the inside [29].

The onset of devolatilisation occurs around 450°C for coal particles and 350°C for biomass particles [33]. As demonstrated in Figure 2.7, the onset of devolatilisation is heavily dependent on the particle heating rate, as high heating rates result

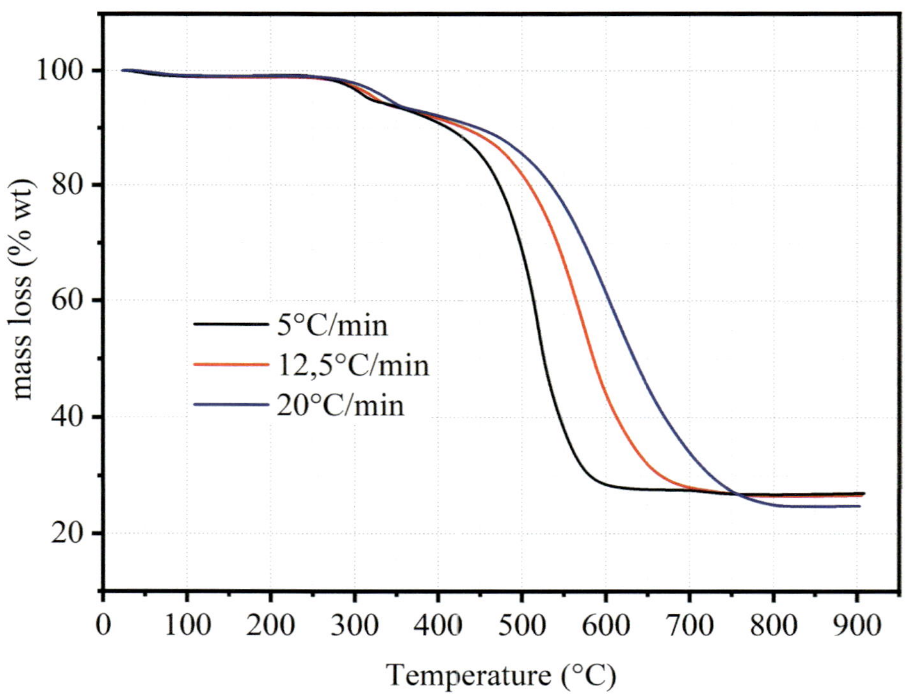

FIGURE 2.7 TGA diagram showing the effect of heating rate on Hwange coal.

in a lag of the devolatilisation process [11]. Sami et al. [34] demonstrated that the devolatilisation of bituminous coal commences at ranges of 430°C for a heating rate of 10°C/s, whilst at a heating rate of 10,000°C/s it commences at temperature ranges of 1,200°C. The initial stages of devolatilisation result in the rapid release of about 80% of the volatiles followed by a slow release of the remaining volatiles [29].

The volatiles that are released initially are mainly made of tar, aliphatics, carbon dioxide, and water, whilst those released later are made of carbon monoxide, hydrogen, HCN, and benzene [30]. The initial volatiles are released because there is a disruption of hydrogen bonds, hence there is vapourisation of these compounds. Only the non-covalently bonded molecules are transported out of the coal particle. As the temperature increases, the bonded molecules are released as well through depolymerisation. Depolymerisation is a process where weak chemical bonds break up to become smaller molecular chains or fragments. When these smaller molecular chains are small enough, they vapourise and are transported outside the coal particle as "volatiles", leaving a solid carbonaceous residue called char. Coal rank also affects the volatile yield and composition as low-rank coals contain more oxygen-containing functional groups that promote the production of water, carbon dioxide, and methyl-based compounds [35]. Bituminous coals generally yield large amounts of tar, whilst low-rank coals produce large amounts of light volatiles. When analysed, light volatiles may contain species such as methane, ethene, ethane, carbon monoxide, hydrogen, and water vapour.

Another parameter of interest that influences volatile yields and composition is maximum temperature and residence time. This is the temperature and time a sample is held at the maximum temperature during the devolatilisation process. Volatile yield increases with higher final temperatures and longer residence times, as elevated temperatures accelerate reaction fronts and promote tar fragmentation, while extended soak periods allow more volatiles to diffuse from coal particles [36].

2.5.3 Homogeneous Gas Combustion

Once devolatilisation is complete, the volatiles undergo homogeneous gas combustion, whilst the char undergoes heterogeneous solid to gas combustion. The volatiles that are released are generally responsible for the flame propagation, especially if sufficient oxidant is available. Since volatiles form a film or cloud around the particle, all the oxygen is consumed by the volatiles through rapid homogeneous combustion before reaching the particle surface. At the initial stage of devolatilisation, the homogeneous combustion occurs relatively further from the particle surface. As the devolatilisation slows down, the flame front moves closer to the coal particle. As soon as relatively all the volatiles have evolved, the flame becomes attached to the char surface resulting in the heterogeneous char combustion. This simplified process is illustrated in Figure 2.8.

The homogeneous gas-phase combustion can progress using a one-step, two-step, or multiple-step reaction mechanism. The common mechanisms used by researchers are the Westbrook and Dryer [37] two-step mechanism, Jones and Lindstedt [38] four-step mechanism. The simplest of these mechanisms is the global one-step reaction mechanism, which forms a simple equation without any intermediate species.

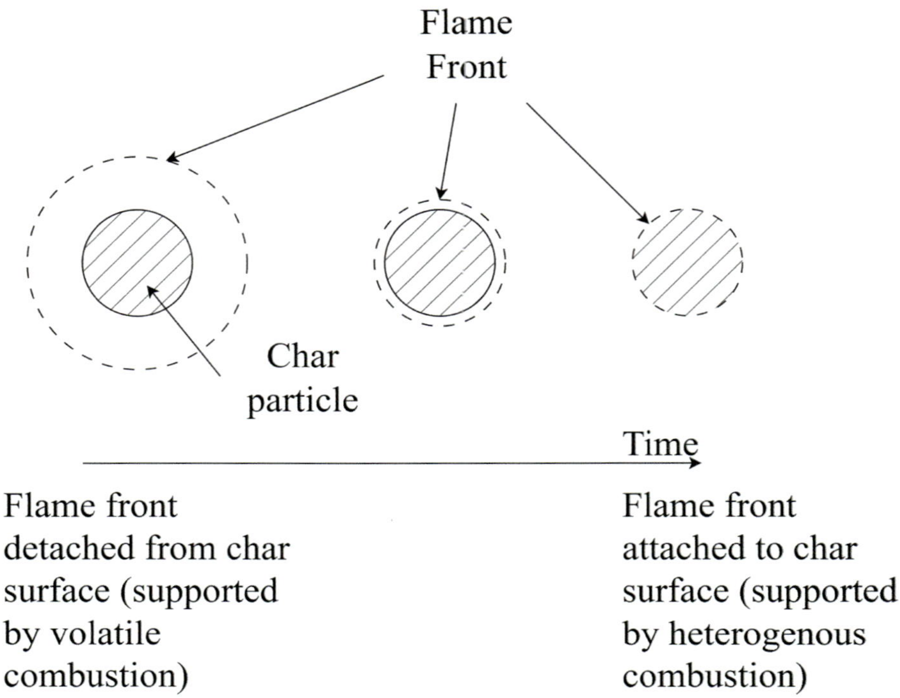

FIGURE 2.8 Flame behaviour during homogeneous volatile combustion.

Using the general lumped form of volatiles, Equation (2.19) can be used to represent the mechanism [39].

$$C_xH_yO_z + (2x + \frac{y}{2} - \frac{z}{2})O_2 \rightarrow xCO_2 + \frac{y}{2}H_2O \tag{2.19}$$

For a more accurate approach, the two-step reaction mechanism is employed, taking carbon monoxide as the intermediate species. Equations (2.20) and (2.21) can then be used to represent the mechanisms involved in the global two-step reaction mechanism [40, 41].

$$C_xH_yO_z + (\frac{x}{2} + \frac{y}{4} - \frac{z}{2})O_2 \rightarrow xCO + \frac{y}{2}H_2O \tag{2.20}$$

$$CO + 0.5O_2 \rightarrow CO_2 \tag{2.21}$$

The four-step reaction mechanism, as given by Equations (2.22) to (2.25), can be called upon to improve on accuracy [40, 41].

$$C_xH_yO_z + (\frac{x}{2} + \frac{y}{4} - \frac{z}{2})O_2 \rightarrow xCO + \frac{y}{2}H_2O \tag{2.22}$$

$$CO + 0.5O_2 \rightarrow CO_2 \tag{2.23}$$

$$CO + H_2O \rightarrow CO_2 + H_2 \tag{2.24}$$

$$H_2 + 0.5O_2 \rightarrow H_2O \tag{2.25}$$

2.5.4 Heterogeneous Char Combustion

Char oxidation is a surface reaction process between the solid char and gaseous oxygen. As with fluid flow principles, a boundary layer is formed between a solid body and gaseous oxidant or products. Various reactions occur within the boundary layer depending on the mechanism that is adopted, as illustrated in Figure 2.9. On a macro scale, for the char and volatiles to interact with the oxidant and subsequent release of products, the following stages can be used to describe this process:

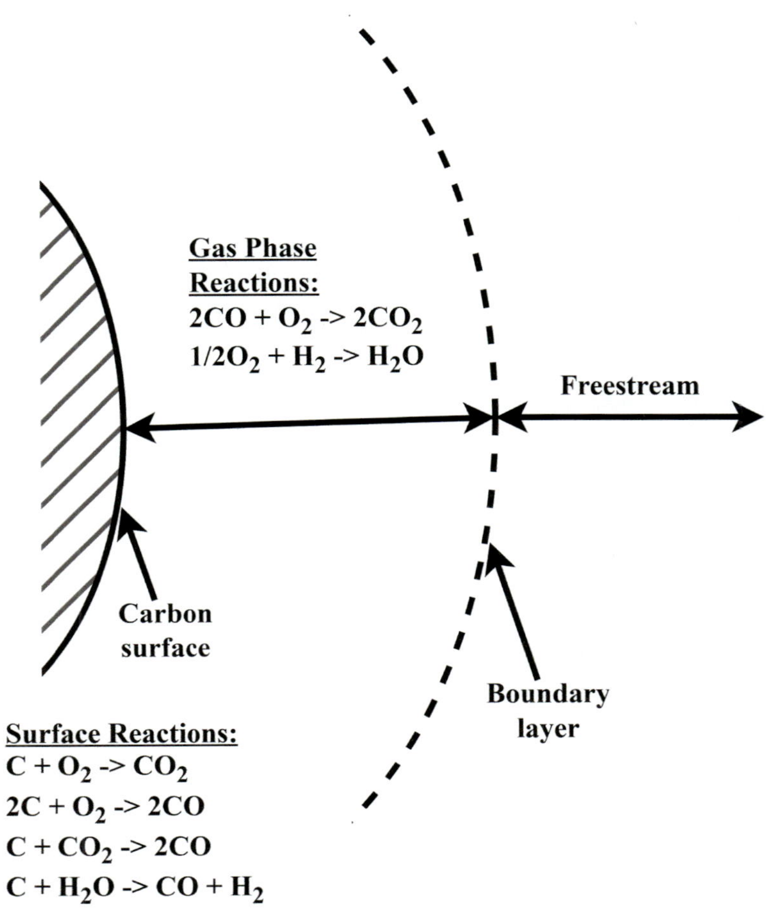

FIGURE 2.9 Heterogeneous chemical reactions that occur within the boundary layer.

- transport of the oxygen molecule by convection or diffusion to the fuel surface,
- adsorption of the oxygen molecule to the fuel surface,
- various elementary reactions take place,
- desorption of product molecules from the fuel surface, and
- transport of the product molecules by convection or diffusion from the fuel surface.

Char combustion progresses at a much slower pace than devolatilisation and homogeneous combustion, which means parameters related to burnout such as the index and time are limited by the heterogeneous char combustion reaction. To a certain extent, heterogeneous char combustion only starts once devolatilisation is virtually complete though this is not set in stone for cases where occasional jets of volatiles are released as in such cases devolatilisation and homogeneous combustion can proceed in parallel [29].

According to the one film model, the solid fuel surface is where the flame is at its strongest. The maximum temperature in the one-film model happens at the char surface, and there is no flame in the gas phase as illustrated in Figure 2.10. In the two-film model, the CO generated at the surface reacts with entering O_2 at a flame sheet that is located a certain distance from the surface as illustrated in Figure 2.11. In the continuous film model, the flame zone is dispersed throughout the boundary layer.

The structure of the char particle has a tremendous effect on the progression of heterogeneous char combustion. As explained previously, char is a porous substance; hence, highly porous char is more reactive, whilst mineral ash deposits tend to block pores and retard char reactions [19, 20, 50]. The pores within a char particle are randomly oriented, with small pores acting as a surface for char reactivity and large pores acting as channels for reactant and product transport [30].

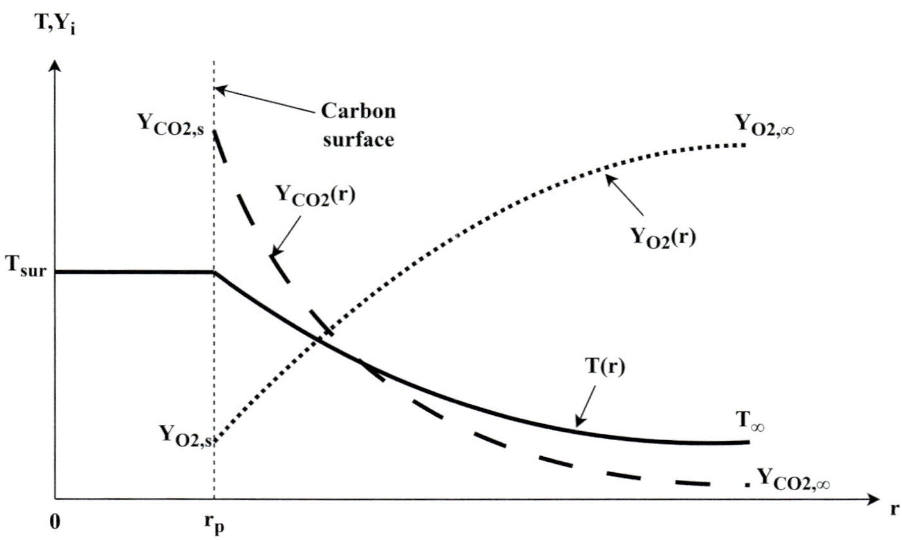

FIGURE 2.10 One-film model representative of heterogeneous char combustion.

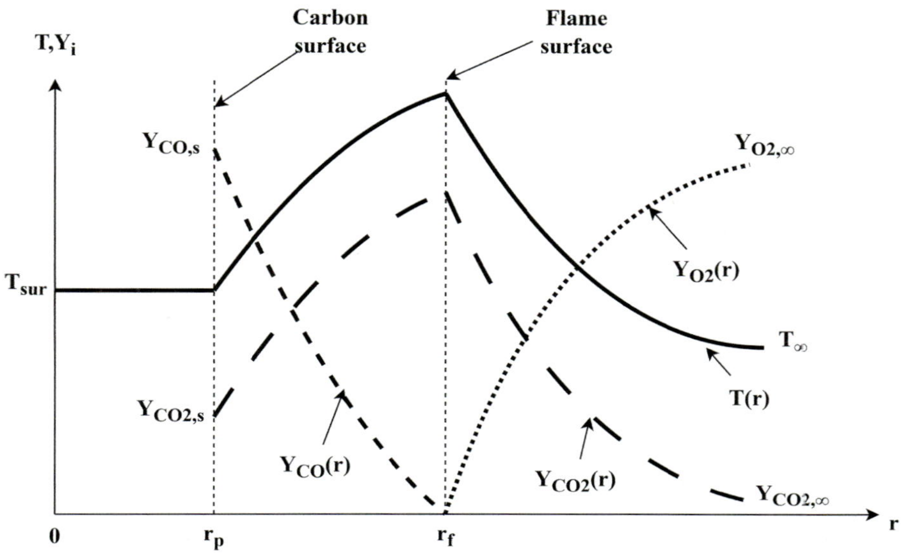

FIGURE 2.11 Two-film models representative of heterogeneous char combustion.

The classification of char is quite subjective as it depends on its application and intended use. The main properties that are of interest during combustion are porosity, shape, swelling ratio, and residual mass ratio shown in Table 2.6. These parameters allow us to describe the actual char properties exhibited during the heterogeneous char combustion stage. Group A char is highly porous, whilst Group C chars are typically dense. During combustion, there are chars that exhibit no visible changes; then, there are those that exhibit some particle fusion but do not swell. These particles are generally dense in nature. The other types of char swell to become an opaque sphere with many visible bubbles, whilst others swell to become transparent with just a few bubbles. When the bubbles rupture, they result in irregularly shaped chars as well [43–45].

2.5.5 Ash Formation

Ash formation is the last stage at which non-combustible constituents such as aluminium, iron, potassium, silicon, and other elements react with oxygen or amongst

TABLE 2.6
Classification of Char Obtained from Coal [43, 44]

Char Groups	Group A	Group B	Group C
Porosity, %	>70	40–70	<40
Average wall thickness, μm	<5	>5	>5
Shape	Spherical subspherical	Subspherical	Angular
Typical swelling ratio	>1.3	<1.0	<0.9
Typical residual mass ratio	0.1–0.5	0.1–0.5	1.0

TABLE 2.7
Slagging and Fouling Characteristics of Typical Combustion Applications

Fuel	Furnace or Testing Conditions	Slagging, Fouling Index, and Ash Characteristics	Reference
Coal, wheat straw, and petroleum coke	Thermogravimetric analyser (TGA), 30°C to 700°C	• Coal and coke had high SiO_2, Al_2O_3, Fe_2O_3 • Blending with biomass reduced total ash content, improved ash fusibility, lowered risk of deposit formation	Kumar and Nandi [48]
Coal and biomass (wood, bark, straw)	Pulverised fuel, fluidised bed, and grate-fired boilers	• Biomass ash is rich in alkali metals (K_2O, Na_2O) and low in SiO_2 and Al_2O_3	Karampinis et al. [49]
Coal and biomass (torrefied palm kernel shell)	540-MW pulverised coal boiler	• Biomass blending increased the flue gas temperature and deposition rate • Ash with higher K_2O and Cl content • Adding an aluminosilicate neutralises alkalis and reduces chlorine-induced corrosion	Pronobis et al. [46]

themselves to form solid residues. At high furnace temperatures, some of these mineral components melt and form slag. Ash represents the fixed inorganic fraction left after carbon is burnt [46]. Coal contains a variety of mineral matter such as silicates, carbonates, sulphites, and oxides existing in both crystalline and amorphous forms. During the process of combustion, minerals undergo processes such as coalescence, fragmentation, fusion, vapourisation, and condensation, producing ash particles of diverse size and composition depending on their temperature and residence time [47]. The slagging index (SI) can be used to quantify the effect of ash on a combustion system, as represented in Equation (2.26).

Table 2.7 summarises a few applications with regard to ash formation in industrial furnaces and boilers.

$$SI = \frac{Fe_2O_3 + CaO + MgO}{SiO_2 + Al_2O_3} \tag{2.26}$$

2.6 HEAT TRANSFER DURING SOLID FUEL COMBUSTION

To understand the process of heat transfer during solid fuel combustion, we typically use spherical coordinates. During drying and devolatilisation, which are endothermic processes, heat is transferred from the furnace environment to the particle surface through convection and radiation, then through conduction within the fuel particle, as illustrated in Figure 2.12.

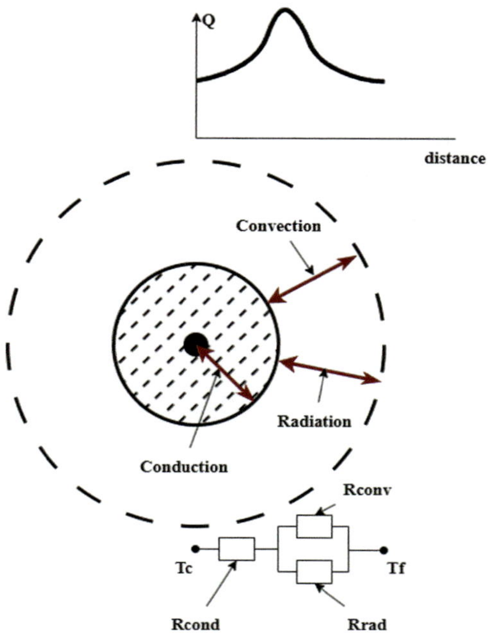

FIGURE 2.12 Particle heating process based on spherical coordinates.

The transfer of heat within the fuel particle is represented by the conduction Equation (2.27) [50]:

$$q = k_c A_p \left[\frac{T_{sur} - T_c}{r_p} \right] \quad (2.27)$$

where:

- q represents the specific heat transfer,
- k_c represents the thermal conductivity,
- A_p represents the surface area of the fuel particle,
- T_{sur} represents the temperature at the particle surface,
- T_c represents the temperature at the centre of the particle, and
- r_p represents the radius of the particle.

During drying, T_{sur} is assumed to be within the range of 180°C and as the cooling effect of steam continues to decrease, this surface temperature continues to increase as well. The overall heat balance experienced by a fuel particle during combustion can be represented using Equation (2.28) [50]:

$$m_p c_p \frac{dT_p}{dt} = h_{conv} A_p \left(T_f - T_p \right) + \epsilon \sigma A_p \left(T_f^4 - T_p^4 \right) + \left(Q_{vap} + Q_{vol} + Q_c \right) \quad (2.28)$$

where:

h_{conv} represents the heat transfer coefficient,
ϵ represents the emissivity,
σ represents the absorptivity,
c_p represents the specific heat capacity at constant pressure,
T_f represents the gas temperature,
T_p represents the particle temperature,
Q_{vap} represents the heat of vapourisation,
Q_{vol} represents the heat of volatilisation, and
Q_c represents the heat of combustion.

2.7 MASS TRANSFER DURING SOLID FUEL COMBUSTION

The movement of species during solid combustion plays a pivotal role in the understanding of how to model combustion processes. Mass transfer regulates the rate at which reactants enter the fuel and products exit the combustion zone [30]. During drying, the moisture must be transferred from within the particle on to the furnace which relates to the diffusion model [8, 51]. This allows the droplet evaporation rate equation to be applied when modelling drying. Salomatov et al. [52] showed experimentally that the moisture that diffuses from within the particle goes to the surface, forming a liquid film around the particle, which is technically a surface layer of water with the same diameter. This moisture film has a tremendous effect on ignition delay. Different species participating in the combustion process have varying diffusivities (requiring multiple Stefan-Maxwell equations to relate these fluxes) which makes the process a multi-component diffusion problem. Heterogeneous char combustion is generally diffusion limited as it depends on the transport of oxidant to the particle surface and products from the particle [29].

2.8 SUMMARY

In summary, this chapter established the theoretical and practical foundations for understanding coal and biomass co-combustion, integrating experimental characterisation with computational modelling principles. The discussion bridged classical combustion theory with modern CFD applications, showing how physical, chemical, and thermal parameters govern combustion efficiency, emissions, and stability.

- Proximate and ultimate analysis showed that blending sawdust (biomass) with bituminous coal decreases fixed carbon (from 53.97% to 15.62%) and heating value (from 22.51 MJ/kg to 18.07 MJ/kg), whilst increasing volatile matter and oxygen content. This was shown to enhance ignition though at the expense of combustion stability.
- Biomass exhibits higher porosity and larger mean particle diameters, which tend to improve diffusion properties. This allows for better ignition and burnout of biomass particles.

- Devolatilisation initiates at lower temperatures for sawdust (biomass) (~350°C) and for bituminous coal (~450°C), which allows for faster volatile release and better ignition properties.
- Incorporating accurate physical and chemical data into CFD improves the prediction of flame behaviour, temperature profiles, and emission trends; hence better combustion systems can be designed.

REFERENCES

1. International Energy Agency. *World Energy Outlook 2024*; Paris, 2024.
2. International Energy Agency. *World Energy Outlook*; Paris, France, 2017, 1–15, doi:10.1016/0301-4215(73)90024-4.
3. Finance, B.N.E. The Clean Energy Country Competitiveness Index 2016.
4. Zhu, Q. *Coal Sampling and Analysis Standards*; London, 2014.
5. ASTM International Practice for Proximate Analysis of Coal and Coke 2021.
6. International Organization for Standardization Coal and Coke – Proximate Analysis 2024.
7. ASTM International Test Methods for Proximate Analysis of Coal and Coke by Macro Thermogravimetric Analysis 2024.
8. Li, L.; Bai, X.; Qu, C.; Zhou, K.; Sun, Y. Effects of Moisture on the Ignition and Combustion Characteristics of Lignite Particles: Modeling and Experimental Study. *ACS Omega* **2022**, *7*, 34912–34920, doi:10.1021/acsomega.2c03462.
9. Cao, Y.; Xiao, Y.; Wang, Z.P.; Li, Q.W.; Shu, C.M.; Jiang, X.R.; Wu, S.L. Recent Progress and Perspectives on Coal Dust Sources, Transport, Hazards, and Controls in Underground Mines. *Process Safety and Environmental Protection* **2024**, *187*, 159–194, doi:10.1016/j.psep.2024.04.095.
10. ANSYS. *ANSYS FLUENT User's Guide*; *ANSYS,* 2021, 1–170.
11. Marangwanda, G.T.; Madyira, D.M.; Chihobo, C.H. Determination of Cocombustion Kinetic Parameters for Bituminous Coal and Pinus Sawdust Blends. *ACS Omega* **2022**, *7*, 32108–32118, doi:10.1021/acsomega.2c03342.
12. Wnorowska, J.; Ciukaj, S.; Kalisz, S. Thermogravimetric Analysis of Solid Biofuels with Additive under Air Atmosphere. *Energies (Basel)* **2021**, *14*, 2257, doi:10.3390/en14082257.
13. Mishra, R.K.; Mohanty, K. Pyrolysis Kinetics and Thermal Behavior of Waste Sawdust Biomass Using Thermogravimetric Analysis. *Bioresource Technology* **2018**, *251*, 63–74, doi:10.1016/j.biortech.2017.12.029.
14. Marangwanda, G.T.; Madyira, D.M.; Babarinde, T.O. Coal Combustion Models: A Review. *Journal of Physics: Conference Series* **2019**, *1378*, 032070, doi:10.1088/1742-6596/1378/3/032070.
15. Maisyarah, A.; Shiun, J.; Nasir, F.; Hashim, H. Ultimate and Proximate Analysis of Malaysia Pineapple Biomass from MD2 Cultivar for Biofuel Application. *Chemical Engineering Transactions* **2018**, *63*, 127–132, doi:10.3303/CET1863022.
16. ASTM International Test Method for Determining Specific Heat Capacity by Differential Scanning Calorimetry 2024.
17. ASTM International Test Method for Gross Calorific Value of Coal and Coke 2019.
18. Bhuiyan, A.A.; Blicblau, A.S.; Islam, A.K.M.S.; Naser, J. A Review on Thermo-Chemical Characteristics of Coal/Biomass Co-Firing in Industrial Furnace. *Journal of the Energy Institute* **2018**, *91*, 1–18, doi:10.1016/j.joei.2016.10.006.

19. Chen, X.; Shi, X.; Zhang, Y.; Zhang, Y.; Ma, Q. Numerical Simulation Study on Coal Spontaneous Combustion: Effect of Porosity Distribution. *Combustion Science and Technology* **2023**, *195*, 472–493, doi:10.1080/00102202.2021.1961760.
20. Wang, Q.; Liu, K.; Wang, S. Effect of Porosity on Ignition and Burning Behavior of Cellulose Materials. *Fuel* **2022**, *322*, 124158, doi:10.1016/j.fuel.2022.124158.
21. Marangwanda, G.T.; Madyira, D.M.; Ndungu, P.G.; Chihobo, C.H. Combustion Characterisation of Bituminous Coal and Pinus Sawdust Blends by Use of Thermo-Gravimetric Analysis. *Energies (Basel)* **2021**, *14*, 7547, doi:10.3390/en14227547.
22. Marangwanda, G.T.; Madyira, D.M. Evaluating Combustion Ignition, Burnout, Stability, and Intensity of Coal–Biomass Blends Within a Drop Tube Furnace Through Modelling. *Energies* **2025**, *18*, 1322, doi:10.3390/en18061322.
23. Wardach-Święcicka, I.; Kardaś, D. Modeling of Heat and Mass Transfer during Thermal Decomposition of a Single Solid Fuel Particle. *Archives of Thermodynamics* **2013**, *34*, 53–71, doi:10.2478/aoter-2013-0010.
24. Hosseini Rahdar, M.; Nasiri, F.; Lee, B. A Review of Numerical Modeling and Experimental Analysis of Combustion in Moving Grate Biomass Combustors. *Energy & Fuels* **2019**, *33*, 9367–9402, doi:10.1021/acs.energyfuels.9b02073.
25. Okolo, G.N.; Everson, R.C.; Neomagus, H.W.J.P.; Roberts, M.J.; Sakurovs, R. Comparing the Porosity and Surface Areas of Coal as Measured by Gas Adsorption, Mercury Intrusion and SAXS Techniques. *Fuel* **2015**, *141*, 293–304, doi: 10.1016/j.fuel.2014.10.046.
26. Yin, C.; Rosendahl, L.A.; Condra, T.J.; Kær, S.K. Use of Numerical Modeling in Design for Co-Firing Biomass in Wall-Fired Burners. *Chemical Engineering Science* **2004**, *59*, 3281–3292, doi:10.1016/j.ces.2004.04.036.
27. Al-Abbas, A.H.; Naser, J. Effect of Chemical Reaction Mechanisms and NOx Modeling on Air-Fired and Oxy-Fuel Combustion of Lignite in a 100-KW Furnace. *Energy & Fuels* **2012**, *26*, 3329–3348, doi:10.1021/ef300403a.
28. Sankar, G.; Kumar, D.S.; Balasubramanian, K.R. Computational Modeling of Pulverized Coal Fired Boilers – A Review on the Current Position. *Fuel* **2019**, *236*, 643–665, doi:10.1016/j.fuel.2018.08.154.
29. Turns, S.; Haworth, D. *An Introduction to Combustion: Concepts and Applications*; McGraw-Hill series in mechanical engineering; 4th ed.; McGraw-Hill: New York, 2021.
30. Tillman, D. *The Combustion of Solid Fuels and Wastes,*1st ed; Academic Press: London, 2014.
31. Zhang, H.; Cen, K.; Yan, J.; Ni, M. The Fragmentation of Coal Particles during the Coal Combustion in a Fluidized Bed. *Fuel* **2002**, *81*, 1835–1840, doi:10.1016/S0016-2361(02)00111-4.
32. Gao, M.; Xiao, Y.; Miao, Z.; Pel, L.; Wan, K.; He, Q.; Xue, S. Characteristics of Moisture Transfer and Surface Crack Development of a Single Lignite Particle Driven by Humidity Difference. *ACS Omega* **2021**, *6*, 18702–18710, doi:10.1021/acsomega.1c01519.
33. Khasraw, D.; Spooner, S.; Hage, H.; Meijer, K.; Li, Z. Devolatilisation Characteristics of Coal and Biomass with Respect to Temperature and Heating Rate for HIsarna Alternative Ironmaking Process. *Fuel* **2021**, *284*, 119101, doi:10.1016/j.fuel.2020.119101.
34. Sami, M.; Annamalai, K.; Wooldridge, M. Co-Firing of Coal and Biomass Fuel Blends. *Progress in Energy and Combustion Science* **2001**, *27*, 171–214, doi:10.1016/S0360-1285(00)00020-4.
35. Zhangfa, W. *Fundamentals of Pulverised Coal Combustion*; IEA Clean Coal Centre: London, 2005.
36. Xu, W.C.; Tomita, A. The Effects of Temperature and Residence Time on the Secondary Reactions of Volatiles from Coal Pyrolysis. *Fuel Processing Technology* **1989**, *21*, 25–37, doi:10.1016/0378-3820(89)90012-X.

37. Westbrook, C.K.; Dryer, F.L. Simplified Reaction-Mechanisms for the Oxidation of Hydrocarbon Fuels in Flames. *Combustion Science and Technology* **1981**, *27*, 31–43.
38. Jones, W.P.; Lindstedt, R.P. Global Reaction Schemes for Hydrocarbon Combustion. *Combustion and Flame* **1988**, *73*, 233–249, doi:10.1016/0010-2180(88)90021-1.
39. Ma, L.; Jones, J.M.; Pourkashanian, M.; Williams, A. Modelling the Combustion of Pulverized Biomass in an Industrial Combustion Test Furnace. *Fuel* **2007**, *86*, 1959–1965, doi:10.1016/j.fuel.2006.12.019.
40. Yin, C.; Kaer, S.K.; Rosendahl, L.A.; Hvid, S.L. Modelling of Pulverized Coal and Biomass Co-Firing in a 150 KW Swirling-Stabilized Burner and Experimental Validation. *The Proceedings of the International Conference on Power Engineering (ICOPE)* **2009**, *2009.2*, 305–310, doi:10.1299/jsmeicope.2009.2._2-305_.
41. Álvarez, L.; Yin, C.; Riaza, J.; Pevida, C.; Pis, J.J.J.; Rubiera, F. Biomass Co-Firing under Oxy-Fuel Conditions: A Computational Fluid Dynamics Modelling Study and Experimental Validation. *Fuel Processing Technology* **2014**, *120*, 22–33, doi:10.1016/j.fuproc.2013.12.005.
42. Bakkaloğlu, E.; Ceylan, S.; Topcu, Y. Thermokinetics and Synergistic Effect Analysis of Peat-Lignite Coal Co-Pyrolysis. *Energy Sources, Part A: Recovery, Utilization and Environmental Effects* **2024**, *46*, 3124–3135, doi:10.1080/15567036.2024.2310143.
43. Wu, H.; Bryant, G.; Benfell, K.; Wall, T. An Experimental Study on the Effect of System Pressure on Char Structure of an Australian Bituminous Coal. *Energy and Fuels* **2000**, *14*, 282–290, doi:10.1021/EF990066J.
44. Wall, T.F.; Liu, G.S.; Wu, H.W.; Roberts, D.G.; Benfell, K.E.; Gupta, S.; Lucas, J.A.; Harris, D.J. The Effects of Pressure on Coal Reactions during Pulverised Coal Combustion and Gasification. *Progress in Energy and Combustion Science* **2002**, *28*, 405–433, doi:10.1016/S0360-1285(02)00007-2.
45. Benfell, K.E.; Liu, G.S.; Roberts, D.G.; Harris, D.J.; Lucas, J.A.; Bailey, J.G.; Wall, T.F. Modeling Char Combustion: The Influence of Parent Coal Petrography and Pyrolysis Pressure on the Structure and Intrinsic Reactivity of Its Char. *Proceedings of the Combustion Institute* **2000**, *28*, 2233–2241, doi:10.1016/S0082-0784(00)80633-5.
46. Pronobis, M.; Wejkowski, R.; Kalisz, S.; Ciukaj, S. Conversion of a Pulverized Coal Boiler into a Torrefied Biomass Boiler. *Energy* **2023**, *262*, 125442, doi:10.1016/j.energy.2022.125442.
47. Werther, J. Potentials of Biomass Co-Combustion in Coal-Fired Boilers. In *Proceedings of the 20th International Conference on Fluidized Bed Combustion*; Springer Berlin Heidelberg: Berlin, Heidelberg, 2009; pp. 27–42.
48. Kumar, P.; Nandi, B.K. Impact of Wheat Straw and Petroleum Coke Blending on Combustion Behavior of High Ash Coal. *Energy Sources, Part A: Recovery, Utilization, and Environmental Effects* **2023**, *45*, 3125–3137, doi:10.1080/15567036.2023.2193152.
49. Karampinis, E.; Grammelis, P.; Agraniotis, M.; Violidakis, I.; Kakaras, E. Co-firing of Biomass with Coal in Thermal Power Plants: Technology Schemes, Impacts, and Future Perspectives. *WIREs Energy and Environment* **2014**, *3*, 384–399, doi:10.1002/wene.100.
50. Marangwanda, G.T.; Madyira, D.M.; Babarinde, T.O. Combustion Models for Biomass: A Review. *Energy Reports* **2020**, *6*, 664–672, doi:10.1016/j.egyr.2019.11.135.
51. Han, Q.; Cui, C.; Jiang, S.; Deng, C.; Jin, Z. Effect of Water Evaporation on the Inhibition of Spontaneous Coal Combustion. *ACS Omega* **2022**, *7*, 6824–6833, doi:10.1021/acsomega.1C06418.
52. Salomatov, V.; Kuznetsov, G.; Syrodoy, S.; Gutareva, N. Mathematical and Physical Modeling of the Coal–Water Fuel Particle Ignition with a Liquid Film on the Surface. *Energy Reports* **2020**, *6*, 628–643, doi:10.1016/j.egyr.2020.02.006.

3 Fundamentals of Computational Fluid Dynamics Modelling

3.1 INTRODUCTION

Computational fluid dynamics (CFD) is a computer-based tool for modelling the behaviour of systems that involve fluid flow, heat transfer, and other related physical processes. It works by solving the equations of fluid flow (in a special form) over a region of interest, with specified (known) conditions on the boundary of that region [1]. CFD grew brick by brick from mathematical and physics-based fundamentals that allow users to solve algebraic equations, particularly through discretisation of partial differential equations. The development of CFD, as illustrated in Figure 3.1, shows how in the 1920s, pioneers of CFD such as Richardson proposed ideas that could be used to predict weather patterns. This was basically breaking down the atmosphere into influential cells, which is some form of discretisation. Further along the way, in the 1950s, researchers such as Von Neumann proposed numerical methods to predict shock behaviour, a critical parameter in weapons design during that time.

The 1980s experienced the most advancement with regard to CFD development coupled with computational advancement [2]. Powerful algorithms such as the SIMPLE, still utilised in modern CFD platforms, were developed during this time [3]. Commercialisation of CFD began around the mid-1980s, leading to the exponential growth of algorithms and CFD capabilities [1]. Continued improvement in computational power enabled more complex applications, as highlighted in Figure 2.1 in the previous chapter, which shows the research trends over the past decades. At present, CFD trends are pointing towards high performance computers (HPC), graphic processing units (GPU), cloud computing, and artificial intelligence since more data is currently being handled with a need for accuracy.

Advances in computing power, together with powerful graphics and interactive 3D manipulation of models, have made the process of creating a CFD model and analysing results much less labour-intensive, reducing time and, hence, cost. Advanced solvers contain algorithms which enable robust solutions of the flow field in a reasonable time. In the context of coal-biomass co-combustion, CFD modelling enables detailed interpretation of temperature fields, flow behaviour, particle trajectories, and pollutant formation without the prohibitive cost of full-scale experiments. As a result of these factors, CFD is now an established industrial design tool in aviation,

 DOI: 10.1201/9781003647942-3

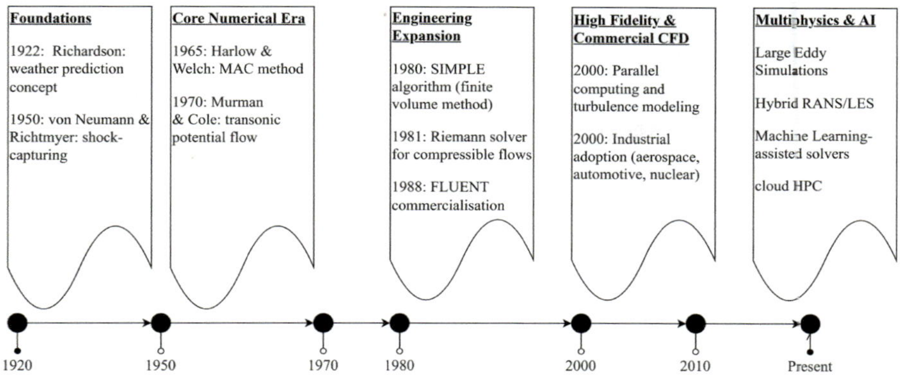

FIGURE 3.1 Development of CFD over the years.

motor design, structural analysis, fluid flow, and HVAC systems, as it helps to reduce design time scales and improve processes throughout the engineering world. CFD provides a cost-effective and accurate alternative to scale model testing, with variations on the simulation being performed quickly, offering obvious advantages. As of 2025, the prominent commercial software providers of CFD are ANSYS, SIEMENS, and COMSOL, only to mention just a few. Some companies have in-house CFD architectures which serve their specific needs, such as those found in Formula 1 and certain thermal power plants, particularly nuclear-fired plants.

CFD is considered a physics-based modelling approach, since it utilises validated algebraic equations in its calculations [4]. This is quite different from statistical modelling, which is generally governed by input and output parameters and corresponding relationships without necessarily considering the limitations of such applications [5, 6]. For example, the Bernoulli equation is devised to determine the expected flow rates, pressure, fluid velocity and head by use of validated algebraic equations. In comparison, statistical analysis is designed to extrapolate or interpolate the expected flow rates, pressure, fluid velocity, and head values based on historical input parameters, output parameters, and existing relationships. Machine learning (ML) on its own is also an extension of statistical modelling since it tries to develop relationships based on existing data. When ML is infused with physics-based models such as CFD, most of its shortcomings are addressed [7].

This chapter presents the numerical foundations that govern CFD modelling, as shown in the flowchart presented in Figure 3.2. This will be done by focusing on what happens at stages four and five, which is the introduction of governing equations, then the discretisation process, as well as the solving of these discretised equations.

3.2 GOVERNING EQUATIONS

The process of breaking down a physical domain, for example, the path of fluid flow between inlet and outlet, into smaller sections is similar to the mathematical process of creating a differential equation. The individual smaller pieces are then

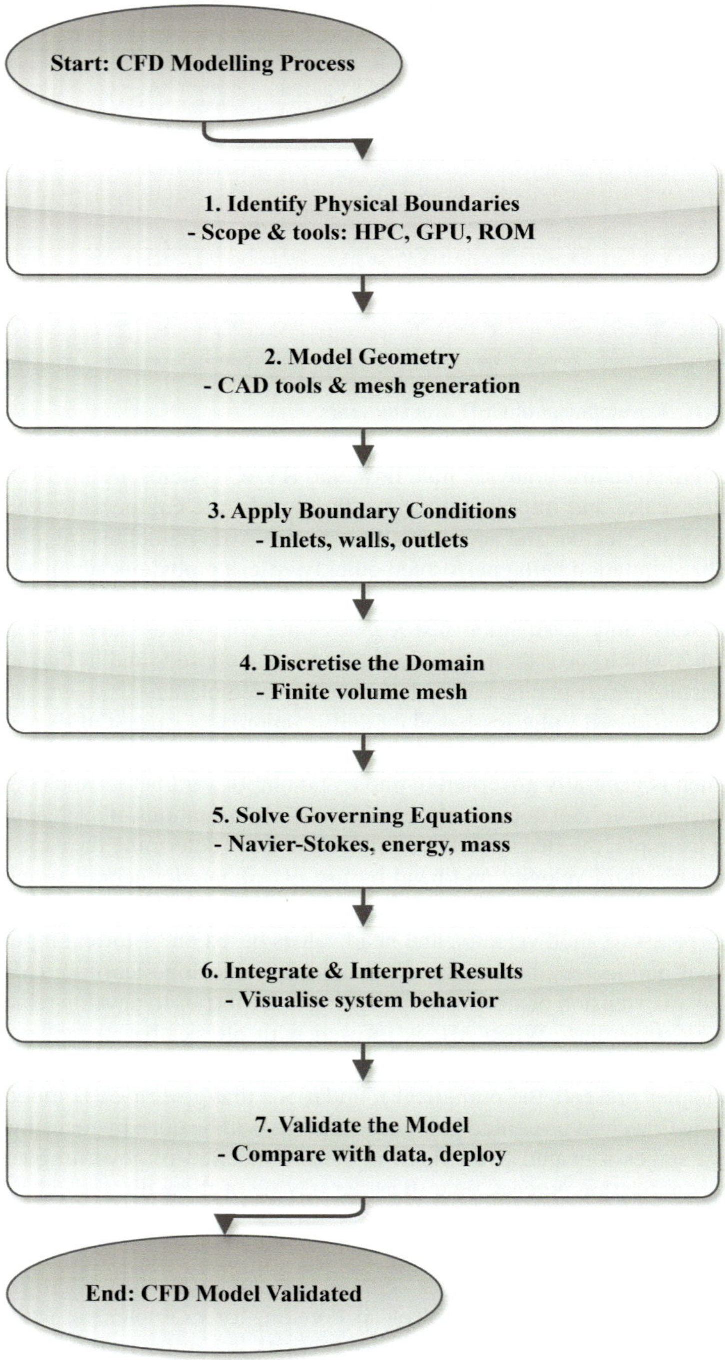

FIGURE 3.2 CFD modelling workflow.

analysed individually, acknowledging that each piece is influenced directly by the neighbouring pieces [2]. The conservation equation in its differential form, as represented by Equation (3.1), is used to analyse the interactions these smaller particles go through. This conservation equation can be applied to the mass, momentum, and energy principles [1]. For modelling purposes, a differential equation is formulated and discretised using the finite volume element methods to obtain an equation that can be numerically solved.

$$\frac{\partial}{\partial t}(\rho.\phi)+\underbrace{\frac{\partial}{\partial x_i}(\rho.c_i.\phi)}_{\text{convection}}=\underbrace{\frac{\partial}{\partial x_i}\left(\Gamma.\frac{\partial\phi}{\partial x_i}\right)}_{\text{diffusion}}+S_\phi \tag{3.1}$$

where:

ρ represents the density,
φ represents the variable (mass, specific enthalpy, temperature, etc.),
x_i represents the direction (x_1: x direction, x_2: y direction, x_3: z direction for Cartesian coordinates),
c_i represents the velocity (u: x component, v: y component, w: z component),
Γ represents the diffusion coefficient of the variable, and
S_φ represents the source or sink parameter of the variable.

Since coal and biomass combustion involves complex gas-to-solid interactions, the solution of Equation (3.1) makes apparent multiple variables related to velocity, pressure, temperature, and chemical species resulting in nonlinear partial differential equations that are difficult to solve. Combustion of a simple fuel requires at least 40 elementary reactions to form the products, meaning coal and biomass require a substantial amount of representative elementary reactions [8]. Considering that every reaction will be feeding into another reaction, representing all these reactions through the conservation equation becomes a mammoth task. To simplify this task, combustion modelling tools are called upon that consider only the important physics.

3.2.1 Continuity Equation

To model the continuity equation, which is basically the conservation of mass equation, the source and diffusion terms reduce to zero, which means the terms representing the instantaneous change in density and local change in velocity are the only ones that remain relevant. Equation (3.2) summarises the continuity equation as applied to CFD [1].

$$\frac{\partial}{\partial t}(\rho)+\frac{\partial}{\partial x_i}(c_i)=0 \tag{3.2}$$

3.2.2 Momentum Equation

The momentum equation accounts for the forces acting on a fluid particle as it travels through the control volume. In its complete form, it is represented by Equations (3.3) to (3.5). The tensor stresses that act on the fluid particle are represented by the matrix (3.6). Surface forces are not simple to analyse since they consist of both normal and tangential components.

$$\frac{\partial \sigma_{xx}}{\partial x} + \frac{\partial \tau_{yx}}{\partial y} + \frac{\partial \tau_{xz}}{\partial z} = \rho \frac{Du}{Dt} \tag{3.3}$$

$$\frac{\partial \sigma_{yy}}{\partial y} + \frac{\partial \tau_{yx}}{\partial x} + \frac{\partial \tau_{yz}}{\partial z} = \rho \frac{Dv}{Dt} \tag{3.4}$$

$$\frac{\partial \sigma_{zz}}{\partial z} + \frac{\partial \tau_{zx}}{\partial x} + \frac{\partial \tau_{zy}}{\partial y} - \rho g = \rho \frac{Dw}{Dt} \tag{3.5}$$

$$\begin{pmatrix} \sigma_{xx} & \tau_{xy} & \tau_{xz} \\ \tau_{yx} & \sigma_{yy} & \tau_{yz} \\ \tau_{zx} & \tau_{zy} & \sigma_{zz} \end{pmatrix} \tag{3.6}$$

where:

δ represents normal stress (pressure),
τ represents shear stress (viscous), and

$\frac{D}{Dt}$ represents the material derivative of a variable.

When special assumptions are applied to the momentum equation, it reduces to a set of Navier-Stokes equations as represented by Equations (3.7) to (3.9) [9].

$$\frac{\partial}{\partial t}(\rho u) + \frac{\partial}{\partial x_i}(\rho c_i u) = -\frac{\partial P}{\partial x} + \frac{\partial}{\partial x_i}\left(\mu \frac{\partial u}{\partial x_i}\right) + S_{\phi,u} \tag{3.7}$$

$$\frac{\partial}{\partial t}(\rho v) + \frac{\partial}{\partial x_i}(\rho c_i v) = -\frac{\partial P}{\partial y} + \frac{\partial}{\partial x_i}\left(\mu \frac{\partial v}{\partial x_i}\right) + S_{\phi,v} \tag{3.8}$$

$$\frac{\partial}{\partial t}(\rho w) + \frac{\partial}{\partial x_i}(\rho c_i w) = -\frac{\partial P}{\partial z} + \frac{\partial}{\partial x_i}\left(\mu \frac{\partial w}{\partial x_i}\right) + S_{\phi,w} \tag{3.9}$$

where:

P represents the pressure and
μ represents the dynamic viscosity.

3.2.3 Energy Equation

Combustion is highly dependent on the energy equation, as temperature changes influence the behaviour of multiple parameters such as heat transfer, enthalpy of reactions, and chemical energy release. Equation (3.10) illustrates the variation of internal energy during the combustion process [10].

$$\frac{\partial}{\partial t}(\rho c_v T) + \frac{\partial}{\partial x_i}(\rho c_i c_v T) = -P\frac{\partial c_i}{\partial x_i} + \frac{\partial}{\partial x_i}\left(k_c \frac{\partial T}{\partial x_i}\right) + \Psi + S_{c_v T} \tag{3.10}$$

where:

c_v represents the specific heat capacity at constant volume (J/kg),
T represents the temperature (K),
k_c represents the thermal conductivity, and
Ψ represents the dissipation function.

3.2.4 Species Transport Equations

The movement of gaseous species to and from the combusting particles' solid surfaces during combustion must be accurately modelled in CFD. The reactants are transported to the surface whilst the products are transported from the surface. Combustion of solid fuels such as coal or biomass with an oxidant is considered a multiphase process as solid phase and gaseous phase species are in constant interaction. Discrete phase model (DPM) or particle phase model is the overall term used to describe all the interacting models that capture particle to particle and continuum to particle interaction during the combustion process [11]. Usually, after each iteration, the parameters related to size (diameter, density, length, etc.) of solid fuel particles reduce and will need to be captured accordingly. During each stage of combustion (drying, devolatilisation, char combustion, volatile combustion, and ash formation), specific laws are required; thus, the conservation equations are written accordingly [12]. Using discrete phase principles, the transport equation of each particle is modelled successfully, as represented in Equation (3.11) [13]. The diffusion flux of a species (J_i) is dependent on the characteristics of the flow based on the Schmidt and Lewis numbers.

$$\frac{\partial}{\partial t}(\rho X_i) + \frac{\partial}{\partial x_i}(\rho c_i X_i) = -\frac{\partial}{\partial x_i}(J_i) + R_i + S_i \tag{3.11}$$

where:

X_i represents the local mass fraction of a species,
R_i represents the rate of production of a species, and
S_i represents the source term of a species.

3.3 NUMERICAL DISCRETISATION

Analytical solutions to the Navier-Stokes equations exist for only the simplest of flows under ideal conditions. To obtain solutions for real flows, a numerical approach must be adopted whereby the equations are replaced by algebraic approximations that can be solved using a numerical method. The most accurate numerical computations in fluid dynamics come from solving the Navier-Stokes equations. In three dimensions, the conservation of momentum is written in each of the coordinate directions, giving a total of five equations to be solved. These equations are highly coupled and non-linear [13].

3.3.1 Finite Volume Discretisation

During modelling by use of an element-based finite volume method, the first step involves discretising the spatial domain using a certain meshing technique. The mesh is used to construct finite volumes, which are used to conserve relevant quantities such as mass, momentum, and energy. The conservation equations for mass, momentum, and energy are integrated over each control volume, and Gauss' Divergence Theorem is applied to convert volume integrals involving divergence and gradient operators to surface integrals [14].

The finite volume method ensures that the discretisation is conservative, i.e., mass, momentum, and energy are conserved in a discrete sense. It also allows for application on irregular meshes consisting of arbitrary polyhedra in three dimensions or arbitrary polygons in two dimensions. This increased flexibility can be used to a great advantage in generating grids about arbitrary geometries.

3.3.2 Steady One-Dimensional Convection and Diffusion

To illustrate the discretisation, this section develops the case for a steady one-dimensional convection and diffusion case. In the absence of sources, the steady convection and diffusion of a property ϕ in each one-dimensional flow field along the x direction is governed by Equation (3.12). The flow must also satisfy the continuity as given in Equation (3.13).

$$\frac{d}{dx}(\rho u \phi) = \frac{d}{dx}\left(\Gamma \frac{d\phi}{dx}\right) \tag{3.12}$$

$$\frac{d}{dx}(\rho u) = 0 \tag{3.13}$$

We consider the one-dimensional control volume shown in Figure 3.3. P represents the location of the variable, W represents the variable location on the west, E represents the location on the east, u_w represents the velocity from W to point P, and u_e represents the velocity from point P to E. Integration of transport and continuity equations over a control volume results in Equations (3.14) and (3.15).

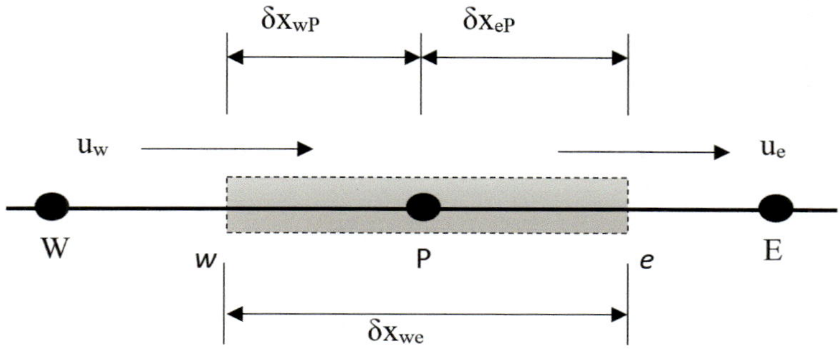

FIGURE 3.3 Control volume around P.

$$(\rho uA\phi)_e - (\rho uA\phi)_w = \left(\Gamma A\frac{\partial\phi}{\partial x}\right)_e - \left(\Gamma A\frac{\partial\phi}{\partial x}\right)_w \tag{3.14}$$

$$(\rho uA)_e - (\rho uA)_w = 0 \tag{3.15}$$

If the mesh is symmetrical enough, the east and west faces which represent the inlet and outlet faces through which the variable is travelling will have the same area; hence $A_w = A_e = A$. By employing the central differencing approach to represent the contribution of the diffusion terms on the right-hand side of the equation, we obtain discretised equations represented in Equations (3.16) and (3.17). The cell face values of the variables F and D are illustrated in Equation (3.18).

$$F_e\phi_e - F_w\phi_w = D_e(\phi_E - \phi_P) - D_w(\phi_P - \phi_W) \tag{3.16}$$

$$F_e - F_w = 0 \tag{3.17}$$

$$F_w = (\rho u)_w; F_e = (\rho u)_e; D_w = \frac{\Gamma_w}{\delta x_{WP}}; D_e = \frac{\Gamma_e}{\delta x_{PE}} \tag{3.18}$$

where:

F represents the convective mass flux per unit area (F=ρu) and

D represents the diffusion conductance at cell faces $D = \frac{\Gamma}{\delta x}$

When applying the central differencing scheme, that is, averaging the influence between W and P as well as that between E and P, the cell face values of property ϕ are simplified into Equations (3.19) and (3.20). Substitution of these terms into Equation (3.16) and subsequent rearrangement simplifies to become (3.21).

$$\phi_w = (\phi_W + \phi_P)/2 \tag{3.19}$$

$$\phi_e = (\phi_P + \phi_E)/2 \tag{3.20}$$

$$\left[\left(D_w + \frac{F_w}{2}\right) + \left(D_e - \frac{F_e}{2}\right) + F_e - F_w\right]\phi_P = \left(D_w + \frac{F_w}{2}\right)\phi_W + \left(D_e - \frac{F_e}{2}\right)\phi_E \tag{3.21}$$

$$a_P\phi_P = a_W\phi_W + a_E\phi_E \tag{3.22}$$

In summary, the central differencing scheme has convective and diffusive parameters represented as $a_W = (D_W + F_W/2)$, $a_E = (D_E - F_E/2)$, and $a_P = (D_W + F_W/2) + (D_E - F_E/2)$. A non-dimensional Peclet number (Pe) is defined as the ratio between the relative strengths of convection and diffusion effects. This allows researchers to study the transportive nature of flows, giving information about the tendency to converge. When the Peclet number is zero, the system has no convective effects, only diffusion. When the Peclet number tends towards ∞, the system has no diffusion, only convective effects. Stable and accurate systems have a Peclet number that is less than two [1].

3.3.3 Discretisation Schemes

The upwind differencing scheme identifies the inadequacies of the central differencing scheme, especially the need to account for directional flow. Either the flow is positive, which implies that $u_w > 0$; $u_e > 0$, then $F_w > 0$; $F_e > 0$, or it's negative. The convective and diffusive parameters are represented as $a_W = (D_W + \max(F_W, 0))$, $a_E = (D_E + \max(-F_E, 0))$, and $a_P = a_W + a_E$ [14].

Combining the upwind differencing scheme and the central differencing scheme produces the hybrid differencing scheme. The hybrid scheme combines the favourable advantages of both schemes by using the central differencing scheme when the Peclet number is less than two and the upwind differencing scheme when the Peclet number is greater than two. The power law scheme introduces a polynomial as a way of reducing errors associated with first order approximations of the equations when using the upwind differencing scheme and the central differencing scheme. Higher order schemes involve more neighbour points and reduce the discretisation errors by bringing in a wider influence. The quadratic upstream interpolation for convective kinetics (QUICK) scheme of Leonard (1979) uses a three-point upstream-weighted quadratic interpolation for cell face values. Since the scheme is based on a quadratic function, its accuracy in terms of the Taylor series truncation error is third order on a uniform mesh [15].

3.3.4 Solving Discretised Equations

The process of discretising the governing equations of fluid flow results in a system of linear algebraic equations. The complexity and size of the set of equations depend on the dimensionality of the problem, the number of grid nodes, and the discretisation practice. The system of equations can be solved by direct methods that include Cramer's rule matrix inversion or Gaussian elimination.

The other method used to solve these equations is the iterative method, as employed by most CFD solvers. This involves the repeated application of relatively simple algorithms, leading to eventual convergence after several repetitions. The simplest of these algorithms is the Thomas algorithm, which is computationally inexpensive and has the advantage that it requires a minimum amount of storage compared to Gauss-Seidel or Jacobi point-by-point iteration methods, which are slow to converge [16, 17].

3.4 TURBULENCE MODELLING

Turbulence is an irregular motion which in general makes its appearance in fluids when they flow past solid surfaces or even when neighbouring streams of the same fluid flow past or over one another. Turbulence develops as an instability of laminar flow. Perhaps the most important feature of turbulence from an engineering point of view is its enhanced diffusivity. Turbulent diffusion greatly enhances the transfer of mass, momentum, and energy. Turbulence consists of fluctuations in the flow field in time and space. It is a complex process, mainly because it is three-dimensional, unsteady, and consists of many scales [2, 18].

Turbulence occurs when the inertia forces in the fluid become significant compared to viscous forces and is characterised by a high Reynolds Number. In principle, the Navier-Stokes equations describe both laminar and turbulent flows without the need for additional information [19]. However, turbulent flows at realistic Reynolds numbers span a large range of turbulent length and time scales and would generally involve length scales much smaller than the smallest finite volume mesh, which can be practically used in a numerical analysis. The Direct Numerical Simulation (DNS) of these flows requires computing power.

To enable the effects of turbulence to be predicted, a large amount of CFD research has concentrated on methods that make use of turbulence models. Turbulence models have been specifically developed to account for the effects of turbulence without recourse to a prohibitively fine mesh and direct numerical simulation. Most turbulence models are statistical turbulence models, with the two exceptions to this being the Large Eddy Simulation (LES) model and the Detached Eddy Simulation (DES) model. Turbulence models seek to solve a modified set of transport equations by introducing averaged and fluctuating components. For compressible flows, the averaging is actually weighted by density (Favre-averaging) [20, 21].

To enable the effects of turbulence to be predicted, a large amount of CFD research has concentrated on methods that make use of turbulence models. Turbulence models have been specifically developed to account for the effects of turbulence without recourse to a prohibitively fine mesh and direct numerical simulation. Most turbulence models are statistical turbulence models, with the two exceptions to this being the Large Eddy Simulation (LES) model and the Detached Eddy Simulation (DES) model. Turbulence models seek to solve a modified set of transport equations by introducing averaged and fluctuating components. For compressible flows, the averaging is actually weighted by density (Favre-averaging).

Turbulence is inherently three-dimensional and time-dependent. Thus, an enormous amount of information is required to completely describe combustion and turbulent flow. Since resolving all eddies is computationally prohibitive, different modelling approaches are used. An ideal turbulence model should introduce the minimum amount of complexity while capturing the essence of the relevant physics.

3.4.1 Reynolds-Averaged Equations (RANS)

Variables undergoing turbulence are easily analysed by treating them as a parameter that combines fluctuating and average parameters, $\phi = \overline{\phi} + \phi''$ (variable = average + fluctuation) [12]. Substituting the averaged quantities into the original transport equations results in the Reynolds averaged equations. When substituted into the Navier-Stokes equation and mathematical operations are performed, the Reynolds-Averaged Navier-Stokes (RANS) equations are formed. Equation (3.23) represents the Reynolds averaged continuity equation, whilst Equations (3.24) to (3.26) represent the Reynolds averaged momentum equations along the x, y, and z directions respectively [21].

$$\frac{\partial}{\partial x_i}(\overline{c}) = 0 \tag{3.23}$$

$$\frac{\partial \overline{u}}{\partial t} + \frac{\partial(\overline{u}\overline{c}_i)}{\partial x_i} = -\frac{1}{\rho}\frac{\partial P}{\partial x} + \frac{1}{\rho}\frac{\partial}{\partial x_i}\left(\mu\frac{\partial(\overline{u})}{\partial x_i}\right) + \left[-\frac{\partial \overline{u'^2}}{\partial x} - \frac{\partial \overline{u'v'}}{\partial y} - \frac{\partial \overline{u'w'}}{\partial z}\right] \tag{3.24}$$

$$\frac{\partial \overline{v}}{\partial t} + \frac{\partial(\overline{v}\overline{c}_i)}{\partial x_i} = -\frac{1}{\rho}\frac{\partial P}{\partial y} + \frac{1}{\rho}\frac{\partial}{\partial x_i}\left(\mu\frac{\partial(\overline{v})}{\partial x_i}\right) + \left[-\frac{\partial \overline{u'v'}}{\partial x} - \frac{\partial \overline{v'^2}}{\partial y} - \frac{\partial \overline{v'w'}}{\partial z}\right] \tag{3.25}$$

$$\frac{\partial \overline{w}}{\partial t} + \frac{\partial(\overline{w}\overline{c}_i)}{\partial x_i} = -\frac{1}{\rho}\frac{\partial P}{\partial z} + \frac{1}{\rho}\frac{\partial}{\partial x_i}\left(\mu\frac{\partial(\overline{w})}{\partial x_i}\right) + \left[-\frac{\partial \overline{u'w'}}{\partial x} - \frac{\partial \overline{v'w'}}{\partial y} - \frac{\partial \overline{w'^2}}{\partial z}\right] \tag{3.26}$$

where:

$\overline{c}_i$ represents the average velocity, $(\overline{u}, \overline{v}, \overline{w})$
x_i represents the coordinates (e.g. x, y, z),
ρ represents the density
μ represents the dynamic viscosity

$\overline{u'^2}$, $\overline{u'v'}$, $\overline{u'w'}$, $\overline{v'^2}$, and $\overline{w'^2}$ are the Reynolds stresses.

The momentum and scalar transport equations contain turbulent flux terms additional to the molecular diffusive fluxes. These are the Reynolds stresses, $\rho\overline{u_i u_j}$. They reflect the fact that convective transport due to turbulent velocity fluctuations will act to enhance mixing caused by thermal fluctuations at the molecular level. When the averaging is applied to the energy equation, the Reynolds-averaged energy equation is formed. Equation (3.27) represents the averaged energy equation based on the total enthalpy [1].

$$\frac{\partial \rho \overline{h_{tot}}}{\partial t} + \nabla.\left(\overline{\rho h_{tot} c_i}\right) = \nabla.\left(\lambda \text{grad}\overline{T} - \overline{\rho h.u_j}\right) + \nabla.\left[\overline{c_i}\left(\tau_{ij} - \overline{\rho u_i u_j}\right)\right] + S_E \tag{3.27}$$

Where,

h_{tot} represents the total enthalpy $\overline{h} + \frac{1}{2}\overline{c_i}^2 + k$

k represents the turbulent kinetic energy.

3.5 GRADIENT DIFFUSION HYPOTHESIS (BOUSSINESQ'S HYPOTHESIS)

The eddy viscosity hypothesis assumes that the Reynolds stresses can be related to the mean velocity gradients and eddy (turbulent) viscosity by the gradient diffusion hypothesis given in Equation (3.28).

$$-\rho\overline{u_i' u_j'} = \mu_t \left(\frac{\partial \overline{u_i}}{\partial x_j} + \frac{\partial \overline{u_j}}{\partial x_i}\right) \tag{3.28}$$

where:

μ_t represents the eddy viscosity.

The eddy viscosity or turbulent viscosity must be modelled appropriately to solve the turbulence equation. Generally, to solve the turbulent viscosity there exist the zero equation, one equation, and two equation models, with the main difference being the manner through which the eddy viscosity is calculated.

The zero equation model uses the characteristic velocity and length scale in a single algebraic expression as given by Equation (3.29).

$$-\rho\overline{u_i' u_j'} = \mu_t \left(\frac{\partial \overline{u_i}}{\partial x_j} + \frac{\partial \overline{u_j}}{\partial x_i}\right) \tag{3.29}$$

The partial differential equations are used to compute the mean fields, whilst the algebraic equations are used for quantities related to turbulence. Models classified under the zero equation models are the simplest and can give good results for some applications. Unfortunately, when the flow separates or recirculates, these models tend to lose their accuracy.

The one-equation model calculates the turbulence characteristic velocity as the square root of turbulence kinetic energy. The most well-known one-equation model for aerospace applications is the Spalart and Allmaras [22]. One-equation models are mostly employed in the nuclear and aeronautics industries. Nevertheless, one-equation models result in significant diffusion, especially for 3D vortical flow regions [18].

3.6 TWO EQUATION TURBULENCE MODELS

Two equation models use, in addition to the Navier-Stokes equations, two transport equations to represent the turbulence properties. The first one is usually that for the turbulence kinetic energy (k). The second one can be that for the dissipation rate of turbulence kinetic energy (ε), dissipation rate (ω), length scale, time scale, product of kinetic energy and length scale, and product of kinetic energy and time scale. Both the k-epsilon (k-ε) and the k-omega two-equation turbulence models use this variable. The two-equation models are the most widely used in industry [18, 23].

Due to their ability to provide a reasonable balance between computational accuracy and numerical effort, two-equation turbulence models are extremely popular. In two-equation models, the turbulence velocity scale is computed from the turbulent kinetic energy, which is provided from the solution of its transport equation. The turbulent length scale is estimated from two properties of the turbulence field, usually the turbulent kinetic energy and its dissipation rate. The dissipation rate of the turbulent kinetic energy is provided from the solution of its transport equation [24].

3.6.1 K-Epsilon Model

K is the turbulence kinetic energy and is defined as the variance of the fluctuations in velocity. It has dimensions of (L^2T^{-2}); for example, m^2/s^2. Epsilon (ε) is the turbulence eddy dissipation (the rate at which the velocity fluctuations dissipate) and has dimensions of k per unit time (L^2T^{-3}); for example, m^2/s^3. For general-purpose simulations, the k-ε model offers a good compromise in terms of accuracy and robustness. The k-ε model introduces two new variables into the system of equations, with the continuity equation remaining the same. However, the momentum equation makes apparent the effective velocity of the fluid as illustrated in Equation (3.31).

$$\frac{\partial \overline{\rho}}{\partial t} + \frac{\partial}{\partial x_i}(\rho \overline{c_i}) = 0 \tag{3.30}$$

$$\frac{\partial \overline{\rho u_j}}{\partial t} + \frac{\partial}{\partial x_j}\left(\rho \overline{u_i u_j}\right) = -\frac{\partial p'}{\partial x_i} + \frac{\partial}{\partial x_j}\left[\mu_{eff}\left(\frac{\partial \overline{u_i}}{\partial x_j} + \frac{\partial \overline{u_j}}{\partial x_i}\right)\right] + S_M \tag{3.31}$$

$$\mu_t = C_\mu \rho \frac{k^2}{\varepsilon} \tag{3.32}$$

where:

p' represents the modified pressure,
μ_{eff} represents the effective viscosity ($\mu_{eff} = \mu + \mu_t$),
S_M represents momentum source, and
C_μ represents a constant.

The k-ε model assumes that the turbulence viscosity (μ_t) is linked to the turbulence kinetic energy and dissipation by Equation (3.32). The values of k and ε come directly from the differential transport equations for the turbulence kinetic energy Equation (3.33) and turbulence dissipation rate Equation (3.34).

$$\frac{\partial(\overline{\rho}k)}{\partial t} + \frac{\partial}{\partial x_j}\left(\overline{\rho u_j k}\right) = \frac{\partial}{\partial x_j}\left[\left(\mu + \frac{\mu_t}{\sigma_k}\right)\frac{\partial k}{\partial x_j}\right] + P_K - \rho\varepsilon \tag{3.33}$$

$$\frac{\partial(\overline{\rho}\varepsilon)}{\partial t} + \frac{\partial}{\partial x_j}\left(\overline{\rho u_j \varepsilon}\right) = \frac{\partial}{\partial x_j}\left[\left(\mu + \frac{\mu_t}{\sigma_\varepsilon}\right)\frac{\partial \varepsilon}{\partial x_j}\right] + \frac{\varepsilon}{k}(c_{\varepsilon 1} P_K - c_{\varepsilon 2}\rho\varepsilon \tag{3.34}$$

where:

P_k represents shear production of turbulence.

The constants C_μ, σ_k, σ_ε, $C_{\varepsilon 1}$, and $C_{\varepsilon 2}$ are constants with values equal to 0.09, 1.01, 1.31, 1.44, and 1.44, respectively, for the k-ε turbulence model [22].

3.6.2 K-Omega Model

One of the advantages of the k-omega (ω), also known as the Wilcox formulation, is the near-wall treatment for low-Reynolds number computations. The model does not involve the complex non-linear damping functions required for the k-ε model and is therefore more accurate and more robust [25]. A low-Reynolds k-ε model would typically require a near-wall resolution of y<+ 0.2, while a low-Reynolds number k-ω model would require at least y<+2. In industrial flows, even y<+2 cannot be guaranteed in most applications, and for this reason, a new near-wall treatment was developed for the k-ω models [26]. It allows for a smooth shift from a low-Reynolds number form to a wall function formulation. The k-ω model gained popularity mainly because it can be integrated to the wall without using any damping functions. This allows the model to be accurate and robust for a wide range of boundary layer flows with large pressure gradients [22]. When employed in aerospace and turbo-machinery communities, several sub-models/options of k-ω can be employed to account for compressibility effects, transitional flows, and shear-flow corrections

[18]. The main problem with the Wilcox model is its well-known strong sensitivity to free stream conditions [27].

The k-ω model has the turbulent kinetic energy and turbulent frequency equations represented as Equations (3.35) and (3.36). This model assumes that the turbulence viscosity is linked to the turbulence kinetic energy and turbulent frequency through Equation (3.37).

$$\frac{\partial(\bar{\rho}k)}{\partial t}+\frac{\partial}{\partial x_i}\left(\overline{\rho u_i k}\right)=\frac{\partial}{\partial x_i}\left[\left(\mu+\frac{\mu_t}{\sigma_k}\right)\frac{\partial k}{\partial x_i}\right]+P_K - \beta'\rho k\omega \tag{3.35}$$

$$\frac{\partial(\bar{\rho}\omega)}{\partial t}+\frac{\partial}{\partial x_i}\left(\overline{\rho u_j \omega}\right)=\frac{\partial}{\partial x_i}\left[\left(\mu+\frac{\mu_t}{\sigma_\omega}\right)\frac{\partial \omega}{\partial x_i}\right]+\alpha\frac{\omega}{k}P_K - \beta\rho\omega^2 \tag{3.36}$$

$$\mu_t=\rho\frac{k}{\omega} \tag{3.37}$$

where:

k represents turbulent kinetic energy and
ω represents turbulent frequency.

The constants β', α, β, σ_k, and σ_ω are constants with values equal to 0.09, 5/9, 0.075, 2, and 2 respectively for the k-ω turbulence model [22].

3.6.3 The Shear Stress Transport

The shear stress transport (SST) model accounts for the transport of turbulent shear stresses. This allows it to give highly accurate predictions about the onset and magnitude of flow separation under adverse pressure gradients. In essence, the SST model formulation is based on the k-ω model. The proper transport behaviour is obtained by a limiting coefficient to the formulation of the eddy-viscosity as given in Equation (3.38) [28].

$$\frac{\mu_t}{\rho}=\frac{a_1 k}{\max\left(a_1\omega, SF_2\right)} \tag{3.38}$$

where:

a_1 is a model constant,
F_2 represents the blending function, and
S represents an invariant measure of the strain rate.

3.6.4 RNG k-ε Model

The RNG k-ε model is based on renormalisation group analysis of the Navier-Stokes equations by Yakhot et al. [29] to account for the effects of smaller scales of motion. It is similar in form to the standard k-ε model but includes various refinements, which include an additional term to the dissipation equation that significantly improves the accuracy for rapidly strained flows. The effect of swirl on turbulence is included in the RNG model as well, enhancing accuracy for swirling flows [30]. While the standard k-ε model is a high-Reynolds-number model, the RNG model provides an analytically derived differential formula for effective viscosity that accounts for low-Reynolds-number effects. Effective use of this feature does, however, depend on an appropriate treatment of the near-wall region. Equation (3.39) illustrates the RNG k-ε turbulence model. The transport equations for turbulence generation and dissipation are the same as those for the standard k-ε model, but the model constants differ, as the constant $C_{\varepsilon 1}$ is replaced by the function $C_{\varepsilon 1RNG}$ given in Equation (3.40). The model has shown good results for applications with complex shear flows, high strain rates, swirl, separation, combusting flows, rotating cavities, as well as indoor air simulations [31–33].

$$\frac{\partial(\bar{\rho}\varepsilon)}{\partial t}+\frac{\partial}{\partial x_j}\left(\overline{\rho u_i \varepsilon}\right)=\frac{\partial}{\partial x_i}\left[\left(\mu+\frac{\mu_t}{\sigma_{\varepsilon RNG}}\right)\frac{\partial \varepsilon}{\partial x_i}\right]+\frac{\varepsilon}{k}\left(C_{\varepsilon 1RNG}P_K - C_{\varepsilon 2RNG}\rho\varepsilon + C_{\varepsilon 1RNG}P_{\varepsilon b}\right) \tag{3.39}$$

$$C_{\varepsilon 1RNG} = 1.42 - f_\eta \tag{3.40}$$

$$f_\eta = \frac{n\left(1-\dfrac{\eta}{4.38}\right)}{\left(1+\beta_{RNG}\eta^3\right)} \tag{3.41}$$

$$\eta = \sqrt{\frac{P_k}{\rho C_{\mu RNG}}} \tag{3.42}$$

The constants $C\mu_{RNG}$, $\sigma_{\varepsilon 1RNG}$, β_{RNG}, $\sigma_{\varepsilon RNG}$, and $C_{\varepsilon 2RNG}$ are constants with values equal to 0.085, 0.7179, 0.012, 0.7179, and 1.68 respectively for the RNG k-ε turbulence model [22].

In summary, the standard k-ε coefficients are empirically derived and valid for fully turbulent flows only. The model is robust and easy to implement; hence, it is suitable for initial iterations, initial screening of alternative designs, and parametric studies. The standard k-ε model performs poorly for complex flows that involve severe pressure gradients, separation, and strong streamline curvature. On the other hand, RNG k-ε coefficients are analytically derived with significant changes to the eddy dissipation equation. This improves the ability of the RNG k-ε model to address all the shortcomings of the k-ε model. The SST k-ω model combines the k-ω model for use near walls and the standard k-ε model for regions away from walls using a blending function. This helps in the modelling of low Reynolds flows as well as within the boundary layer.

3.7 LARGE EDDY SIMULATION (LES)

Large eddy simulations (LES) resolve large-scale turbulent structures that have been spatially filtered. To put this into context, when coal particles go through sieve analysis, there are particles that are collected at the bottom of the tray and those that remain within the sieves. Essentially, if one focuses on analysing coal particles that are captured within the sieves, it means a filter has been applied to the analysis. Similarly, LES is a method that only focuses on turbulent structures that obey a certain filtering rule (spatial filtering) and ignores the smaller turbulent structures. This brings to light the difference between LES and Reynolds-averaged equations, as both tend to have similar equation structures but different interpretations. Reynolds-averaged equations employ averaging for all turbulent scales as represented in Equations (3.23) to (3.26), whilst LES only focuses on spatially filtered turbulent scales. Equation (3.43) shows the equation that is produced after spatial filtering [34].

$$\phi = \tilde{\phi} + \phi'' \tag{3.43}$$

where:

ϕ represents the variable,
$\tilde{\phi}$ represents the large turbulent structures that have been filtered, and
ϕ'' represents the small turbulent structures that have passed through the filter.

The difference between LES and Reynolds averaged equations becomes apparent when one tries to manipulate the parameter with a double prime. For Reynolds averaged equations, the average of the fluctuating variable is zero, but for LES, the average of the filtered sub-grid variable is not zero. The overall momentum equation as applied in LES is represented by (3.44).

$$\frac{\partial \bar{\rho}\tilde{\varphi}_i}{\partial t} + \frac{\partial \left(\bar{\rho}\tilde{\varphi}_i \tilde{u}_j\right)}{\partial x_j} = -\frac{\partial \bar{p}}{\partial x_i} + \frac{\partial}{\partial x_j}\left(\Gamma_{ij}\right) - \frac{\partial}{\partial x_j}\left(\bar{\rho}\tau_{ij}\right) \tag{3.44}$$

where:

Γ_{ij} represents the viscous stress tensor and
τ_{ij} represents the sub-grid stresses in LES.

3.8 DIRECT NUMERICAL SIMULATION (DNS)

Direct Numerical Simulation (DNS) resolves all turbulent scales directly from the Navier-Stokes equations without modelling. This involves numerically solving Equations (3.2) up to (3.10) in their full three-dimensional and time-dependent nature. This makes DNS computationally expensive; hence, most applications are restricted to simple geometries and low Reynolds numbers [18]. Generally, to successfully set up a DNS model, the computational domain's dimensions must be sufficient to include the largest turbulence scales. Also, the dissipation length scale, sometimes referred to as the Kolmogorov micro-scale, must be captured by the grid resolution [35, 36].

The Kolmogorov micro-scale and Kolmogorov micro-timescale determine how many grid points are needed for a particular DNS. The higher the Reynolds number, the finer the mesh must be [37]. Based on the Kolmogorov equation, the numerical time step should be proportional to $Re^{1/2}$ whilst the domain should be proportional to $Re^{3/4}$ [1]. To enhance numerical stability of the algorithm, the DNS time step is generally less than the Kolmogorov micro timescale [18].

3.9 SUMMARY

In summary, this chapter establishes the mathematical and physical principles that are used to establish CFD modelling. Of particular interest is how CFD is developed so it can be applied to combustion modelling.

- To perform numerical modelling, conservation equations in their differential form of mass, momentum, energy, and species transport are first developed. These equations are then discretised using a preferred scheme: central difference, upwind, hybrid, or QUICK. Algebraic equations are then formed, which can be solved by iterative methods, as in the case of CFD. Once solved, the solution is integrated over the whole domain to give a representation of the physical phenomenon that is occurring.
- The conservation equations needed to capture the turbulence exhibited by a fluid are developed through various techniques that are well-developed in the literature. To solve these equations, there exist Reynolds-averaged models, LES, and DNS models, as summarised in Table 3.1. Reynolds-averaged models such as k-ε, k-ω, SST, and RNG provide a balance between computational cost and engineering accuracy, while LES and DNS offer higher resolution.
- In general, LES filters large eddies and models sub-grid scales, making it ideal for transient combustion applications. DNS resolves all turbulent scales directly from the Navier-Stokes equations, offering high accuracy.

TABLE 3.1
Examples of Turbulence Modelling in Combustion Applications

Application	Turbulence Model	Reference
Aerodynamic parameters of large aspect biomass; laminar flame speed and gas temperature of coal combustion	DNS	Wang et al. [38]; Hara et al. [39]
Combustion characteristics and NOx formation for swirl-stabilised coal flame; impact of devolatilisation and char on flame structure; full-loop combustion modelling of municipal solid waste and coal	LES	Meller et al. [40]; Rabacal et al. [41]; Xie et al. [42]
Turbulent flame analysis	One equation model (Spalart-Allmaras)	Belmrabet et al. [43]
Comparison of AI and CFD for auto ignition mechanisms; residence time in a multi-jet reactor; performance of enclosed methane air flames	Two equation models (k-ε; realisable k-ω; k-ω SST)	Nsamou et al. [44]; Ju et al. [45]; Jamlkhoo and Moghiman [46]

REFERENCES

1. Versteeg, H.K.; Malalasekera, W. *An Introduction to Computational Fluid Dynamics: The Finite Volume Method*; 2nd ed.; Pearson Education Ltd: Harlow, England, 2007.
2. Anderson, J.D. *Computational Fluid Dynamics: The Basics With Applications*; Holman, J., Lloyd, J., Eds.; McGraw-Hill Book Co Ltd: New York, 1995.
3. Patankar, S. V. *Numerical Heat Transfer and Fluid Flow*; CRC Press: Boca Raton, 2018.
4. Marangwanda, G.; Musonda, I. *Development of a Boiler Healthcare Monitoring System for Predictive Maintenance of Boiler Tube Failures;* New York, 2026.
5. Mao, R.; Lan, Y.; Liang, L.; Yu, T.; Mu, M.; Leng, W.; Long, Z. Rapid CFD Prediction Based on Machine Learning Surrogate Model in Built Environment: A Review. *Fluids* **2025**, *10*, 193, doi:10.3390/fluids10080193.
6. Zimont, V.L.; Massoudi, M.; Birouk, M. Conceptual Limitations of the Probability Density Function Method for Modeling Turbulent Premixed Combustion and Closed-Form Description of Chemical Reactions' Effects. *Fluids* **2021**, *6*, 142, doi:10.3390/fluids6040142.
7. Lu, X.; Wang, Y. Surrogate Modeling for Radiative Heat Transfer Using Physics-Informed Deep Neural Operator Networks. *Proceedings of the Combustion Institute* **2024**, *40*, 105282, doi:10.1016/j.proci.2024.105282.
8. Turns, S. *An Introduction to Combustion: Concepts and Applications*; McGraw-Hill series in mechanical engineering; 2nd ed.; McGraw-Hill: New York, 1996.
9. Marangwanda, G.T.; Madyira, D.M. Evaluating Combustion Ignition, Burnout, Stability, and Intensity of Coal–Biomass Blends Within a Drop Tube Furnace Through Modelling. *Energies* **2025**, *18*, 1322, doi:10.3390/en18061322.
10. Marangwanda, G.T.; Madyira, D.M.; Babarinde, T.O. Combustion Models for Biomass: A Review. *Energy Reports* **2020**, *6*, 664–672, doi:10.1016/j.egyr.2019.11.135.

11. Kumar, H.; Mohapatra, S.K.; Singh, R.I. Three-Dimensional CFD Modelling of a Fluidised Bed Combustor Fuelled by Biomass and Coal. *Materials Research Innovations* **2015**, *19*, 118–124, doi:10.1179/1432891715Z.0000000001636.
12. Al-Abbas, A.H.; Naser, J. Effect of Chemical Reaction Mechanisms and NOx Modeling on Air-Fired and Oxy-Fuel Combustion of Lignite in a 100-KW Furnace. *Energy & Fuels* **2012**, *26*, 3329–3348, doi:10.1021/ef300403a.
13. Álvarez-Bermúdez, C.; Chapela, S.; Varela, L.G.; Gómez, M.Á. CFD Simulation of an Internally Cooled Biomass Fixed-Bed Combustion Plant. *Resources* **2021**, *10*, 77, doi:10.3390/resources10080077.
14. Kaushik, A. Critical Evaluation of Four Differencing Schemes for A Steady Convection-Diffusion Problem. *Research Journal of Mathematical and Statistical Sciences*, **2016**, *4*, 6–12.
15. Ayalew, M.; Aychluh, M.; Suthar, D.L.; Purohit, S.D. Quadratic Upwind Differencing Scheme in the Finite Volume Method for Solving the Convection-Diffusion Equation. *Mathematical and Computer Modelling of Dynamical Systems* **2023**, *29*, 265–285, doi :10.1080/13873954.2023.2282974.
16. Messa, G.V.; Yang, Q.; Adedeji, O.E.; Chára, Z.; Duarte, C.A.R.; Matoušek, V.; Rasteiro, M.G.; Sean Sanders, R.; Silva, R.C.; Souza, F.J. De Computational Fluid Dynamics Modelling of Liquid–Solid Slurry Flows in Pipelines: State-of-the-Art and Future Perspectives. *Processes* **2021**, *9*, doi:10.3390/pr9091566.
17. Ferziger, J.H.; Perić, M.; Street, R.L. *Computational Methods for Fluid Dynamics*; 4th ed.; Springer International Publishing: Cham, 2020.
18. Argyropoulos, C.D.; Markatos, N.C. Recent Advances on the Numerical Modelling of Turbulent Flows. *Applied Mathematical Modelling* **2015**, *39*, 693–732, doi:10.1016/J.APM.2014.07.001.
19. Saldana, M.; Gallegos, S.; Gálvez, E.; Castillo, J.; Salinas-Rodríguez, E.; Cerecedo-Sáenz, E.; Hernández-Ávila, J.; Navarra, A.; Toro, N. The Reynolds Number: A Journey from Its Origin to Modern Applications. *Fluids* **2024**, *9*, 299, doi:10.3390/fluids9120299.
20. Caron, C.; Lauret, P.; Bastide, A. Machine Learning to Speed up Computational Fluid Dynamics Engineering Simulations for Built Environments: A Review. *Build Environ* **2025**, *267*, 112229, doi:10.1016/j.buildenv.2024.112229.
21. Gorman, J.; Bhattacharyya, S.; Cheng, L.; Abraham, J.P.; Gorman, J.; Bhattacharyya, S.; Cheng, L.; Abraham, J.P. *Applications of Computational Fluid Dynamics Simulation and Modeling*; Bhattacharyya, S., Ed.; IntechOpen: London, 2022.
22. ANSYS. *ANSYS FLUENT User's Guide*; ANSYS, 2021, 1–170.
23. Kumar, H.; Mohapatra, S.K.; Singh, R.I. Review on CFD Modelling of Fluidized Bed Combustion Systems Based on Biomass and Co-Firing. *Journal of The Institution of Engineers (India): Series C* **2018**, *99*, 449–474, doi:10.1007/s40032-017-0361-2.
24. Kim, J.; Lee, S.; Tahmasebi, A.; Jeon, C.H.; Yu, J. A Review of the Numerical Modeling of Pulverized Coal Combustion for High-Efficiency, Low-Emissions (HELE) Power Generation. *Energy and Fuels* **2021**, *35*, 7434–7466, doi:10.1021/acs.energyfuels.1c00343.
25. Menter, F.R. *Improved Two-Equation k-Turbulence Models for Aerodynamic Flows*; California, 1992.
26. Rumsey, C.L.; Spalart, P.R. Turbulence Model Behavior in Low Reynolds Number Regions of Aerodynamic Flowfields. *AIAA Journal* **2012**, *47*, 982–993, doi:10.2514/1.39947.
27. Swanson, R.C. *Solving Two-Equation Turbulence Models With a Perspective on Solving Transport Equations*; Virginia, 2021.

28. Balasubramanian, R.; Barrows, S.; Chen, J.P. Investigation of Shear-Stress Transport Turbulence Model for Turbomachinery Applications. *46th AIAA Aerospace Sciences Meeting and Exhibit* **2008**, doi:10.2514/6.2008-566.
29. Yakhot, V.; Orszag, S.A.; Thangam, S.; Gatski, T.B.; Speziale, C.G. Development of Turbulence Models for Shear Flows by a Double Expansion Technique. *Physics of Fluids A: Fluid Dynamics* **1992**, *4*, 1510–1520, doi:10.1063/1.858424.
30. Turutoglu, C.; Cadirci, S. A New Variant of Renormalization Group Turbulence Model for Flows in Axial Compressor Cascade. *Physics of Fluids* **2025**, *37*, doi:10.1063/5.0274264.
31. Marangwanda, G.T. *Optimisation of Hwange Bituminous Coal and Pinus Sawdust Co-Combustion Using CFD Analysis*; University of Johannesburg: Johannesburg, 2024.
32. Sankar, G.; Kumar, D.S.; Balasubramanian, K.R. Computational Modeling of Pulverized Coal Fired Boilers – A Review on the Current Position. *Fuel* **2019**, *236*, 643–665, doi:10.1016/j.fuel.2018.08.154.
33. Hamdi, T.; Hamdi, F.; Molima, S.; Hernandez, J.J.; Chrigui, M. Computational Analysis on the Effect of Methanol Energy Ratio on the Spray and Combustion Pattern of a Dual-Fuel Compression Ignition Engine. *Journal of Energy Resources Technology, Part A: Sustainable and Renewable Energy* **2025**, *1*, doi:10.1115/1.4068210.
34. Rutland, C.J. Large-Eddy Simulations for Internal Combustion Engines – A Review. *International Journal of Engine Research* **2011**, *12*, 421–451, doi:10.1177/1468087411407248.
35. Kasagi, N.; Shikazono, N. Contribution of Direct Numerical Simulation to Understanding and Modelling Turbulent Transport. *Proceedings – Royal Society of London, A* **1995**, *451*, 257–292, doi:10.1098/rspa.1995.0125.
36. Rogallo, R.S.; Moin, P. Numerical Simulation of Turbulent Flows. *Annual Review of Fluid Mechanics* **1984**, *16*, 99–137, doi:10.1146/annurev.fl.16.010184.000531.
37. Launder, B.E..; Sandham, N.D. *Closure Strategies for Turbulent and Transitional Flows*; Launder, B.E., Sandham, N.D., Eds.; Cambridge University Press: Cambridge, 2002.
38. Wang, J.; Ma, L.; Jiang, M.; Fang, Q.; Yin, C.; Tan, P.; Zhang, C.; Chen, G. Direct Numerical Simulation of the Drag, Lift, and Torque Coefficients of High Aspect Ratio Biomass Cylindrical Particles. *Physics of Fluids* **2024**, *36*, doi:10.1063/5.0186621.
39. Hara, T.; Muto, M.; Kitano, T.; Kurose, R.; Komori, S. Direct Numerical Simulation of a Pulverized Coal Jet Flame Employing a Global Volatile Matter Reaction Scheme Based on Detailed Reaction Mechanism. *Combustion and Flame* **2015**, doi:10.1016/j.combustflame.2015.07.027.
40. Meller, D.; Engelmann, L.; Stein, O.T.; Kempf, A.M. Exhaust Gas Recirculation (EGR) Analysis of a Swirl-Stabilized Pulverized Coal Flame with Focus on NOx Release Using FPV-LES. *Fuel* **2023**, *343*, 127939, doi:10.1016/J.FUEL.2023.127939.
41. Rabaçal, M.; Costa, M.; Vascellari, M.; Hasse, C.; Rieth, M.; Kempf, A.M. A Large Eddy Simulation Study on the Effect of Devolatilization Modelling and Char Combustion Mode Modelling on the Structure of a Large-Scale, Biomass and Coal Co-Fired Flame. *Journal of Combustion* **2018**, *2018*, 1–15, doi:10.1155/2018/7036425.
42. Xie, J.; Zhong, W.; Shao, Y.; Liu, Q.; Liu, L.; Liu, G. Simulation of Combustion of Municipal Solid Waste and Coal in an Industrial-Scale Circulating Fluidized Bed Boiler. *Energy & Fuels* **2017**, *31*, 14248–14261, doi:10.1021/acs.energyfuels.7b02693.
43. Belmrabet, T.; Talice, M.; Delussu, G.; Mulas, M.; Hanchi, S. Combustion Analysis Using Roe's Scheme and the Spalart-Allmaras Model. *AIAA Journal* **2012**, *47*, 2726–2737, doi:10.2514/1.44331.

44. Nsamou, F.C.; Noume, H.; Wandji Nyamsi, W.; Bomba, V.; Mouangue, R. Optimization of Auto-Ignition Time for a Skeletal Mechanism in a Piloted Jet Flame Using Artificial Neural Networks. *Journal of Energy Resources Technology, Part A: Sustainable and Renewable Energy* **2025**, *1*, doi:10.1115/1.4069064.
45. Ju, J.; Duan, X.; Sarkodie, B.; Hu, Y.; Jiang, H.; Li, C. Numerical Simulation of Flow Field and Residence Time of Nanoparticles in a 1000-Ton Industrial Multi-Jet Combustion Reactor. *Chinese Journal of Chemical Engineering* **2022**, *51*, 86–99, doi:10.1016/j.cjche.2021.12.008.
46. Jamalkhoo, M.H.; Moghiman, M. The Development and Validation of an Improved K-ω SST Turbulence Model for the Simulation of Non-Premixed Confined Methane/Air Flames. *Iranian Journal of Science and Technology – Transactions of Mechanical Engineering* **2025**, *49*, 533–547, doi:10.1007/s40997-024-00823-5.

4 Modelling of Chemical Phenomena during Combustion

4.1 INTRODUCTION

Modelling chemical interactions during combustion forms the basis of developing relevant CFD models. Since fuels that undergo combustion can exist as solid, liquid, or gas, combustion models are generally specific to a certain phase of fuel. Emphasis will be placed on how to model the chemistry behind the combustion of solid fuels, particularly coal and biomass blends. The chemistry behind solid fuel combustion is classified under drying, devolatilisation, char combustion, and volatile combustion. Extra chemistry appears during the formation of important emissions from the combustion process. The models that can be called upon to predict devolatilisation are discussed in detail, giving clarification on the compromises that are taken to balance accuracy and computational demand. Evidence shows that for modest applications where the yield rate is important, the single-rate kinetic devolatilisation model is enough, whilst the functional group devolatilisation vapourisation cross-linking (FG-DVC) is called upon when the composition of the volatiles is of interest. The chapter highlights the chemical mechanisms responsible for char and volatile combustion, taking special attention to explore the eddy-dissipation concept and the multiple surface reaction models – models that are good when representing solid fuel combustion. The last section explores the mechanisms such as the extended Zeldovich that are responsible for the formation of NOx, SOx, and particulate matter. A comprehensive summary on the industrial cases with their modelling challenges and abatement techniques accompanies each discussion.

4.2 OVERVIEW OF SOLID FUEL COMBUSTION MODELS

Combustion of solid fuels is based on the fact that a fuel particle goes through distinct stages when it enters the combustion zone, namely: drying, devolatilisation, char combustion, volatile combustion, and ash formation [1]. This chapter discusses the chemical phenomenon that governs combustion in detail. Drying entails the removal of moisture (surface and inherent) when the particle is heated up to around 105°C in the absence of oxygen. Devolatilisation involves the removal of light volatiles and heavy volatiles when the heating continues further in the absence of oxygen. The quantitative (mass of volatiles released and the rate of release) and qualitative

DOI: 10.1201/9781003647942-4

(chemical composition of volatiles) analysis of volatiles is rather complicated, making this stage one of the most important ones in the combustion experimental process. As for CFD, volatile combustion sub-models use mechanisms that are mainly based on the results of the devolatilisation stage [2].

Char combustion requires knowledge of the reactions that occur within the boundary layer and those occurring within the free stream. Fortunately, the experimental analysis of this stage is less cumbersome when information about the fuel is adequate [1]. However, CFD requires specific input parameters about particle shape or porosity, which tend to be overlooked by some of the researchers. Similarly, because of the vast biomass species, it is difficult to exhaust all the possible fuel blend compositions between coal and biomass whilst testing various operating conditions [3].

Combustion of solid fuels such as coal or biomass with an oxidant such as air is considered a multiphase process since the solid phase and gaseous phase interact with each other. Discrete phase model (DPM) or particle phase model (PPM) is the overall term used to describe all the interacting models that capture particle-to-particle and continuum-to-particle interaction during the combustion process. Usually, after each iteration, the solid fuel particle changes its density, which needs to be updated accordingly. During each stage of combustion (drying, devolatilisation, char combustion, volatile combustion, and ash formation), specific laws are required; thus, the conservation equations are written accordingly [4]. The following sub-sections discuss these laws in detail and critically analyse them.

4.3 DRYING MODEL

Drying models take an important role, especially for cases where there is a considerable amount of moisture in the fuel, i.e., wet coal and untreated biomass. Drying is a water vapourisation process which is affected by the local pressure and temperature within the combustion chamber [5]. Coal contains less surface moisture but more inherent moisture, which makes it difficult to remove all the water molecules at this stage. However, some drying models assume complete removal of moisture, or rather the fuel properties are given on a dry basis [6]. To model particle drying in CFD, a logic argument code is employed to test particle temperature (T_p) with respect to water vapourisation temperature (T_{vap}), $T_p < T_{vap}$. The testing law only activates the release of H_2O molecules; hence, no volatiles are released. Typically, the modelled energy conservation is given by Equation (4.1) [7]:

$$m_p.c_{p,p}.\frac{dT_p}{dt} = h_{conv}.A_p.(T_\infty - T_p) + \varepsilon_p.A_p.\sigma.\left(T_w^4 - T_p^4\right) - Q_v \quad (4.1)$$

where:

m_p represents particle mass (kg),
$c_{p,p}$ represents particle specific heat capacity at constant pressure (J/kg.K),
T_p represents particle temperature (K),
h_{conv} represents convective heat transfer coefficient (W/m².K),
A_p represents particle surface area (m²),

$T\infty$ represents continuous phase local temperature (K),
ε_p represents particle surface emissivity,
σ represents the Stefan-Boltzmann constant,
T_w represents wall temperature (K), and
Q_v represents heat transfer due to vapourisation (W).

After the particle has lost its surface water, diffusion occurs due to a water vapour concentration gradient between the particle surface and bulk fluid. Of importance are the equations used to represent the flux of water vapour and diffusion of water vapour, as simplified for CFD by Ma et al. [7] and Tabet and Gökalp [8], given by Equations (4.2) and (4.3). Modelling of the diffusion coefficient is dependent on the particle diameter which makes it easier to model solid fuel particles that are assumed to be spherical during combustion [9]. As suggested by Yin [10] whilst studying the combustion of brown coal in a 609-MW furnace, adding free moisture to the primary oxidiser stream tends to improve the accuracy of the model.

$$N_w = k_w.\left(\frac{P_{sat}}{R.T_p} - X_w.\frac{P}{\mathfrak{R}.T}\right) \tag{4.2}$$

$$\frac{k_w d_p}{D_w} = 2.0 + 0.6 Re_p^{1/2} Sc^{1/3} \tag{4.3}$$

where:

N_w represents the flux of water vapour,
P represents the pressure of the bulk gas,
P_{sat} represents saturated water vapour pressure,
X_w represents the molar fraction of the water vapour in the bulk gas,
$\mathcal{R}$ represents gas constant,
k_w represents mass transfer coefficient,
D_w represents the diffusion coefficient of water vapour in bulk gas,
Re_p represents particle Reynolds number, and
Sc represents Schmidt number.

4.4 DEVOLATILISATION MODELS

Devolatilisation happens when a fuel loses its volatile components, which takes place when the solid fuel particle (coal or biomass) is heated in the absence of oxygen. The same energy equation that is modelled during coal drying is called upon with a minor change to recognise the heat of vapourisation for the volatiles during the devolatilisation process. When coal is heated up during particle drying and devolatilisation, it softens up and swells, which affects the devolatilisation process to a larger extent, unlike for biomass [11, 12]. To account for this effect, a swelling coefficient (C_{sw}) is introduced which compares the size of the particle before and after swelling as a function of particle diameter and volatile mass fractions, as shown in Equation (4.4) [9].

$$\frac{d_p}{d_{p,0}} = 1 + (C_{sw} - 1)\frac{\left(1 - f_{w,0}\right).m_{p,0} - m_p}{f_{v,0}.\left(1 - f_{w,0}\right).m_{p,0}} \tag{4.4}$$

where:

C_{sw} represents the swelling coefficient,

$f_{w,0}$ represents initial mass fraction of boiling/evaporating material (if wet combustion is modelled),

$m_{p,0}$ represents initial particle mass,

$f_{v,0}$ represents initial mass fraction of volatiles in the bulk gas,

d_p represents particle diameter, and

$d_{p,0}$ represents initial particle diameter.

The term $\frac{\left(1 - f_{w,0}\right)m_{p,0} - m_p}{f_{v,0}\left(1 - f_{w,0}\right)m_{p,0}}$ is the ratio of the mass that has evolved during devolatilisation to the total volatile mass of the particle.

The reason for the large number of devolatilisation models is due to a lot of assumptions made during CFD modelling of this stage of combustion [8]. Non-phenomenological devolatilisation models (simple models) focus on the volatile yield rate, whilst phenomenological devolatilisation models (complex models) focus on both the yield rate and the composition [5]. The use of complex models is limited in industrial applications because they require a lot of computational power to execute; thus, no validations have been sufficiently carried out for most scenarios. The following sub-sections present the main devolatilisation models as summarised in Figure 4.1.

4.4.1 Constant Rate Model

This model assumes that volatiles can be released at a constant rate as represented by a constant rate chemical reaction as shown in Equation (4.5) [8]. The average rate constant with respect to bituminous coal devolatilisation is equivalent to twelve (12 s^{-1}) as suggested by Pillai [5].

$$-\frac{dm_p}{dt} = A_0.f_{v,0}.\left(1 - f_{w,0}\right).m_{p,0} \tag{4.5}$$

where:

A_0 represents rate constant and

$f_{v,0}$ represents initial mass fraction of volatiles in the bulk gas.

4.4.2 Single-Rate Kinetic Model

From a chemical kinetics background, this model takes the first-order reaction approach to represent the devolatilisation process as shown by Equations (4.6) and (4.7) [5]. The equation is modelled as a function of the volatiles remaining in the particle.

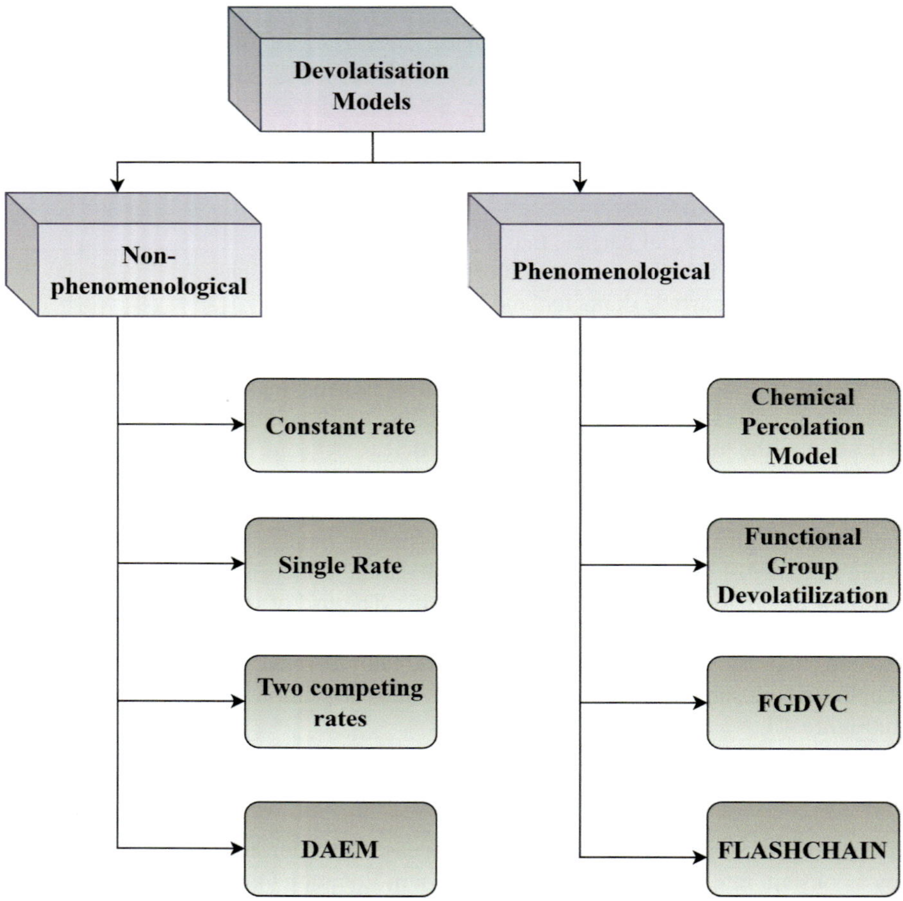

FIGURE 4.1 Main devolatilisation models for solid fuels.

$$-\frac{dm_p}{dt} = -k.[(1-f_{v,0}).(1-f_{w,0}).m_{p,0}] \tag{4.6}$$

$$k = A.e^{-\left(\frac{E_a}{\Re T}\right)} \tag{4.7}$$

where:

k represents the kinetic rate,
A represents the pre-exponential factor,
E_a represents activation energy, and
T represents absolute temperature.

As supported by Tabet et al. [8], this model offers the best compromise for determining volatile yield rate and computational requirements; hence, its wide use for most industrial applications. Different exponents relative to each coal and biomass species have been proposed by researchers in the quest to improve the model accuracy [13]. From a theoretical perspective, the pre-exponential factor and activation energy constants can be obtained by means of the functional group-depolymerisation vapourisation cross-linking code (FG-DVC) model [14]. However, the best compromise is obtained by carrying out thermogravimetric experiments on the fuel samples [15]. The main limitation of the single-rate kinetic model is its failure to predict volatile compositions [16, 17]. A compromise of determining the volatile compositions experimentally before modelling using a thermogravimetric analyser coupled to a mass spectrometer can be employed. Similarly, a theoretical approach to volatile composition deduction can be employed based on the proximate and ultimate analysis of the fuel [18]. Table 4.1 summarises few applications that have used the single-rate kinetic model to model their combustion processes.

4.4.3 Two Competing Reactions Model

This model appreciates that volatiles can be classified; hence, they must be treated differently. Because coal contains both heavy volatiles and light volatiles, the devolatilisation rates are treated differently. The light volatile matter, which has a higher H/C ratio, is modelled such that it is released at low temperatures, whilst the heavy volatile matter, with a low H/C ratio, is released at high temperatures. This model was refined by Kaboyashi [23] as shown in Equation (4.8), so it can be adapted for CFD. Table 4.2 summarises some of the relevant applications of the two competing reactions model.

TABLE 4.1
Single-Rate Devolatilisation Model Applications

Case	Fuel	Outcome	Reference
Co-combustion of coal with olive waste	Coal and olive waste	Validation was with ± 5 to 10% for burnout	Álvarez et al. [19]
Co-combustion of coal and pine sawdust in a 1-MW down-fired furnace	Coal and pine sawdust	No validation was reported	Backreedy et al. [20]
Pyrolysis of biomass	Pine sawdust, oak, almond shells, gumtree	No validation was reported	Di Blasi [21]
Co-combustion of coal and biomass	Coal and grass	Validation using temperature and species mass fractions at furnace outlet data	Pallarés et al. [22]

TABLE 4.2
Two Competing Rate Devolatilisation Model Applications

Case	Fuel	Outcome	Reference
Combustion of coal and coke in a drop tube furnace	coal	Used species mass fractions, gas velocity, and temperature measurements for validation	Zhuo and Shen [25]
Combustion of coal in 600-MW powerplant	coal	Validation showed max deviation of 5% for temperature and 4 % for O_2 species mass fractions	Wang et al. [26]
Combustion in a 630-MW tangentially fired furnace	coal	At the furnace exit, validation using O_2 mass fraction showed a 5.7% deviation and 3.3% for NOx	Chang et al. [27]
Combustion in a swirl burner in a 60-kW furnace	Lignite with oxyfuel	Used another previously done model that uses large eddy simulations for validation	Sadiki et al. [28]
Combustion of biomass in a blast furnace	Charcoal	The burnout rates used for both spherical and cylindrical shaped biomass was within 10% agreement	Liu and Shen [29]

$$\frac{m_v(t)}{m_{p,0} - m_{p,daf}} = \int_0^t \left(\alpha_1.k_1 + \alpha_2.k_2\right).e^{-\int_0^t (k_1 + k_2)dt} dt \tag{4.8}$$

$$k_1 = A_1.e^{-\frac{E_{a1}}{\Re T}}, k_2 = A_2.e^{-\frac{E_{a2}}{\Re T}} \tag{4.9}$$

where:

k_1 and k_2 represent the competing kinetic rates that control devolatilisation at low temperatures and high temperatures respectively,

$m_v(t)$ represents the volatile yield at a certain time,

α_1 and α_2 represent the mass stoichiometric factor of volatiles evolved at low temperatures and high temperatures, and

$m_{p,daf}$ represents the mass of dry ash-free particles.

It is important to note that $m_{p,0} - m_{p,daf} = (1 - f_{w,0})m_{p,0} - m_a$. By separating the devolatilisation process, the model tends to be more accurate than the single-rate kinetic model from a theoretical perspective. However, the main limitations associated with the single-rate kinetic model are shared with this model as well [16, 17]. The approach is further complicated when trying to quantify the heavy volatiles against the light volatiles during experimental analysis, which leaves room for a lot of assumptions, thus affecting the credibility of the modelled results. An extension to this model is the universal devolatilisation process model, which uses theoretical calculations specific to the two competing reactions model to determine the competing kinetic parameters [24].

4.4.4 Distributed Activation Energy Model

The distributed activation energy model (DAEM) when applied to devolatilisation assumes that the pyrolysis process is comprised of independent reactions that occur parallel to each other. What is important is that these reactions have unique properties, such as order of reaction, as well as activation energy. This approach makes the model more accurate than the single-rate kinetic model and the two competing kinetic reactions, though the main limitations are shared [30]. This means, at a heating rate of β the relationship between temperature, activation energy, and pre-exponential factor is represented by Equation (4.10).

$$\ln\frac{\beta}{T^2} = \ln\left(\frac{A\mathcal{R}}{E_a}\right) + 0.6075 - \frac{E_a}{\Re T} \tag{4.10}$$

The constant 0.6075 is obtained through simplification of the single-rate kinetic model. The DAEM model has been known to produce good results, especially with complex reactions such as biomass pyrolysis as well as low temperature degradation [31–34]. A variation to this model is the universal devolatilisation process model (UDPM), which proposes use of extra constants so as to make the model more accurate.

4.4.5 Chemical Percolation Devolatilisation Model

The chemical percolation devolatilisation (CPD) model characterises the devolatilisation behaviour of rapidly heated solid fuels based on the physical and chemical transformations that take place within the fuel structure. This model can be classified as a phenomenological model since it is not based on the Arrhenius equation. This model was initially proposed specifically to study coal species behaviour. The approach is based on taking coal as a lattice that has aliphatic bridge connections (carbon atoms forming open chains). These bridges break down during the devolatilisation process but not at the same temperature [13]. The main advantage of this over the constant rate, single kinetic, two-competing rates, and DAEM models is its ability to predict volatile compositions during the modelling process. This means after performing CFD modelling, the researcher will have information regarding the molecular weight and amount of specific species generated during the devolatilisation process [8].

Use of the CPD model with CFD modelling requires input of the initial fraction of bridges in lattice, initial fraction of intact bridges, lattice coordination number, cluster molecular weight, and side chain molecular weight. Such parameters are determined by sophisticated mass spectrometry, nuclear magnetic resonance, and infrared spectrometry experiments. Biomass chemical chains are divided such that they are represented as containing lignin, cellulose, and hemicellulose, with each having its own specific CPD input parameters [35].

4.4.6 Functional Group Devolatilisation Model

As with the previous CPD model, the Functional Group Devolatilisation Model (FGD) can predict the volatile compositions. The difference with the CPD model is in how the coal structure is treated. Inasmuch as the CPD model prefers coal having a lattice structure with aliphatic bridge connections, the FGD model prefers representing the coal structure as being based on elemental and functional groups. Ma et al. [36] employed this model with satisfactory results for a 0.5-MW pulverised furnace test facility.

4.4.7 Functional Group Depolymerisation Vapourisation and Crosslinking Devolatilisation Model

The functional group depolymerisation, vapourisation, and crosslinking devolatilisation model (FG-DVC) combines the functional group model for gas evolution and the statistical depolymerisation, vapourisation, and cross-linking model. A subroutine based on the FG model is able to calculate the tar and char compositions throughout the devolatilisation process [37]. The other subroutine based on the DVC model can deduce the molecular weight of the different fragments from devolatilisation [13]. Theoretically, light fragments evolve into tar whilst the heavy fragments become metaplastic. This process of evolving involves depolymerisation, cross-linking, and transport [38]. This model showed a considerable deviation between 5% and 10% during validation by Alvarez et al. [19].

4.4.8 FLASHCHAIN Model

The FLASHCHAIN Model is a phenomenological model that can predict the volatile contents with good accuracy. The principle of this model is based on four generic structural components used to characterise coal, which are labile bridges, char links, aromatic nuclei, and peripheral groups. More emphasis is placed on labile bridges because their evolution determines the tar and gas yields, which in turn determine the volatile yield rate and composition [13].

4.5 HETEROGENEOUS CHAR COMBUSTION MODELS

As soon as the fuel particle becomes free of volatiles, the remaining solid substance, comprising mainly of char and mineral elements, softens into a porous substance. Most CFD models assume that the char particle is spherical in shape as a way of simplifying the modelling process. Heterogeneous char combustion is a surface process which makes the effect of particle shape pronounced at this stage. The global reactions between carbon, oxygen, carbon dioxide, and water result in the formation of carbon monoxide within the boundary layer as represented by Equations (4.11) to (4.13) [1, 39].

	Enthalpy of Reaction	
$C + \frac{1}{2}O_2 \rightarrow CO$	+122.9 kJ/kmol	(4.11)
$C + CO_2 \rightarrow CO$	−172.0 kJ/kmol	(4.12)
$C + H_2O \rightarrow CO + H_2$	−131.0 kJ/kmol	(4.13)

The task of bringing heterogeneous char combustion models to correctly predict industrial combustion is hindered in certain instances by the assumptions that are employed. Industrial applications find it difficult to achieve 100% combustion, which is the default setting for most combustion modelling. As such, validation of heterogeneous char combustion models becomes difficult to a large extent [13].

4.5.1 Diffusion Limited Rate Model

This model assumes a surface reaction that proceeds at a rate determined by the diffusion of gaseous oxidant from bulk flow to particle surface [5]. Baum and Street [40] refined the equations to be compatible with CFD as represented in Equation (4.14).

$$\frac{dm_p}{dt} = -A_p.D_0.\frac{X_{ox}.T_\infty.\rho}{S_b.(T_p + T_\infty)} \tag{4.14}$$

where:

m_p represents particle mass,
A_p represents particle surface area,
D_0 represents diffusion rate coefficient of oxygen in the carrier gas,
X_{ox} represents mass fraction of oxygen in the carrier gas,
$T\infty$ represents the temperature of the carrier gas, and
S_b represents the stoichiometric coefficient of oxygen during char oxidation (S_b = 1 for complete combustion and S_b = 0.5 for incomplete combustion).

This model assumes a constant particle diameter throughout char combustion, which then becomes one of its limitations, since particle diameter must decrease as char combusts, releasing carbon dioxide or monoxide [8].

4.5.2 Kinetics Diffusion-Limited Model

As a way of improving the accuracy of the diffusion-limited model, char reaction is assumed to depend on both diffusion of gaseous oxidant towards the particle surface alongside chemical reaction kinetics at the particle surface. In other words, the char surface reaction rate is controlled partly by the diffusion rate (physical control) and partly by reaction kinetics (chemical control). The chemical control is due to the chemical reaction kinetics which might also include pore diffusion. Baum and Street [40] alongside Field [41] helped develop this model for CFD applications. The overall equation that captures the diffusion rate and chemical kinetic rate is captured in Equation (4.15):

$$\frac{dm_p}{dt} = -A_p . p_{ox} . \frac{D_0 . k}{D_0 + k} . \phi_{eh} \tag{4.15}$$

where:

P_{ox} represents partial pressure of oxygen in the carrier gas,
D_0 represents diffusion rate coefficient of oxygen in the carrier gas,
k represents the chemical reaction kinetic rate constant, and
Φ_{eh} represents enhancement factor dependent on particle shape.

As a step further in improving model accuracy, the kinetics-diffusion limited model is divided into the Field char oxidation model and the Gibb char oxidation model depending on the application. The Field char oxidation model performs better when the char particle is considered spherical, hence it is mostly adapted for coal combustion. The Field char oxidation model assumes a stagnant boundary layer around the char particle, which gives important information related to the diffusion rate when oxygen diffuses through the boundary layer only [41]. Equation (4.15) reduces to become Equation (4.16) [5]:

$$\frac{dm_p}{dt} = -A_p . \left(D_o^{-1} + k^{-1}\right)^{-1} . Y_{ox} . \frac{P}{P_{atm}} \tag{4.16}$$

where

Y_{ox} represents the mole fraction of oxygen in the carrier gas,
P represents local pressure in the boundary layer, and
P_{atm} represents atmospheric pressure.

The Gibb char oxidation model is more adapted for irregular shapes and porous particles; hence, it is better suited for biomass combustion. This model has parameters that capture the char particle void fraction, volume/internal surface area ratio, effective internal diffusion coefficient of oxygen within the pores, and the product ratio of the carbon atoms/oxygen molecules involved in the oxidation processes [5].

4.5.3 Smith Intrinsic Char Reaction Model

Modification of the kinetics-diffusion char model so as to incorporate change in diffusion rates due to the existence of pores within a char particle results in the development of a model called the intrinsic char reaction model as propounded by Smith [36]. The chemical rate is then expressed as a function of intrinsic chemical and pore diffusion rates as demonstrated in Equation (4.17):

$$k = \eta.\frac{d_p}{6}.\rho_p.A_g.k_i \tag{4.17}$$

where

η represents the ratio of the actual combustion rate to the rate attainable if no pore diffusion resistance existed,

ρ_p represents apparent density of char,

A_g represents specific internal surface area of the char particle, and

k_i represents intrinsic reactivity.

4.5.4 Multiple Surface Reaction Model

This model takes the approach of assuming the char particle depletes based on the stoichiometric coefficients obtained from the chemical reactions given in Equations (4.11) to (4.13). This model has demonstrated good predictions for char combustion under excess air; thus, it is preferred for most applications. It is, however, limited when dealing with rich mixtures that promote the gasification of char with H_2O or CO. This is because this model has difficulties recognising reactions that occur within the boundary layer [42]. This model is also limited with regard to combustion scenarios that alternate between non-premixed, premixed, and partially premixed combustion [5, 43].

4.5.5 Multistep Semi-Global Kinetics Char Combustion Model

This model is built upon the basis of the multiple surface reaction model by placing more emphasis on intermediate reactions that result in radical formation. Hurt and Carlo [44] proposed this model after studying the carbon-oxygen mechanism as applied to CFD. As such, this model is good at recognising the limiting reactions that affect the progression of the combustion process [45, 46].

4.5.6 Boundary Layer Models

These models try to divide the zones of reaction around the char particle into different sections. The simplest is the single film model, which assumes that CO is the primary product of oxidation within the boundary layer represented as a single film around the particle (boundary layer). However, the burnout is usually overpredicted with the single film. As a step further, Burke and Schumann [47] adopted the double-film model for CFD which automatically assumed the existence of two distinct regions within the boundary layer. Specific reactions take place within each region. Makino and Law [48] developed the continuous film model which considers oxidation of C and CO within the boundary layer to CO_2 which comes at the expense of computational power and complexity. This in turn makes the continuous film model unattractive for engineering modelling applications [22, 49].

4.6 HOMOGENEOUS VOLATILE MATTER COMBUSTION MODELS

The volatiles released during devolatilisation create a source for gaseous combustion. Volatiles carry almost 50% of the energy for coal and more for biomass. Ignition, local temperature, flame stability, species distribution, and pollutant formation are all influenced by homogeneous volatile combustion. In CFD simulations, the volatile gases are taken as one single "artificial" species, $C_xH_yO_z$, or a few artificial species. Since homogeneous combustion is a reaction between gases, homogeneous combustion models are applicable to any type of fuel, and the usual reactions are presented by Equations (4.18) to (4.20) [7].

Enthalpy of Reaction

$$H_2 + \frac{1}{2}O_2 \rightarrow H_2O \quad 242.0\,\text{kJ/mol} \tag{4.18}$$

$$CO + \frac{1}{2}O_2 \rightarrow CO_2 \quad 283.0\,\text{kJ/mol} \tag{4.19}$$

$$CO + H_2O \rightarrow CO_2 + H_2 \quad 41.1\,\text{kJ/mol} \tag{4.20}$$

$$C_xH_yO_z + \left(2x + \frac{y}{2} - \frac{z}{2}\right)O_2 \rightarrow xCO_2 + \frac{y}{2}H_2O \tag{4.21}$$

For homogeneous volatile matter combustion models to be employed, knowledge of the chemical reaction mechanism to be used needs to be considered as well. The simplest of these mechanisms is the global one-step reaction mechanism which forms a simple equation without any intermediate species. Using the general lumped form of volatiles, Equation (4.21) can be used to represent the mechanism [7].

For a more accurate approach, the global two-step reaction mechanism is employed, taking carbon monoxide as the intermediate species. Equations (4.22) and (4.23) can then be used to represent the mechanisms involved in the global two-step reaction mechanism [19, 50].

$$C_xH_yO_z + \left(\frac{x}{2} + \frac{y}{4} - \frac{z}{2}\right)O_2 \rightarrow xCO + \frac{y}{2}H_2O \tag{4.22}$$

$$CO + 0.5O_2 \rightarrow CO_2 \tag{4.23}$$

However, another mechanism was proposed by Jones and Lindstedt [51] which is a four-step mechanism (JL4) to improve on the accuracy as given by Equations (4.24) to (4.27) [19, 50].

$$C_xH_yO_z + \left(\frac{x}{2} + \frac{y}{4} - \frac{z}{2}\right)O_2 \rightarrow xCO + \frac{y}{2}H_2O \tag{4.24}$$

$$CO + 0.5O_2 \rightarrow CO_2 \tag{4.25}$$

$$CO + H_2O \rightarrow CO_2 + H_2 \tag{4.26}$$

$$H_2 + 0.5O_2 \rightarrow H_2O \tag{4.27}$$

These different reaction mechanisms are then applied to the different homogeneous reaction models depending on the complexity of the problem and the required accuracy.

4.6.1 Eddy-Dissipation Model

The eddy-dissipation model (EDM) assumes that the chemical kinetic rate is directly proportional to the time required to mix reactants at a molecular level. Due to the turbulent nature of combustion, the eddy properties are responsible for giving information concerning mixing time. The simplified chemical kinetic rate equation is given in Equation (4.28) [5, 52].

$$k \propto \frac{k_t}{\varepsilon_t} \tag{4.28}$$

where

k_t represents turbulent kinetic energy and

ε_t represents turbulent dissipation energy.

One of the major drawbacks of the eddy-dissipation model stems from the lack of information concerning the constants used, meaning more time is spent on validation before using the model for each application process [5, 52].

4.6.2 Finite Rate Chemistry Model

This model computes the chemical reaction rates of individual species going through reactions during the combustion process [8]. The finite rate kinetics model assumes turbulence and chemical reaction rates do not interact. Since combustion is based on multiple mechanisms that require a lot of information, a lot of processing power is also required to solve those mechanisms [53].

4.6.3 Eddy-Dissipation Concept Model

The eddy-dissipation concept (EDC) is an extension of the eddy-dissipation model. As with the eddy-dissipation model, the EDC model depends mainly on the eddies formed (fine structures) as mixing occurs at a micro scale [54]. The fine structures are demarcated by a parameter from the Kolmogorov scale which implies that reactions take place in the fine structures, whilst the larger structures are considered inert until they break up into smaller eddies. The only major setback of the EDC model is the need to work under high-Reynolds-number conditions so as to assure better breakup of the flame [53, 55].

4.6.4 Equilibrium Approach Model

This model uses a probability density function in the form of lookup tables to describe the turbulence-chemistry interactions during the combustion process. As noted by Tabet et al. [8], no individual species transport equations are solved, but rather the mean and variance of the species mixture fraction are solved.

4.6.5 Mixed-Is-Reacted Approach Model

As the name suggests, once the fuel and the oxidant come into contact, products are formed. This model is also referred to as the equilibrium model because the mass fractions of the reactants (fuel and oxidant) and products are obtained by solving the conservation equation, considering the variables to be the mean as well as the variance of the mixture fraction for each fuel. This model works best for applications with high Damköhler numbers (Da>>1), meaning turbulence occurs very quickly compared to the actual reaction [22].

4.7 EMISSION MODELS

Emissions from solid fuel combustion come in different forms; hence, there is interest from most environmental studies and climate change advocates. Particulate matter, carbon-based species, SOx, and NOx are the main products produced during the combustion of coal or biomass species. Since the char and volatile combustion reactions result in the formation of CO_2 and CO, which have already been discussed, this section focuses on NOx, SOx, and particulate formation models.

4.7.1 NOx Pollutant Formation

Oxides of nitrogen, which are nitrogen monoxide (NO), nitrogen dioxide (NO_2), nitrous oxide (N_2O), and other less common compounds like nitrogen tetroxide (N_2O_4) are clustered together and referred to as NOx. Nitrogen monoxide, also known as nitric oxide, is the most common NOx found just after combustion, though it quickly combines with the atmosphere to form NO_2 and acid rain. Although NO is considered a pollutant, it is also important to acknowledge that most living organisms require nitrogen for balanced growth and functioning, humans included. As a result, the notion of pollution only comes into play when the NO reacts to form other harmful compounds [56].

The major mechanisms of NOx formation are thermal NOx, fuel NOx, and prompt NOx. Thermal NOx is formed when nitrogen and oxygen combine at relatively high temperatures and oxygen-rich environments. Fuel NOx is mainly due to the nitrogen inherent in the fuel before combustion, which makes HCN radicals more pronounced in the NO formation equation. Prompt NOx is formed when atmospheric nitrogen reacts with hydrocarbons derived from the fuel. This reaction occurs at low temperatures and/or fuel-rich conditions [16]. Modelling of NOx emissions includes the two most important mechanisms responsible for NOx formation, which are fuel NOx and thermal NOx, whilst prompt NOx formation is disregarded since the air-to-fuel ratios employed during combustion are lean enough to avoid fuel-rich pockets.

Thermal NOx is the main contributor to NOx pollutant formation. The formation rate is considered a function of temperature and residence time of nitrogen at high temperatures. As such, thermal NOx increases exponentially with temperature. The classic chemical reaction equation depicting thermal NOx is modelled by the Zeldovich mechanism, as given by Equations (4.29) and (4.30). Naturally, from reaction kinetics theory, there exists a number of intermediate reactions when nitrogen reacts with oxygen to form NOx. However, an extended Zeldovich mechanism, Equation (4.31) to (4.33), was coined to further refine the representative reactions [56].

$$N_2 + 2O_2 \rightarrow 2NO_2 \tag{4.29}$$

$$N_2 + O_2 \rightarrow 2NO \tag{4.30}$$

$$N_2 + O \leftrightarrow N + NO \tag{4.31}$$

$$N + O_2 \leftrightarrow O + NO \tag{4.32}$$

$$N + OH \leftrightarrow H + NO \tag{4.33}$$

Abatement techniques used in industry with regard to thermal NOx are based on trying to reduce the flame peak temperature, residence time of the flame zone, and amount of excess air available. Experiments have shown that thermal NOx varies linearly with residence time up to about 5 seconds, varies linearly with N2 concentration, varies exponentially with temperature, and varies with the square root of O_2 concentration, as given by Equation (4.34). Equation (4.34) is only valid for a residence time below 5 seconds as well as a temperature range of 1800°C–2000°C [56, 57].

$$\left[NO\right] = C_1.t.e^{C_2/T}.\left[N_2\right].\left[O_2\right]^{1/2} \tag{4.34}$$

where:

[NO] represents nitrogen oxide concentration,

C_1 represents constant = 5.2×10^{17},

t represents residence time,

C_2 represents constant = -7.23×10^5

T represents temperature,

$[N_2]$ represents nitrogen dioxide concentration, and

$[O_2]$ represents oxygen concentration

Fuel NOx is due to the oxidation of nitrogen contained within the fuel, normally bound chemically to the structure as represented by Equation (4.35). De Soete [57] propounded mechanisms that are responsible for the formation of fuel NOx.

$$\text{Fuel N} + O_2 \rightarrow NO, NO_2, CO_2, H_2O, \text{trace elements} \tag{4.35}$$

The nitrogen bound within the volatile matter is released during devolatilisation whilst the one bound within the char remains to be released during char oxidation. The common assumption during fuel NOx modelling is the conversion of volatile-N to mostly HCN radicals and the rest to NH_3. Depending on whether the region is rich or lean, the HCN radicals and NH_3 will react to form either NO or N_2 respectively. Clean fuels do not have any fuel NOx unlike coal and biomass which usually contribute to fuel NOx. Modelling of fuel NOx is heavily dependent on the properties of the fuel. Techniques used to reduce NOx formation include fuel switching, additives, fuel pre-treatment, oxidiser switching, low excess air, staged combustion, flue gas recirculation, pulsed combustion, reburning, burner out of service, and post combustion treatment techniques [56].

4.7.2 SOx Pollutant Formation

The SOx formation mechanism is based on the findings of Kramlich et al. [58] who proposed 20 reversible reactions and 12 intermediate species that include S, S_2, SH, SO, SO_2, H_2S, H, H_2, OH, H_2O, O, and O_2. Introduction of biomass waste does not necessarily imply a reduction in SOx emissions mainly because other forms of biomass waste contain a sulphur composition higher than that for the coal used. Fortunately, pre-treatment of the biomass waste gives a good reduction of combustion emissions whilst tackling the menace caused by the disposal of biomass waste. Dealing with SOx modelling is quite sensitive towards the distribution between the sulphur left in char against the sulphur that would have vapourised during the devolatilisation stage. As such, pyritic sulphur is assumed to convert into organic sulphur which then appears in the char and volatiles. This assumption simplifies the modelling process by grouping all the important mechanisms [59].

Zhou et al. [60] demonstrated a decrease in emissions for both NOx and SOx when coal was co-combusted with corn stalk or sawdust. Several authors also reported NOx reduction, with Akhtar et al. [61] pointing out how the volatiles released at low temperatures for biomass combustion are responsible for the reduction in NOx emissions. This was related to the decrease in overall combustion temperature; thus, theoretically, thermal NOx is reduced [1]. Although co-combustion of coal and biomass generally results in NOx reduction and SOx reduction, power plant operators are also concerned about the variation in combustion parameters such as ignition and burnout indices [62].

4.8 SUMMARY

Several modelling approaches for coal biomass co-firing were discussed in this chapter. For devolatilisation, the single-rate kinetic rate model is preferred when the objective is to capture volatile yield rates without overwhelming the model with input requirements. However, when the composition of the released volatiles is critical, phenomenological models such as the CPD, FGD, FG-DVC, and FLASHCHAIN are better positioned to address such tasks which is at the expense of computational power and extensive fuel characterisation. The simplest heterogeneous char oxidation model is the diffusion limited rate model which provides only coarse estimates of reaction rates, while the kinetic diffusion limited model proposed by Field and Gibb improves the accuracy by accounting for both transport and surface kinetics. Since the oxidiser has to diffuse to the surface of the solid fuel particle, using boundary layer models tends to improve the accuracy of heterogeneous char combustion models. Homogeneous volatile matter combustion models increase complexity as the number of reactions that need to be modelled increases. The eddy-dissipation model is efficient but sensitive to the constant that is employed. The eddy-dissipation concept is ideal when the task is to model multiple reactions. Thermal NOx depends strongly on flame temperature and residence time, while fuel NOx pathways are governed by the transformation of fuel-bound nitrogen into intermediates such as HCN and NH_3. Sulphur oxide emissions hinge on how sulphur partitions between volatiles and char during combustion. Importantly, co-firing with biomass can reduce both NOx and SOx formation, especially if the biomass has lower nitrogen and sulphur content.

REFERENCES

1. Turns, S. *An Introduction to Combustion: Concepts and Applications*; McGraw-Hill series in mechanical engineering; 2nd ed.; McGraw-Hill: New York, 1996.
2. Marangwanda, G.T.; Madyira, D.M.; Babarinde, T.O. Coal Combustion Models: A Review. *Journal of Physics: Conference Series* **2019**, *1378*, 032070, doi:10.1088/1742-6596/1378/3/032070.
3. Álvarez, A.; Pizarro, C.; García, R.; Bueno, J.L.; Lavín, A.G. Determination of Kinetic Parameters for Biomass Combustion. *Bioresource Technology* **2016**, *216*, 36–43, doi:10.1016/j.biortech.2016.05.039.
4. Al-Abbas, A.H.; Naser, J. Effect of Chemical Reaction Mechanisms and NOx Modeling on Air-Fired and Oxy-Fuel Combustion of Lignite in a 100-KW Furnace. *Energy & Fuels* **2012**, *26*, 3329–3348, doi:10.1021/ef300403a.
5. ANSYS. *ANSYS FLUENT User's Guide;* ANSYS: Pennsylvania, 2021, 1–170.
6. Ghenai, C.; Janajreh, I. CFD Analysis of the Effects of Co-Firing Biomass with Coal. *Energy Conversion and Management* **2010**, *51*, 1694–1701, doi:10.1016/j.enconman.2009.11.045.
7. Ma, L.; Jones, J.M.; Pourkashanian, M.; Williams, A. Modelling the Combustion of Pulverized Biomass in an Industrial Combustion Test Furnace. *Fuel* **2007**, *86*, 1959–1965, doi:10.1016/j.fuel.2006.12.019.
8. Tabet, F.; Gökalp, I. Review on CFD Based Models for Co-Firing Coal and Biomass. *Renewable and Sustainable Energy Reviews* **2015**, *51*, 1101–1114, doi:10.1016/j.rser.2015.07.045.
9. Yin, C.; Rosendahl, L.A.; Condra, T.J.; Kær, S.K. Use of Numerical Modeling in Design for Co-Firing Biomass in Wall-Fired Burners. *Chemical Engineering Science* **2004**, *59*, 3281–3292, doi:10.1016/j.ces.2004.04.036.
10. Yin, C. Effects of Moisture Release and Radiation Properties in Pulverized Fuel Combustion: A CFD Modelling Study. *Fuel* **2016**, *165*, 252–259, doi:10.1016/J.fuel.2015.10.057.
11. Supramono, D.; Sitorus, A.F.; Nasikin, M. Synergistic Effect on the Non-Oxygenated Fraction of Bio-Oil in Thermal Co-Pyrolysis of Biomass and Polypropylene at Low Heating Rate. *Processes* **2020**, *8*, 57, doi:10.3390/pr8010057.

12. Sadhukhan, A.K.; Gupta, P.; Saha, R.K. Modeling and Experimental Investigations on the Pyrolysis of Large Coal Particles. *Energy and Fuels* **2011**, *25*, 5573–5583, doi:10.1021/ef201162c.
13. Sankar, G.; Kumar, D.S.; Balasubramanian, K.R. Computational Modeling of Pulverized Coal Fired Boilers – A Review on the Current Position. *Fuel* **2019**, *236*, 643–665, doi:10.1016/j.fuel.2018.08.154.
14. Badzioch, S.; Hawksley, P.G.W. Kinetics of Thermal Decomposition of Pulverized Coal Particles. *Industrial & Engineering Chemistry Process Design and Development* **1970**, *9*, 521–530, doi:10.1021/i260036a005.
15. Marangwanda, G.T.; Madyira, D.M.; Ndungu, P.G.; Chihobo, C.H. Combustion Characterisation of Bituminous Coal and Pinus Sawdust Blends by Use of Thermo-Gravimetric Analysis. *Energies (Basel)* **2021**, *14*, 1–20, doi:10.3390/en14227547.
16. Choi, C.R.; Kim, C.N.; Ryul, C.; Nyung, C. Numerical Investigation on the Flow, Combustion and NOx Emission Characteristics in a 500MWe Tangentially Fired Pulverized-Coal Boiler. *Fuel* **2009**, *88*, 1720–1731, doi:10.1016/j.fuel.2009.04.001.
17. Pérez-Jeldres, R.; Cornejo, P.; Flores, M.; Gordon, A.; García, X. A Modeling Approach to Co-Firing Biomass/Coal Blends in Pulverized Coal Utility Boilers: Synergistic Effects and Emissions Profiles. *Energy* **2017**, *120*, 663–674, doi:10.1016/j.energy.2016.11.116.
18. Marangwanda, G.T.; Madyira, D.M.; Chihobo, H.C.; Babarinde, T.O. Modelling Co-Combustion of Bituminous Coal and Pine Sawdust: Thermal Behaviour. *Fuel Communications* **2021**, *9*, 100035, doi:10.1016/j.jfueco.2021.100035.
19. Álvarez, L.; Yin, C.; Riaza, J.; Pevida, C.; Pis, J.J.; Rubiera, F. Biomass Co-Firing under Oxy-Fuel Conditions: A Computational Fluid Dynamics Modelling Study and Experimental Validation. *Fuel Processing Technology* **2014**, *120*, 22–33, doi:10.1016/j.fuproc.2013.12.005.
20. Backreedy, R.I.; Fletcher, L.M.; Jones, J.M.; Ma, L.; Pourkashanian, M.; Williams, A. Co-Firing Pulverised Coal and Biomass: A Modeling Approach. *Proceedings of the Combustion Institute* **2005**, *30 II*, 2955–2964, doi:10.1016/j.proci.2004.08.085.
21. Di Blasi, C. Modeling Chemical and Physical Processes of Wood and Biomass Pyrolysis. *Progress in Energy and Combustion Science* **2008**, *34*, 47–90, doi:10.1016/j.pecs.2006.12.001.
22. Pallarés, J.; Gil, A.; Cortés, C.; Herce, C. Numerical Study of Co-Firing Coal and Cynara Cardunculus in a 350 MWe Utility Boiler. *Fuel Processing Technology* **2009**, *90*, 1207–1213, doi:10.1016/j.fuproc.2009.05.025.
23. Kobayashi, H.; Howard, J.B.; Sarofim, A.F. Coal Devolatilization at High Temperatures. *Symposium (International) on Combustion* **1977**, *16*, 411–425, doi:10.1016/S0082-0784(77)80341-X.
24. Luo, K.; Xing, J.; Bai, Y.; Fan, J. Universal Devolatilization Process Model for Numerical Simulations of Coal Combustion. *Energy and Fuels* **2017**, *31*, 6525–6540, doi:10.1021/acs.energyfuels.7b00970.
25. Zhuo, Y.; Shen, Y. Transient 3D CFD Study of Pulverised Coal Combustion and Coke Combustion in a Blast Furnace: Effect of Blast Conditions. *Fuel* **2023**, *340*, 127468, doi:10.1016/j.fuel.2023.127468.
26. Wang, H.; Jin, H.; Yang, Z.; Deng, S.; Wu, X.; An, J.; Sheng, R.; Ti, S. CFD Modeling of Flow, Combustion and NOx Emission in a Wall-Fired Boiler at Different Low-Load Operating Conditions. *Applied Thermal Engineering* **2024**, *236*, 121824, doi:10.1016/j.applthermaleng.2023.121824.

27. Chang, J.; Wang, X.; Zhou, Z.; Chen, H.; Niu, Y. CFD Modeling of Hydrodynamics, Combustion and NOx Emission in a Tangentially Fired Pulverized-Coal Boiler at Low Load Operating Conditions. *Advanced Powder Technology* **2021**, *32*, 290–303, doi:10.1016/j.apt.2020.12.008.
28. Sadiki, A.; Agrebi, S.; Chrigui, M.; Doost, A.S.; Knappstein, R.; Di Mare, F.; Janicka, J.; Massmeyer, A.; Zabrodiec, D.; Hees, J.; et al. Analyzing the Effects of Turbulence and Multiphase Treatments on Oxy-Coal Combustion Process Predictions Using LES and RANS. *Chemical Engineering Science* **2017**, *166*, 283–302, doi:10.1016/j.ces.2017.03.015.
29. Liu, Y.; Shen, Y. Computational Fluid Dynamics Study of Biomass Combustion in a Simulated Ironmaking Blast Furnace: Effect of the Particle Shape. *Energy & Fuels* **2018**, *32*, 4372–4381, doi:10.1021/acs.energyfuels.7b03150.
30. de Caprariis, B.; De Filippis, P.; Herce, C.; Verdone, N. Double-Gaussian Distributed Activation Energy Model for Coal Devolatilization. *Energy & Fuels* **2012**, *26*, 6153–6159, doi:10.1021/ef301092r.
31. Mishra, R.K.; Mohanty, K. Pyrolysis Kinetics and Thermal Behavior of Waste Sawdust Biomass Using Thermogravimetric Analysis. *Bioresource Technology* **2018**, *251*, 63–74, doi:10.1016/j.biortech.2017.12.029.
32. Chen, R.; Cai, J.; Li, X.; Lyu, Q.; Qi, X. Modelling of Large Biomass and Coal Particle Based on a Novel C-DAEM: A Numerical Study on Heat Transfer and Pyrolysis Behavior. *Energy* **2023**, *283*, 129137, doi:10.1016/j.energy.2023.129137.
33. Wang, J.; Lian, W.; Li, P.; Zhang, Z.; Yang, J.; Hao, X.; Huang, W.; Guan, G. Simulation of Pyrolysis in Low Rank Coal Particle by Using DAEM Kinetics Model: Reaction Behavior and Heat Transfer. *Fuel* **2017**, *207*, 126–135, doi:10.1016/J.FUEL.2017.06.078.
34. Rabaçal, M.; Costa, M.; Vascellari, M.; Hasse, C.; Rieth, M.; Kempf, A.M. A Large Eddy Simulation Study on the Effect of Devolatilization Modelling and Char Combustion Mode Modelling on the Structure of a Large-Scale, Biomass and Coal Co-Fired Flame. *Journal of Combustion* **2018**, *2018*, 1–15, doi:10.1155/2018/7036425.
35. Sheng, C.; Azevedo, J.L.T. Modeling Biomass Devolatilization Using the Chemical Percolation Devolatilization Model for the Main Components. *Proceedings of the Combustion Institute* **2002**, *29*, 407–414, doi:10.1016/S1540-7489(02)80054-2.
36. Ma, L.; Gharebaghi, M.; Porter, R.; Pourkashanian, M.; Jones, J.M.; Williams, A. Modelling Methods for Co-Fired Pulverised Fuel Furnaces. *Fuel* **2009**, *88*, 2448–2454, doi:10.1016/j.fuel.2009.02.030.
37. Chen, Y.; Charpenay, S.; Jensen, A.; Wójtowicz, M.A.; Serio, M.A.; Y., C. Modeling of Biomass Pyrolysis Kinetics. *Symposium International On Combustion* **1998**, *27*, 1327–1334, doi:10.1016/S0082-0784(98)80537-7.
38. Advanced Fuel Research. *Advanced Fuel-Devolatilization Model Functional-Group*; Connecticut, 1980.
39. Wang, Y.; Yan, L. CFD Studies on Biomass Thermochemical Conversion. *International Journal of Molecular Sciences* **2008**, *9*, 1108–1130, doi:10.3390/ijms9061108.
40. Baum, M.M.; Street, P.J. Predicting the Combustion Behaviour of Coal Particles. *Combustion Science and Technology* **1971**, *3*, 231–243, doi:10.1080/00102207108952290.
41. Field, A.M.; Field, M.A. Rate of Combustion Of Size-Graded Fractions of Char from a Low Rank Coal between 1200 K-2000K. *Combustion and Flame* **1969**, *13*, 237–252, doi:10.1016/0010-2180(69)90002-9.
42. Sarroza, A.C.; Bennet, T.D.; Eastwick, C.; Liu, H. Characterising Pulverised Fuel Ignition in a Visual Drop Tube Furnace by Use of a High-Speed Imaging Technique. *Fuel Processing Technology* **2017**, *157*, 1–11, doi:10.1016/j.fuproc.2016.11.002.

43. Zha, Q.; Li, D.; Wang, C.; Che, D. Numerical Evaluation of Heat Transfer and NOx Emissions under Deep-Air-Staging Conditions within a 600 MWe Tangentially Fired Pulverized-Coal Boiler. *Applied Thermal Engineering* **2017**, *116*, 170–181, doi:10.1016/j.applthermaleng.2016.12.088.
44. Hurt, R.H.; Calo, J.M. Semi-Global Intrinsic Kinetics for Char Combustion Modeling. *Combustion and Flame* **2001**, *125*, 1138–1149, doi:10.1016/S0010-2180(01)00234-6.
45. Bhuiyan, A.A.; Naser, J. Numerical Modeling of Biomass Co-Combustion with Pulverized Coal in a Small Scale Furnace. *Procedia Engineering* **2015**, *105*, 504–511, doi:10.1016/j.proeng.2015.05.083.
46. Bhuiyan, A.A.; Blicblau, A.S.; Naser, J. Co-Firing of Biomass and Slagging in Industrial Furnace: A Review on Modelling Approach. *Journal of the Energy Institute* **2017**, *90*, 838–854, doi:10.1016/j.joei.2016.08.010.
47. Burke, S.P.; Schumann, T.E.W. The Mechanism of Combustion of Solid Fuel. In Proceedings of the Proceedings 3rd International Conference Bituminous Coal; 1931; Vol. 2, p. 485.
48. Makino, A.; Law, C.K. Quasi-Steady and Transient Combustion of a Carbon Particle: Theory and Experimental Comparisons. *Symposium (International) on Combustion* **1988**, *21*, 183–191, doi:10.1016/S0082-0784(88)80245-5.
49. Yu, J.; Zhang, M.C. Experimental and Modeling Study on Char Combustion. *Energy & Fuels* **2009**, *23*, 2874–2885, doi:10.1021/ef900018b.
50. Yin, C.; Kaer, S.K.; Rosendahl, L.A.; Hvid, S.L. Modelling of Pulverized Coal and Biomass Co-Firing in a 150 KW Swirling-Stabilized Burner and Experimental Validation. *The Proceedings of the International Conference on Power Engineering (ICOPE)* **2009**, *2009.2*, 305–310, doi:10.1299/jsmeicope.2009.2._2-305_.
51. Jones, W.P.; Lindstedt, R.P. Global Reaction Schemes for Hydrocarbon Combustion. *Combustion and Flame* **1988**, *73*, 233–249, doi:10.1016/0010-2180(88)90021-1.
52. Elfasakhany, A.; Klason, T.; Bai, X.S. Modelling of Pulverised Wood Combustion Using a Functional Group Model. *Combustion Theory and Modelling* **2008**, *12*, 883–904, doi:10.1080/13647830802094344.
53. Shiehnejadhesar, A.; Mehrabian, R.; Scharler, R.; Goldin, G.M.; Obernberger, I. Development of a Gas Phase Combustion Model Suitable for Low and High Turbulence Conditions. *Fuel* **2014**, *126*, 177–187, doi:10.1016/j.fuel.2014.02.040.
54. Farokhi, M.; Birouk, M. Application of Eddy Dissipation Concept for Modeling Biomass Combustion, Part 2: Gas-Phase Combustion Modeling of a Small-Scale Fixed Bed Furnace. *Energy and Fuels* **2016**, *30*, 10800–10808, doi:10.1021/acs.energyfuels.6b01948.
55. Mardani, A.; Ch, H.; Mardani, A. Optimization of the Eddy Dissipation Concept (EDC) Model for Turbulence-Chemistry Interactions under Hot Diluted Combustion of CH4/H2. *Fuel* **2017**, *191*, 114–129, doi:10.1016/j.fuel.2016.11.056.
56. Baukal, C.E.Jr.; Bussman, W. NOx Emissions. In *The John Zink Hamworthy Combustion Handbook*; Baukal Jr., C.E., Ed.; CRC Press: London, 2013; pp. 418–47.
57. De Soete, G.G. Heterogeneous N2O and NO Formation from Bound Nitrogen Atoms during Coal Char Combustion. *Symposium (International) on Combustion* **1991**, *23*, 1257–1264, doi:10.1016/S0082-0784(06)80388-7.
58. Kramlich, J.C.; Malte, P.C.; Grosshandler, W.L. The Reaction of Fuel-Sulfur in Hydrocarbon Combustion. *Symposium (International) on Combustion* **1981**, *18*, 151–161, doi:10.1016/S0082-0784(81)80020-3.
59. Manovic, V.; Grubor, B. Correlation for the Total Sulfur Content in Char after Devolatilization. *Energy and Fuels* **2006**, *20*, 133–137, doi:10.1021/ef050244a.

60. Zhou, C.; Liu, G.; Wang, X.; Qi, C. Co-Combustion of Bituminous Coal and Biomass Fuel Blends: Thermochemical Characterization, Potential Utilization and Environmental Advantage. *Bioresource Technology* **2016**, *218*, 418–427, doi:10.1016/j.biortech.2016.06.134.
61. Akhtar, J.; Yaseen, A.; Munir, S. Effect of Rice Husk Co-Combustion with Coal on Gaseous Emissions and Combustion Efficiency. *Energy Sources, Part A: Recovery, Utilization, and Environmental Effects* **2018**, *40*, 1010–1018, doi:10.1080/15567036.2018.1468513.
62. Gao, H.; Runstedtler, A.; Majeski, A.; Boisvert, P.; Campbell, D. Optimizing a Woodchip and Coal Co-Firing Retrofit for a Power Utility Boiler Using CFD. *Biomass Bioenergy* **2016**, *88*, 35–42, doi:10.1016/j.biombioe.2016.03.006.

5 Modelling of Physical Phenomena during Co-Combustion

5.1 INTRODUCTION

This chapter discusses the modelling of physical phenomena that govern the co-combustion of solid fuels, with particular emphasis on fluid mechanics, particle dynamics, multiphase interactions, and heat transfer. The section on fluid dynamics develops turbulence models as applied to solid fuel combustion, highlighting the RNG k–ε formulation combined with near-wall treatments as the most suitable approach for capturing swirling, high-Reynolds-number flows typical of furnaces. Particle dynamics are then considered, with attention to the drag laws that dictate the motion of pulverised particles entrained in the gas stream. Because solid fuels such as biomass often exhibit irregular shapes, drag correlations that incorporate shape factor and sphericity are presented in detail to improve the accuracy of particle trajectory and heat transfer predictions.

Multiphase modelling, which bridges the interactions between the fluid and solid phases, is introduced through the discrete phase model and illustrated using industrial case studies. This provides insight into how particle-to-fluid coupling affects combustion efficiency and pollutant formation. The chapter then turns to the energy conservation equation, focusing on radiation – the dominant mode of heat transfer in combustion systems. While basic radiation models account for the direction and position of radiative fluxes, greater accuracy is achieved by incorporating spectral effects, since combustion gases such as CO_2 and H_2O each possess distinct radiative properties. By integrating turbulence, particle dynamics, multiphase flow, and radiation models, this chapter establishes the physical modelling framework required for reliable CFD simulations of coal biomass co-firing.

5.2 TURBULENCE MODELS

Turbulence models seek to solve the conservation equations consisting of mass, momentum, energy, and species concentration. To acknowledge the effect of turbulence, modified conservation equations are formed by introducing averaged and fluctuating components for each parameter, as shown in Equation (5.1). This gives rise to the Reynolds-Averaged Navier-Stokes (RANS) equations with extra parameters related to average and fluctuating parameters which are shown in Equations

 DOI: 10.1201/9781003647942-5

(5.2) to (5.4). Various hypotheses have been propounded by various CFD researchers to simplify these extra parameters; hence, the various turbulence models.

$$\varphi = \overline{\varphi} + \varphi'' \qquad (5.1)$$

Momentum (e.g. Reynolds-Averaged Navier-Stokes):

$$\frac{\partial \overline{u}}{\partial t} + \frac{\partial}{\partial x_i}\left(\overline{c_i}\,\overline{u}\right) = -\frac{1}{\rho}\frac{\partial P}{\partial x} + \nu \frac{\partial}{\partial x_i}\left(\frac{\partial \overline{u}}{\partial x_i}\right) - \left[\frac{\partial \overline{u''u''}}{\partial x} + \frac{\partial \overline{u''v''}}{\partial y} + \frac{\partial \overline{u''w''}}{\partial z}\right] \qquad (5.2)$$

$$\frac{\partial \overline{v}}{\partial t} + \frac{\partial}{\partial x_i}\left(\overline{c_i}\,\overline{v}\right) = -\frac{1}{\rho}\frac{\partial P}{\partial y} + \nu \frac{\partial}{\partial x_i}\left(\frac{\partial \overline{v}}{\partial x_i}\right) - \left[\frac{\partial \overline{v''u''}}{\partial x} + \frac{\partial \overline{v''v''}}{\partial y} + \frac{\partial \overline{v''w''}}{\partial z}\right] \qquad (5.3)$$

$$\frac{\partial \overline{w}}{\partial t} + \frac{\partial}{\partial x_i}\left(\overline{c_i}\,\overline{w}\right) = -\frac{1}{\rho}\frac{\partial P}{\partial z} + \nu \frac{\partial}{\partial x_i}\left(\frac{\partial \overline{w}}{\partial x_i}\right) - \left[\frac{\partial \overline{w''u''}}{\partial x} + \frac{\partial \overline{w''v''}}{\partial y} + \frac{\partial \overline{w''w''}}{\partial z}\right] \qquad (5.4)$$

with

φ representing the variable (e.g. temperature (K), pressure (Pa), species mass fraction)

$\overline{\varphi}$ representing the average quantity (e.g. $\overline{u}; \overline{v}; \overline{w}$ are average values for velocity $\overline{c_i}$ in the x, y, and z velocity),

φ'' representing the fluctuating quantity (e.g. $u''; v''; w''$ are fluctuating values for x, y, and z velocity),

x_i representing the direction (x_1: x direction, x_2: y direction, x_3: z direction for Cartesian coordinates),

ρ representing the density (kg/m^3),

P representing the pressure (Pa), and

ν representing the kinematic viscosity (m^2.s).

The simplest hypothesis uses Boussinesq's approximation [1], and the most common hypothesis is Prandtl's approximation [2], which acknowledges the effects of the boundary layer (SST, k-ω). The Prandtl's approximation results in the formation of more sophisticated two-equation turbulence models. With regard to combustion, the k-epsilon (k-ε) two-equation turbulence model is sufficient for most industrial applications, though variations such as the RNG k-ε model and the realisable k-ε model are called upon to increase model convergence and accuracy [3, 4]. The RNG k-ε turbulence sub-model is known to converge much quicker for moderate swirling combustion compared to the standard k-ε turbulence sub-model, mainly due to a renormalisation (R_ε) function that is introduced as given in Equations (5.5) and (5.6) [4]. The main difference with the standard k-ε model is in the constants $C_{1\varepsilon} = 1.42$, $C_{2\varepsilon} = 1.68$.

$$\frac{\partial}{\partial t}\left(\rho k_t\right) + \frac{\partial}{\partial x_i}\left(\rho k_t c_i\right) = \frac{\partial}{\partial x_j}\left[\alpha_k \mu_{eff} \frac{\partial k_t}{\partial x_j}\right] + G_k + G_b - \rho\varepsilon_t - Y_m + S_k. \qquad (5.5)$$

$$\frac{\partial}{\partial t}(\rho\varepsilon_t)+\frac{\partial}{\partial x_i}(\rho\varepsilon_t c_i)=\frac{\partial}{\partial x_j}\left[\alpha_\varepsilon \mu_{eff}\frac{\partial \varepsilon_t}{\partial x_j}\right]+C_{1\varepsilon}\frac{\varepsilon_t}{k}(G_k+C_{3\varepsilon}G_b)-C_{2\varepsilon}\rho\frac{\varepsilon_t{}^2}{k_t}-R_\varepsilon+S_\varepsilon. \tag{5.6}$$

with

k_t representing the turbulence kinetic energy,

ε_t representing the rate of turbulence dissipation,

G_k representing the generation of turbulence kinetic energy due to the mean velocity gradients,

G_b representing the generation of turbulence kinetic energy due to buoyancy,

Y_m representing the contribution of the ratio between fluctuating dilatation in compressible turbulence to the overall dissipation rate,

α_k representing the inverse effective Prandtl number for turbulence kinetic energy,

α_ε representing the inverse effective Prandtl number for rate of turbulence dissipation, and

μ_{eff} representing the effective viscosity.

5.3 MULTIPHASE MODELLING

Understanding multiphase modelling is crucial when analysing solid fuel combustion applications. The main phases that are present during solid fuel combustion are solid particles, gaseous reactants, intermediate, and product species. The overall application range of multiphase systems is immense, ranging from chemical process industries, energy systems, environmental systems, as well as food industry. The main models used to solve multiphase modelling are the Euler and Euler model and the Euler and Lagrange model [5–7].

5.3.1 Eulerian and Eulerian Frameworks

In the Euler and Euler model, the fluid and solid phases are treated as separate continuums that will be interacting with each other. Normal fluid models can then be applied, such as mass, momentum, and energy conservation. Conversion of the solid phase into a continuum is where all the research lies, as various models have been proposed to facilitate this modelling approach, forming an extra equation called the kinetic theory of granular flow equation. The coupling is made apparent through normal fluid interactions related to source and sink terms within a fluid flow. This framework is suitable for systems that have a high concentration of particles. The most common name used for models under this framework is the discrete element models (DEM) [5, 8].

5.3.2 Eulerian and Lagrangian Frameworks

In the Euler and Lagrange model, only the fluid phase is treated as continua, whilst the solid phase is treated as a discrete phase. In the discrete phase, each particle is tracked through space and time using normal Newton's laws. The coupling is made apparent through normal fluid to solid interactions related to boundary layer theory such as drag-and-lift forces. This model is very useful for systems with low concentrations of the solid phase, hence its application in suspension combustion. The most common name used for models under this framework is the discrete phase models (DPM) [5].

5.3.3 Governing Equations

Since the fluid phase is considered as a continuum, the volume-average Navier-Stokes (VANS) equation is used to represent the fluid phase motion as given by Equations (5.7) and (5.8). When treating particles, Equation (5.9) is used; and for low concentration scenarios, the last two terms on the right tend towards zero, leading to the formation of the discrete phase model [5, 9].

$$\frac{\partial}{\partial t}(\varepsilon_f)+\frac{\partial}{\partial x_i}(\varepsilon_f u_f)=0 \tag{5.7}$$

$$\frac{\partial}{\partial t}(\varepsilon_f u_f \rho_f)+\frac{\partial}{\partial x_i}(\varepsilon_f \rho_f u_f u_f)=\frac{\partial}{\partial x_i}(P)+\frac{\partial}{\partial x_i}(\tau)+\varepsilon_f \rho_f g+S_p \tag{5.8}$$

$$\frac{\partial}{\partial t}(u_p)=-\frac{1}{\rho_p}\frac{\partial}{\partial x_i}(P)+F_D(u_f-u_p)+\frac{(\rho_p-\rho_f)}{\rho_p}g-\frac{1}{\rho_p}\frac{\partial}{\partial x_i}(\bar{\bar{\tau}}_s)+a_{other} \tag{5.9}$$

with

ε_f representing the void fraction,

ρ_f representing the fluid density,

u_f representing the average velocity of the fluid unit,

P representing the pressure,

τ representing the viscous stress tensor,

S_p representing the source term generated by the interaction between particles and volumetric fluid,

u_p representing the particle velocity,

ρ_p representing the particle density,

$\bar{\bar{\tau}}_s$ representing the solid stress tensor caused by particle-to-particle interaction,

F_D representing the drag force, and

a_{other} representing the acceleration caused by an external force

5.4 PARTICLE DYNAMICS

Drag force models are mainly centred on determining the particle drag coefficient, which requires a lot of experimental results for validation. Most solid fuels adopt an irregular shape when pulverised, which causes a lot of problems for CFD researchers who are concerned with accuracy. The following subsections are going to discuss the common drag models applied to solid fuel combustion applications.

5.4.1 Schiller-Naumann Model

The oldest and simplest model employed in CFD codes is the Schiller-Naumann model [4] which has its strength with spherical particles and requires less computational space. Equation (5.10) is the original empirical formula, but to promote numerical stability and closure, Equation (5.11) is employed by most CFD codes.

$$C_D = \frac{24}{Re_p}\left(1 + 0.15 Re_p^{\;0.687}\right) \tag{5.10}$$

$$C_D = \max\left(\frac{24}{Re_p}\left(1 + 0.15 Re_p^{\;0.687}\right), 0.44\right) \tag{5.11}$$

with

C_D representing the particle drag coefficient and

Re_p representing the particle Reynolds number.

Meller et al. [10] modelled the combustion of pulverised coal in a furnace that uses a swirl-stabilised burner and used particle velocity measurements to validate their model. The model provides quick results since the pulverised solid fuel particle is assumed to be spherical in shape. Zhang and Zeng [11] modelled the combustion of anthracite and sawmill sludge in a circulating fluidised bed (CFB) furnace using this model.

5.4.2 Morsi and Alexander Model

Morsi and Alexander [12] also proposed their own model which is sensitive to the velocity at which the particles are travelling in the combustion chamber. Redha et al. [13] applied this model to predict the combustion of coal and wheat straw as given by Equation (5.12). The coefficients a_1, a_2, and a_3 are dependent on the Reynolds number of a particle [14–16].

$$C_D = a_1 + \frac{a_2}{Re_p} + \frac{a_3}{Re_p^2} \tag{5.12}$$

Re_p	a_1	a_2	a_3
$0< Re_p <0.1$	0	24	0
$0.1< Re_p <1$	3.69	22.73	0.0903
$1< Re_p <10$	1.222	29.1667	–3.8889
$10< Re_p <100$	0.6167	46.5	–116.67
$100< Re_p <1\,000$	0.3644	98.33	–2778
$1\,000< Re_p <5\,000$	0.357	148.62	–4.75e4
$5\,000< Re_p <10\,000$	0.46	–490.546	57.87e4
$10\,000< Re_p <50\,000$	0.5191	–1662.5	5.4167e4

5.4.3 Haider and Levenspiel Model

A more common approach with irregular-shaped particles like biomass in determining particle drag coefficients was proposed by Haider and Levenspiel [17]. This model uses particle shape factor coefficients to consider the deviation from the usual spherical shape. The correlation developed by Haider and Levenspiel [17], which is best suited to capturing drag properties of non-spherical particles, is given by Equation (5.13):

$$C_D = \frac{24}{Re_p}(1 + b_1 Re_p^{b_2}) + \frac{b_3 Re_p}{b_4 + Re_p} \tag{5.13}$$

$$b_1 = \exp\left(2.3288 - 6.4581\phi_s + 2.4486\phi_s^{\,2}\right)$$

$$b_2 = 0.0964 + 0.5565\phi_s$$

$$b_3 = \exp\left(4.905 - 13.8994\phi_s + 18.4222\phi_s^{\,2} - 10.2599\phi_s^{\,3}\right)$$

$$b_4 = \exp\left(1.4681 + 12.2584\phi_s - 20.7322\phi_s^{\,2} + 15.8855\phi_s^{\,3}\right)$$

$$\phi_s = \frac{s}{S}$$

with

C_D representing the drag coefficient,

Re_p representing the particle Reynolds number within the bulk flow, and

b_1, b_2, b_3, and b_4 representing the coefficients dependent on ϕ_s the particle shape factor.

The particle shape factor measures the relationship between the surface area of a sphere having the same volume as the particle(s) and the actual surface area (S) of the particle. Studies by Bhuiyan and Naser [18] show that modelling biomass particles as cylinders as compared to spheres increases the model performance.

Similarly, Backreedy et al. [19] applied a shape factor to model pinewood particles going through combustion within a 1 MW furnace. These studies demonstrated the importance of including a shape factor that captures the irregular properties of solid fuel particles going through combustion [20–23]. Guo et al. [24] proved that overlooking the shape factor also affects the heat transfer properties of biomass particles going through combustion by use of a lab scale furnace under a dilute oxidising environment.

5.4.4 Hölzer and Sommerfeld Model

Hölzer and Sommerfeld [25] proposed another particle drag model that is dependent on the sphericity of a particle as given in Equation (5.14). Sphericity is a parameter that captures how a particle is circular in shape depending on the orientation of the particle.

$$C_D = \frac{8}{Re_p}\frac{1}{\sqrt{\phi_{s\parallel}}} + \frac{16}{Re_p}\frac{1}{\sqrt{\phi_s}} + \frac{3}{Re_p}\frac{16}{\phi_s^{3/4}} + 0.4210^{0.4(-\log\phi_s)^{0.2}}\frac{1}{\phi_{s\perp}} \tag{5.14}$$

with

$\phi_{s\parallel}$ representing the elongated particle shape factor and

$\phi_{s\perp}$ representing the elongated particle shape factor.

This model compares the fuel particle to a spherical one by involving the orientation of the particle. The strength of this model is also in its tolerance at higher Reynolds numbers [26, 27]. Vollmari et al. [28] employed this approach when he studied the pressure drop in a fluidised bed for various particle shapes such as spherical, cylindrical, and rhomboids. The level of agreement for cylindrical particles is quite commendable except at the inception of fluidisation. Since suspension combustion within a thermal power plant occurs at regimes above the inception of fluidisation, this model can be applied for combustion as well. Inasmuch as this model is accurate, the computational power required for large furnaces tends to make it unattractive for rapid combustion applications [29]. Apart from the Schiller-Naumann drag model, all the other drag models require programming by the user (user-defined functions) for most modelling software [30].

5.5 RADIATION MODELS

The purpose of combustion as derived from most applications is to produce heat which can in turn be used for other downstream processes such as steam generation. The energy conservation equation that is solved during combustion must obey a thermal balance which incorporates all the heat transfer modes and species absolute enthalpy variation during the reactions. Chemical reactions are either exothermic or endothermic which inevitably impact on the heat transfer models of the system. Radiation heat transfer accounts for most of the energy emitted or absorbed during combustion. Since radiation is very important, a separate equation known as the radiative transfer equation (RTE) which captures radiation parameters such as frequency, direction vector, position vector, path length, scattering coefficient, absorption coefficient, spectral radiation intensity, blackbody emission intensity, solid angle, local absolute temperature, and radiation intensity source term is developed. Equation (5.15) summarises the RTE as applied to combustion modelling [31]:

$$\frac{dI(\vec{r},\vec{s})}{ds}+(\alpha+\sigma_s)I(\vec{r},\vec{s})=\alpha n^2\frac{\sigma T^4}{4\pi}+\frac{\sigma_s}{4\pi}\int_0^{4\pi}I(\vec{r},\vec{s'})\Phi(\vec{s}.\vec{s'})d\Omega \tag{5.15}$$

with

I representing the radiation intensity which depends on position $(\vec{r})$ and direction $(\vec{s})$

s representing the path length,

α representing the absorption coefficient,

σ_s representing the scattering coefficient,

n representing the refractive index,

σ representing the Stefan-Boltzmann constant (5.67×10^{-8} W/m^2K^4),

T representing the local temperature,

$\vec{s'}$ representing the scattering direction vector,

Φ representing the phase function, and

Ω representing the solid angle.

Radiation sub-models can be classified based on the data available and information produced, which are either directional or spectral models with Figure 5.1 showing the main models used for solid fuel combustion that are in existence [32]. The following sub-sections discuss in detail the applications of these models in solid fuel combustion.

5.5.1 Discrete Ordinate Model

The discrete ordinate (DO) model reduces the entire solid angle into a finite number of solid angles, also known as discretisation of radiation solid angles. The accuracy of this model depends on the number of solid angles one wishes to form, which also increases the computational power required [33]. The DO model results in false scattering and ray effects, which give rise to a lack of conservation for the radiative energy. Lu and Wang [34] support this finding, though their work was not necessarily on solid fuel combustion but rather on developing an Artificial Neural Network applicable to combustion gases. The DO model is generally considered less computationally demanding; hence its use by various researchers [35].

5.5.2 P1 Model

The P1 model, also known as the differential approximation model, expands the radiation intensity into an orthogonal series of spherical harmonics, which then transforms the RTE into a set of simultaneous partial differential equations [36, 37]. Huang et al. [38] used this model to predict species variations of NOx and CO when

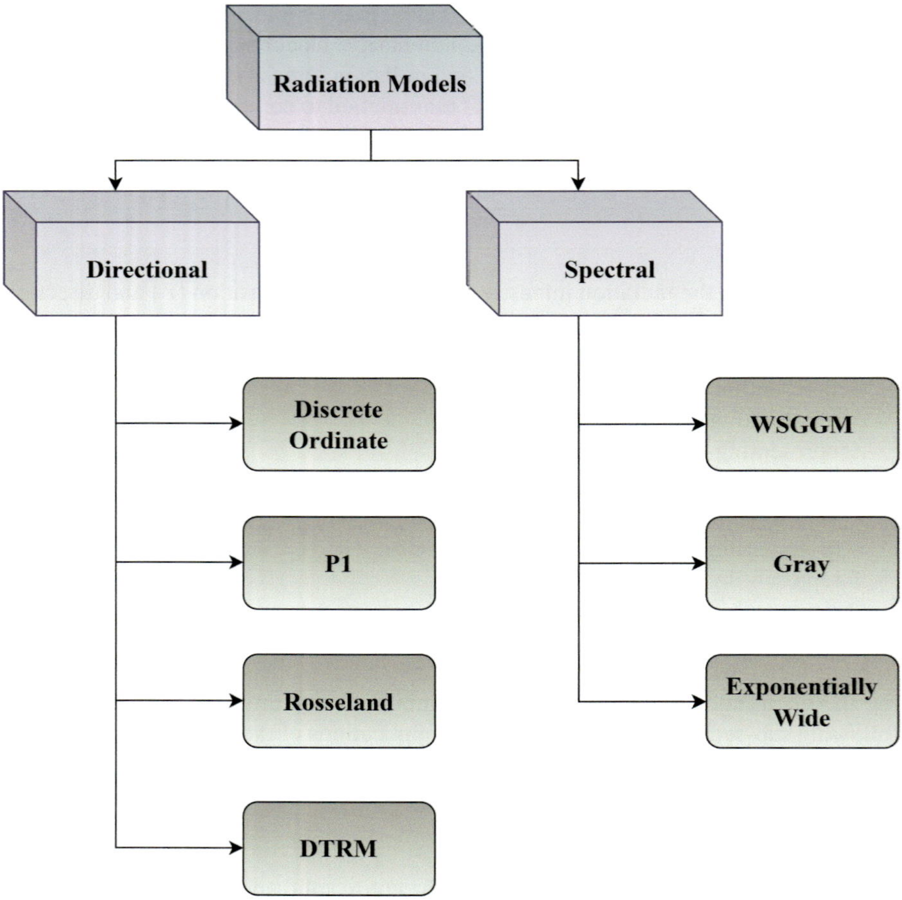

FIGURE 5.1 Main radiation models used for solid fuel combustion.

coal and eucalyptus sawdust are co-combusted within a 1000 MW furnace. The agreement with experimental data at the furnace outlet for temperature, NOx, and CO was within 6%.

5.5.3 Rosseland Model

Further approximations on the P1 model will give the Rosseland model, which appears as a simple conduction equation with all the parameters being temperature dependent. However, the Rosseland model has its limitations near a boundary. As such, it is rarely used to model combustion applications as the thermal boundary layer plays a critical role in combustion modelling [39].

5.5.4 Discrete Transfer Radiation Model

The discrete transfer radiation model (DTRM) is less expensive computationally, as surface radiation is assumed to be a single ray. However, the accuracy of this model can be increased by tracing more rays to represent the surface [32, 40, 41]. With regard to co-firing of coal and biomass, Chowdhury et al. [42] successfully used this model to predict NOx and CO emissions, demonstrating its applicability though with modest accuracy.

5.5.5 Monte Carlo Model

The Monte Carlo model's strength is in its limiting factor with regard to the scattering length of radiation and iterative behaviour when determining the emission, transmission, reflection, absorption, or scattering nature of radiation from a particle. This reduces the computational power by having an equation that does not necessarily calculate radiation beyond the domain [4, 43]. Similarly, with the DTRM, this model's accuracy can be increased by increasing the representative rays [41].

5.5.6 Weighted Sum of Grey Gases Model

Radiation can be described according to its level on the electromagnetic spectrum. CFD can also be manipulated to define models that are dependent on the electromagnetic spectrum. The most common model associated with combustion is the weighted sum of grey gases model (WSGGM). Combustion models become more accurate when both directional and spectral models are used to represent the radiation within a system. As such, the WSGGM is applied together with the Discrete Ordinate model, for example.

This model acknowledges that radiative absorption and emission from a gas is characterised by emissivity as a function of temperature and the product of gas partial pressure and path length. The main emitters of radiation during combustion are CO_2 and H_2O as they are the products that attain the highest temperatures. To simplify the input parameters required, look-up tables have been drafted for the emissivity of CO_2 and H_2O at different temperatures, partial pressures, and characteristic length [44]. The WSGGM is the most common spectral method applied to combustion applications [45–48].

5.5.7 Gray Model

Another spectral model that can be used with combustion gases is the Gray model, though it has proven to be less robust. This model assumes all gas species that are participating in the combustion process possess similar energy levels on the electromagnetic spectrum [4]. This means this model, when used for combustion applications, is just for initial guidance since a lot of gas species with different spectral properties exist during the combustion process.

5.5.8 Exponentially Wide Band Model

Exponentially wide band model (EWBM) has been developed and adapted for CFD based on its ability to predict the emissivity of individual species with known partial pressure and temperature [6, 49]. Yan et al. [50], when studying the possibility of using oxyfuel to combust coal and biomass in a 600-MW furnace, used this model. Inasmuch as this model is relatively accurate, the lack of experimental data with regard to the spectrum of various gases under different combustion conditions limits its application [51]. Holtmeyer et al. [52] also used this model with success for the modelling of the co-combustion process of coal and biomass in a 30-kW combustor under an oxy fuel atmosphere.

5.6 SUMMARY

This chapter has focused on the physical models applied to coal biomass co-firing, covering turbulence, particle dynamics, multiphase flow, and radiation heat transfer. For turbulence, the RNG k–ε formulation is identified as the most suitable for swirling, recirculating combustion flows due to its better convergence and accuracy compared to the standard k–ε model. In multiphase modelling, both Euler-Euler and Euler-Lagrange approaches are important, with the discrete phase model providing better results for dilute pulverised fuel flows commonly used in industrial furnaces. Particle dynamics is represented through drag laws, beginning with the Schiller-Naumann model for spherical particles and extending to correlations such as Haider-Levenspiel and Hölzer-Sommerfeld, which incorporate sphericity and shape factor. These correlations can reduce prediction error in drag coefficients by up to 20% when applied to non-spherical biomass particles, an important improvement given the irregular morphology of biomass particles.

Radiation modelling is described through the radiative transfer equation. Depending on the application, discrete ordinates and P1 models are generally used. The weighted sum of grey gases model is useful when capturing the absorption behaviour of CO_2 and H_2O, which together account for more than 90% of the radiative heat transfer in combustion gases. Overall, the chapter demonstrated that accurate representation of turbulence, particle drag, multiphase coupling, and spectral radiation is essential for credible CFD prediction of combustion behaviour and emissions in coal–biomass systems.

REFERENCES

1. Boussinesq, J. Essai Sur La Theorie Des Eaux Courantes. *Mémoires Academie des Science* **1872**, *23*.
2. Prandtl, F. Bemerkungen Zur Theorie Der Freien Turbulenz. *ZAMM* **1942**, *22*, 241–243.
3. Versteeg, H.K.; Malalasekera, W. *An Introduction to Computational Fluid Dynamics: The Finite Volume Method Approach*; Longman Scientific & Technical: Essex, 1995.
4. ANSYS. *ANSYS FLUENT User's Guide*; ANSY: Pennsylvania; 2021, 1–170.

5. Yang, X.; Xi, T.; Qin, Y.; Zhang, H.; Wang, Y. Computational Fluid Dynamics–Discrete Phase Method Simulations in Process Engineering: A Review of Recent Progress. *Applied Sciences* **2024**, *14*, 3856, doi:10.3390/app14093856.
6. Yan, L.; Yue, G.; He, B. Application of an Efficient Exponential Wide Band Model for the Natural Gas Combustion Simulation in a 300 KW BERL Burner Furnace. *Applied Thermal Engineering* **2016**, *94*, 209–220, doi:10.1016/j.applthermaleng.2015.09.109.
7. Ortner, B.; Peham, L.; Schmidberger, C.; Prieler, R.; Hochenauer, C. CFD Modeling of Sewage Sludge Combustion in an Experimental Entrained Flow Reactor with Integrated Slag Flow Dynamics. *Applied Thermal Engineering* **2025**, *267*, 125779, doi:10.1016/J.applthermaleng.2025.125779.
8. Huang, X.; Jin, X.; Dong, L.; Li, R.; Yang, K.; Li, Y.; Deng, L.; Che, D. CPFD Numerical Study on Tri-Combustion Characteristics of Coal, Biomass and Oil Sludge in a Circulating Fluidized Bed Boiler. *Journal of the Energy Institute* **2024**, *113*, 101550, doi:10.1016/J.joei.2024.101550.
9. Venier, C.M.; Torres, E.; Fouga, G.G.; Rodriguez, R.A.; Mazza, G.; Reyes Urrutia, A. Computational Modeling of Biomass Fast Pyrolysis in Fluidized Beds with Eulerian Multifluid Approach. *Fluids* **2024**, *9*, 301, doi:10.3390/fluids9120301.
10. Meller, D.; Engelmann, L.; Stein, O.T.; Kempf, A.M. Exhaust Gas Recirculation (EGR) Analysis of a Swirl-Stabilized Pulverized Coal Flame with Focus on NOx Release Using FPV-LES. *Fuel* **2023**, *343*, 127939, doi:10.1016/j.fuel.2023.127939.
11. Zhang, Z.; Zeng, Q. Numerical Simulation and Experimental Analysis on Nitrogen and Sulfur Oxides Emissions during the Co-Combustion of Longyan Anthracite and Sawmill Sludge. *Fuel* **2019**, *254*, 115611, doi:10.1016/j.fuel.2019.06.019.
12. Morsi, S.A.A.; Alexander, A.J.J. An Investigation of Particle Trajectories in Two-Phase Flow Systems. *Journal of Fluid Mechanics* **1972**, *55*, 193–208, doi:10.1017/S0022112072001806.
13. Mohammed Redha, A.; Lau, A.; Holuszko, M.; Vakil, A.; Sokhansanj, S. CFD Investigation of Pelletization Effect on Co-firing Coal with Wheat Straw. *Canadian Journal of Chemical Engineering* **2022**, *100*, 237–253, doi:10.1002/cjce.24106.
14. Li, Q.; He, F.; Mei, S.; Xie, J.; Zhang, C.; Deng, Y. Effect of Oxy-Fuel Combustion with Different O2/CO2 Fractions on the Pulverized Coal Combustion Mechanism and Heat Transfer Characteristics in Rotary Kilns: A Numerical Simulation Study. *Chemical Engineering Journal* **2024**, *494*, 153181, doi:10.1016/j.cej.2024.153181.
15. Deniz, C.; Boke, Y.; Aydin, O.; Benim, A. Computational Analysis of Pulverised Coal Co-Firing with Biomass in 150MWe Unit of Tuncbilek Thermal Power Plant. *Isı Bilimi ve Tekniği Dergisi* **2021**, *41*, 37–50, doi:10.47480/isibted.979314.
16. Karampinis, E.; Nikolopoulos, N.; Nikolopoulos, A.; Grammelis, P.; Kakaras, E. Numerical Investigation Greek Lignite/Cardoon Co-Firing in a Tangentially Fired Furnace. *Applied Energy* **2012**, *97*, 514–524, doi:10.1016/j.apenergy.2011.12.032.
17. Haider, A.; Levenspiel, O. Drag Coefficient and Terminal Velocity of Spherical and Nonspherical Particles. *Powder Technology* **1989**, *58*, 63–70, doi:10.1016/0032-5910(89)80008-7.
18. Bhuiyan, A.A.; Naser, J. Thermal Characterization of Coal/Straw Combustion under Air/Oxy-Fuel Conditions in a Swirl-Stabilized Furnace: A CFD Modelling. *Applied Thermal Engineering* **2016**, *93*, 639–651, doi:10.1016/j.applthermaleng.2015.10.024.
19. Backreedy, R.I.; Fletcher, L.M.; Jones, J.M.; Ma, L.; Pourkashanian, M.; Williams, A. Co-Firing Pulverised Coal and Biomass: A Modeling Approach. *Proceedings of the Combustion Institute* **2005**, *30 II*, 2955–2964, doi:10.1016/j.proci.2004.08.085.

20. Black, S.; Szuhánszki, J.; Pranzitelli, A.; Ma, L.; Stanger, P.J.; Ingham, D.B.; Pourkashanian, M. Effects of Firing Coal and Biomass under Oxy-Fuel Conditions in a Power Plant Boiler Using CFD Modelling. *Fuel* **2013**, *113*, 780–786, doi:10.1016/j.fuel.2013.03.075.
21. Madanayake, B.N.; Gan, S.; Eastwick, C.; Ng, H.K. An Investigation into the Use of CFD to Model the Co-Firing of *Jatropha Curcas* Seed Cake with Coal. *International Journal of Green Energy* **2018**, *15*, 605–621, doi:10.1080/15435075.2018.1525559.
22. Thieu Trinh, V.; Kim, S.M.; Kim, K.M.; Lee, B.H.; Jeong, T.Y.; Son, J.S.; Kim, J.M.; Jeon, C.H. In-Depth Numerical Analysis of Combustion and NOx Emission Characteristics in a 125 MWe Biomass Boiler. *Fuel* **2023**, *332*, 125961, doi:10.1016/j.fuel.2022.125961.
23. Ortner, B.; Schmidberger, C.; Prieler, R.; Hochenauer, C. Computationally Inexpensive CFD Modeling of Sewage Sludge and Waste Additive Combustion Experiments in an Entrained Flow Reactor. *Fuel* **2025**, *389*, 134797, doi:10.1016/j.fuel.2025.134797.
24. Guo, N.; García Llamas, Á.D.; Li, T.; Umeki, K.; Gebart, R.; Løvås, T. Computational Fluid Dynamic Simulations of Thermochemical Conversion of Pulverized Biomass in a Dilute Flow Using Spheroidal Approximation. *Fuel* **2020**, *271*, 117495, doi:10.1016/j.fuel.2020.117495.
25. Hölzer, A.; Sommerfeld, M. New Simple Correlation Formula for the Drag Coefficient of Non-Spherical Particles. *Powder Technology* **2008**, *184*, 361–365, doi:10.1016/j.powtec.2007.08.021.
26. Vollmari, K.; Oschmann, T.; Kruggel-Emden, H. Mixing Quality in Mono- and Bidisperse Systems under the Influence of Particle Shape: A Numerical and Experimental Study. *Powder Technology* **2017**, *308*, 101–113, doi:10.1016/j.powtec.2016.11.072.
27. Wang, J.; Xiao, Y.; He, C.; Fang, Q.; Zhang, C.; Yin, C. Dynamic Behavior of Straw Biomass Particles in Gas Flow: A Novel Motion Model and Experimental Validation. *Physics of Fluids* **2025**, *37*, doi:10.1063/5.0268862.
28. Vollmari, K.; Jasevičius, R.; Kruggel-Emden, H. Experimental and Numerical Study of Fluidization and Pressure Drop of Spherical and Non-Spherical Particles in a Model Scale Fluidized Bed. *Powder Technology* **2016**, *291*, 506–521, doi:10.1016/j.powtec.2015.11.045.
29. Pieper, C.; Liedmann, B.; Wirtz, S.; Scherer, V.; Bodendiek, N.; Schaefer, S. Interaction of the Combustion of Refuse Derived Fuel with the Clinker Bed in Rotary Cement Kilns: A Numerical Study. *Fuel* **2020**, *266*, 117048, doi:10.1016/j.fuel.2020.117048.
30. Marangwanda, G.T.; Madyira, D.M. Evaluating Combustion Ignition, Burnout, Stability, and Intensity of Coal–Biomass Blends Within a Drop Tube Furnace Through Modelling. *Energies* **2025**, *18*, 1322, doi:10.3390/EN18061322.
31. Milijašević, M.; Stojiljković, D.; Manić, N. Computational Fluid Dynamics Study of Biomass Moisture Content Impact on Particle Matter Emissions. *Environmental Modeling and Assessment* **2025**, *30*, 193–218, doi:10.1007/s10666-024-09991-9.
32. Wang, Y.; Yan, L. CFD Studies on Biomass Thermochemical Conversion. *International Journal of Molecular Sciences* **2008**, *9*, 1108–1130, doi:10.3390/ijms9061108.
33. Ma, L.; Gharebaghi, M.; Porter, R.; Pourkashanian, M.; Jones, J.M.; Williams, A. Modelling Methods for Co-Fired Pulverised Fuel Furnaces. *Fuel* **2009**, *88*, 2448–2454, doi:10.1016/j.fuel.2009.02.030.
34. Lu, X.; Wang, Y. Surrogate Modeling for Radiative Heat Transfer Using Physics-Informed Deep Neural Operator Networks. *Proceedings of the Combustion Institute* **2024**, *40*, 105282, doi:10.1016/j.proci.2024.105282.
35. Tan, H.P.; Zhang, H.C.; Zhen, B. Estimation of Ray Effect and False Scattering in Approximate Solution Method for Thermal Radiative Transfer Equation. *Numerical Heat Transfer A Applied* **2004**, *46*, 807–829, doi:10.1080/104077890504267.

36. Wu, X.; Hu, F.; Ding, C.; Yang, Y.; Yang, C.; Liao, H.; Lu, K.; Li, B.; Liu, T.; Liu, C.; et al. Progress in Numerical Simulations and Fundamental Characteristics of Pulverized Coal Co-Firing with Ammonia. *International Journal of Hydrogen Energy* **2024**, *82*, 740–758, doi:10.1016/j.ijhydene.2024.07.456.
37. Marangwanda, G.T.; Madyira, D.M.; Babarinde, T.O. Coal Combustion Models: An Overview. *Journal of Physics: Conference Series* **2019**, *1378*, 032070, doi:10.1088/1742-6596/1378/3/032070.
38. Huang, J.; Liao, Y.; Lin, J.; Dou, C.; Huang, Z.; Yu, X.; Yu, Z.; Chen, C.; Ma, X. Numerical Simulation of the Co-Firing of Pulverized Coal and Eucalyptus Wood in a 1000MWth Opposed Wall-Fired Boiler. *Energy* **2024**, *298*, 131306, doi:10.1016/j.energy.2024.131306.
39. Marangwanda, G.T.; Madyira, D.M.; Babarinde, T.O. Combustion Models for Biomass: A Review. *Energy Reports* **2020**, *6*, 664–672, doi:10.1016/j.egyr.2019.11.135.
40. Bhuiyan, A.A.; Naser, J. Computational Modelling of Co-Firing of Biomass with Coal under Oxy-Fuel Condition in a Small Scale Furnace. *Fuel* **2015**, *143*, 455–466, doi:10.1016/j.fuel.2014.11.089.
41. Kim, J.; Lee, S.; Tahmasebi, A.; Jeon, C.H.; Yu, J. A Review of the Numerical Modeling of Pulverized Coal Combustion for High-Efficiency, Low-Emissions (HELE) Power Generation. *Energy and Fuels* **2021**, *35*, 7434–7466, doi:10.1021/acs.energyfuels.1c00343.
42. Chowdhury, M.S.R.; Azad, A.K.; Karim, M.R.; Naser, J.; Bhuiyan, A.A. Reduction of GHG Emissions by Utilizing Biomass Co-Firing in a Swirl-Stabilized Furnace. *Renew Energy* **2019**, *143*, 1201–1209, doi:10.1016/j.renene.2019.05.103.
43. Zhu, Y.L.; Li, S.N.; Fan, J.H.; Kong, W.Q.; Yuan, Y.; Tan, H.P. Numerical Investigation of the Photo-Thermal Characteristics of a Direct Absorption Solar Collector Using Monte Carlo and Finite Volume Methods. *Science China Technological Sciences* **2024**, *67*, 2436–2448, doi:10.1007/s11431-023-2515-5.
44. Pallarés, J.; Gil, A.; Cortés, C.; Herce, C. Numerical Study of Co-Firing Coal and Cynara Cardunculus in a 350 MWe Utility Boiler. *Fuel Processing Technology* **2009**, *90*, 1207–1213, doi:10.1016/j.fuproc.2009.05.025.
45. Tabet, F.; Gökalp, I. Review on CFD Based Models for Co-Firing Coal and Biomass. *Renewable and Sustainable Energy Reviews* **2015**, *51*, 1101–1114, doi:10.1016/j.rser.2015.07.045.
46. Chaiyo, R.; Wongwiwat, J.; Sukjai, Y. Numerical and Experimental Investigation on Combustion Characteristics and Pollutant Emissions of Pulverized Coal and Biomass Co-Firing in a 500 KW Burner. *Fuels* **2025**, *6*, 9, doi:10.3390/fuelS6010009.
47. Ihsan, S.; Prabowo; Widodo, W.A.; Saputra, I.N.A.A.; Hariana Utilization of Palm Frond Waste as Fuel for Co-Firing Coal and Biomass in a Tangentially Pulverized Coal Boiler Using Computational Fluid Dynamic Analysis. *Biomass* **2024**, *4*, 1142–1163, doi:10.3390/biomass4040063.
48. Rekhraj, S.K.; Hasini, H. Numerical Analysis of Flow and Combustion of Coal-Ammonia Blend in Coal-Fired Furnace. *Engineering Research Express* **2024**, *6*, 015511, doi:10.1088/2631-8695/AD299B.
49. Romero, S.S. Modelo de Radiación de Gases Aplicable a Simulaciones CFD En Procesos de Combustión, Polytechnic University of Madrid, 2016.
50. Yan, L.; Liu, Y.; Li, K.; Geng, C.; He, B. Insights into Radiation Property Prediction for Numerical Simulation of Pulverized Coal/Biomass Oxyfuel Combustion. *Clean Energy* **2025**, *9*, 111–123, doi:10.1093/ce/zkae101.

51. Yuan, T.; Wang, C.; Feng, Q.; Sun, Y.; Fan, G.; Chen, M.; Che, D. Computational Fluid Dynamics Simulation on Oxy-Fuel Combustion Performance of a Multiple-Burner Furnace Firing Coking Dry Gas. *Applied Thermal Engineering* **2024**, *252*, 123728, doi:10.1016/j.applthermaleng.2024.123728.
52. Holtmeyer, M.L.; Kumfer, B.M.; Axelbaum, R.L. Effects of Biomass Particle Size during Cofiring under Air-Fired and Oxyfuel Conditions. *Applied Energy* **2012**, *93*, 606–613, doi:10.1016/J.apenergy.2011.11.042.

6 Trends in Modelling Co-Combustion

6.1 INTRODUCTION

The synergy of coal and biomass characteristics rarely results in a linear variation of combustion parameters, hence there is the need to perform extensive experimental work coupled with good modelling techniques [1]. Haykiri-Acma and Yaman [2] noted that blending of fuels results in a synergistic behaviour of kinetic parameters that can be additive or multiplicative. Unfortunately, some researchers have opted to use approximated kinetic parameters during their co-combustion modelling without acknowledging the synergistic behaviour of combustion parameters. Pallarés et al. [3] during their modelling of a 350-MW boiler used devolatilisation kinetic parameters obtained from thermogravimetric analysis (TGA), though they preferred to approximate char combustion kinetic parameters. Williams et al. [4] whilst modelling the co-combustion of American coal with wheat straw employed approximations that are based on coal combustion studies. By use of correlations derived from proximate and ultimate analysis, Bhuiyan et al. [5] approximated the kinetic parameters they used for co-combustion modelling as did some researchers [6, 7].

Some researchers even preferred to employ default values already available in most simulation software whilst disregarding the thresholds through which those kinetic parameters are applicable [8–10]. Consequently, not employing the actual kinetic parameters for co-combustion modelling may not reflect the actual combustion process [11].

In contrast, as with most TGA studies of fuel blends, determination of kinetic parameters is considered an important step in co-combustion modelling studies [12, 13]. Zhang et al. [14] insisted on determining the actual kinetic parameters of Victorian brown coal blends from TGA analysis. Chen et al. [15] also performed similar studies with success whilst determining the kinetic parameters of coal-cornstalk blends. Using correct kinetic parameters during co-combustion modelling, an extension of chemical reaction modelling, improves its prediction capabilities considerably [8, 9].

Since the blending of coal and biomass results in synergy of combustion parameters, it is very important to determine if the synergy is positive or negative. Boumanchar et al. [16] concluded that co-combustion of coal and biomass is a positive synergy process, whilst Xing et al. [17] recorded negative effects at lower temperature ranges and positive effects at higher temperature ranges. The discrepancy is caused by the different methods authors use when they represent synergy as a numerical value. Unfortunately, deduction of co-combustion synergy has no

DOI: 10.1201/9781003647942-6

standardised equation which results in various interpretations dependent on the combustion parameter under investigation. What is common amongst TGA experimental findings for co-combustion is the existence of interaction between the coal and biomass characteristics [18–20].

6.2 EFFECT OF CO-COMBUSTION ON COMBUSTION PARAMETERS

Combustion performance parameters that are important for engineers are ignition, burnout, stability, and intensity. These combustion parameters are dependent on the chemical reaction properties of the fuels under study, which are basically the kinetic parameters as derived from the Arrhenius equation. CFD offers rapid analysis techniques that can be used to determine the onset of ignition, termination of combustion, relative stability, and intensity of the combustion process.

6.2.1 Co-Combustion Modelling and Ignition

By use of CFD, Alvarez et al. [21] used contours of temperature to monitor ignition by visualising the zones with high temperature, as shown in Figure 6.1. Bhuiyan et al. [5] and Pérez-Jeldres et al. [22] also employed the same procedure of monitoring the contours of temperature whilst investigating co-combustion of coal and biomass blends. Ignition point identification is relatively subjective in all of these studies, since it is based on visualisations rather than technical parameters. By use

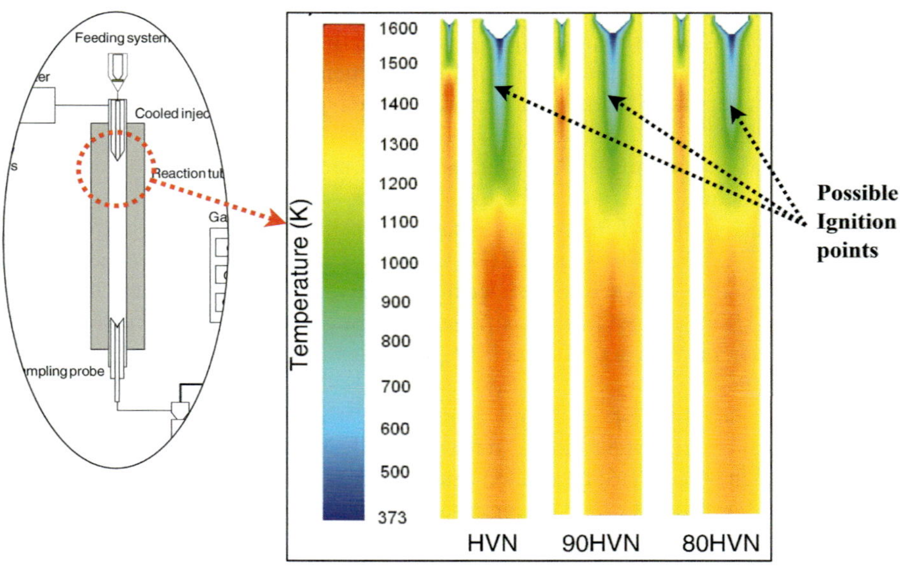

FIGURE 6.1 Temperature contours of Spanish coal and sawdust blends.

of visualisations, Riaza et al. [23] even alluded that Spanish coal and biomass blends ignite above 915 K, which is too high when compared to experimental data.

By comparison, Zhou et al. [24] demonstrated an average ignition temperature of 523 K for Chinese coal and sawdust blends after performing TGA analysis. Identification of the ignition point by use of TGA analysis is basically based on monitoring the mass loss and rate of mass loss curves, as shown in Figure 6.2. The points at which rapid mass loss began to occur shifted towards the left as blending increased, demonstrating how a blended sample is able to ignite at a lower temperature. By incorporating TGA analysis principles into co-combustion modelling with regard to ignition identification, a more technical parameter is deduced.

6.2.2 Co-Combustion Modelling and Burnout

Burnout is a subject of various combustion modelling studies, as well as it is generally linked to combustion efficiency [25]. As with ignition, TGA analysis can help reduce the subjective nature of using contours of temperature to monitor burnout. Figure 6.2 demonstrates that a fuel blend is considered burnt out when the mass loss curve approaches a local inflection point at the right.

Complete combustion is linked to burnout through equations of char oxidation to form CO_2. Some co-combustion modelling studies prefer to monitor burnout using the char content at the furnace exit, whilst others track the char content of each particle as it travels through the furnace [26–28].

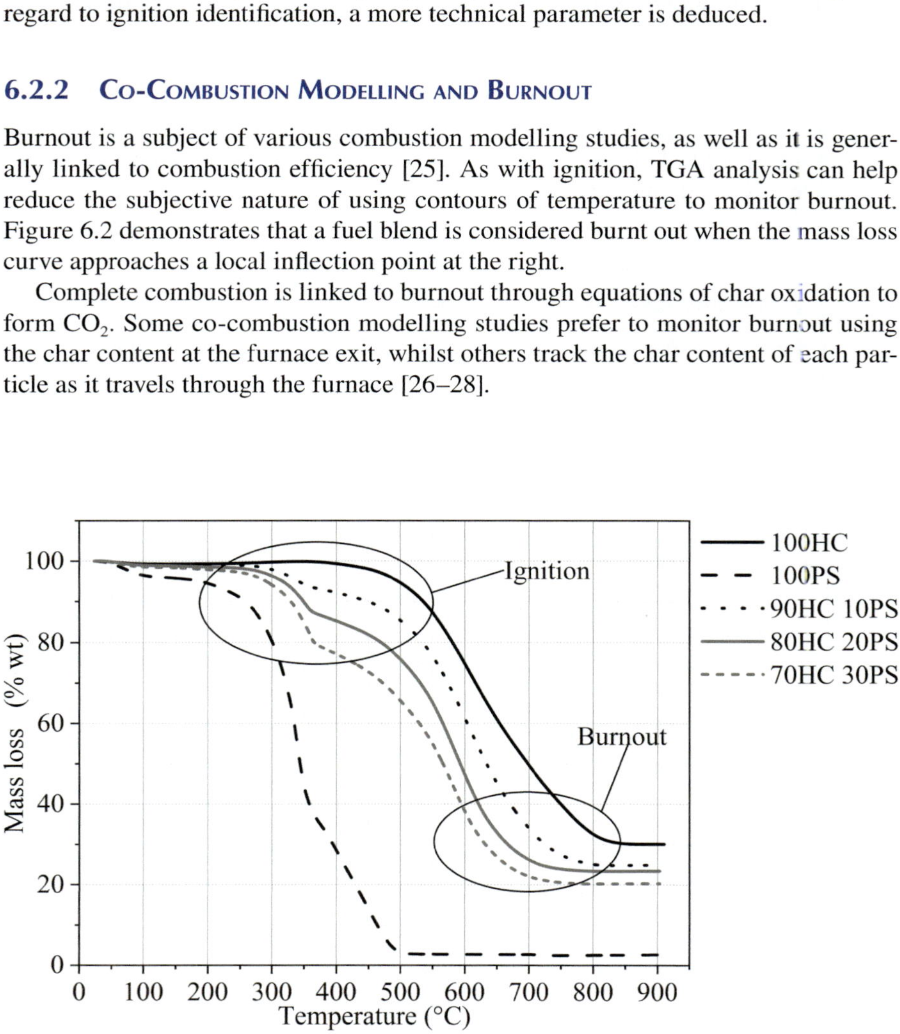

FIGURE 6.2 Mass Loss against temperature for thermal degradation of coal and sawdust blends.

6.2.3 Co-Combustion Modelling and Stability

Combustion stability is an equally important parameter when compared to ignition and burnout. Some researchers have reported on combustion stability from a technical point of view when performing co-combustion studies. Combustion stability is only mentioned in general terms without any reference to a measured parameter by Yin et al. [9] whilst modelling coal and straw co-combustion. A similar reference is made by Tan et al. [29] and Chang et al. [30] without any mention of a measurable parameter. The physical significance of combustion stability can be analysed by measuring the flame behaviour through capturing chemiluminescent images [31]. Of interest to co-combustion are the chemiluminescent images of OH radicals since they only occur within the high reaction zone, which encompasses the flame [32]. However, the chemiluminescent images are also susceptible to subjective interpretation, thus there is the need to use a parameter derived from TGA analysis to represent combustion stability [33]. By extension, a parameter on combustion intensity can also be deduced and investigated.

6.2.4 Case: Experimental Evaluation of Coal and Pinus Sawdust Co-Combustion

To illustrate the experimental determination of ignition, burnout, stability, and intensity indexes, this section uses experimental data of Hwange coal-Pinus sawdust blends. The notation that is employed to represent these fuel blends is 100HC (100% Hwange coal), 90HC10PS (90% Hwange coal and 10% Pinus sawdust by mass), 80HC20PS (80% Hwange coal and 20% Pinus sawdust by mass), 70HC30PS (70% Hwange coal and 30% Pinus sawdust by mass), and 100PS (100% Pinus sawdust by mass). Table 6.1 summarises the proximate and elementary composition of the fuel blends used to deduce the ignition, burnout, stability, and intensity indexes of Hwange coal and Pinus sawdust. Figure 6.3 summarises the diameter distribution of the individual fuel particles based on the Rosin Rammler distribution. Hwange coal has a mean diameter of 189.97 μm and a spread parameter of 3.714, whilst Pinus sawdust has a mean diameter of 202.89 μm and a spread parameter of 5.754. Pulverised coal used in thermal power plants has a mean diameter of 100 μm in most cases [34]. It is however difficult to attain such diameters with regard to biomass.

The sample TG-DTG curves of the Hwange coal and Pinus sawdust, as obtained using a TGA experimental analyser, are given in Figures 6.4 and 6.5. The atmosphere that produces such trends is oxidative, whilst the heating rates are 5, 12.5, and 20°C/min.

Figure 6.4 highlights an example of a TG-DTG curve obtained when a 100HC sample is under a heating rate of 12.5°C and an air atmosphere. Likewise, Figure 6.5 highlights an example of a TG-DTG curve obtained when an 80HC20PS sample goes under a heating rate of 12.5°C/min and an air atmosphere. As noted by Zhou et al. [24], thermal degradation of coal makes apparent four stages on the TG-DTG curves, which are low temperature decomposition stage (closely related to inherent

TABLE 6.1
Chemical Properties of Hwange Coal, Pine Sawdust, Blends

	Proximate Analysis* (weight %)				Ultimate Analysis* (weight%)				
	**Fixed Carbon	Volatile Matter	Ash	HHV (MJ/kg)	C	H	**O	N	S
100HC	53.97	23.10	22.93	22.51	58.671	2.946	13.245	1.613	0.593
90HC 10PS	48.21	29.91	21.88	21.54	58.489	3.367	14.246	1.448	0.570
80HC 20PS	46.35	31.82	21.83	21.18	56.954	3.734	15.632	1.296	0.555
70HC 30PS	46.02	33.74	20.24	21.37	56.885	4.093	17.100	1.143	0.535
100PS	15.62	80.68	3.70	18.07	49.504	6.035	40.404	0.358	0

*On a dry basis, **By difference.

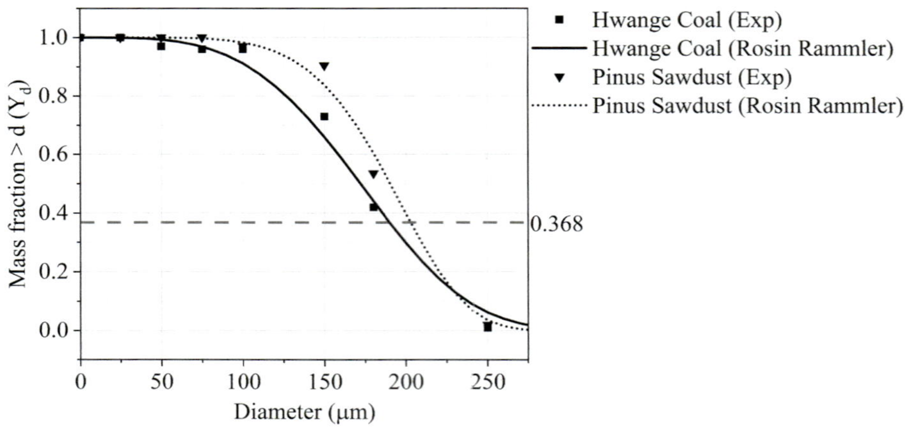

FIGURE 6.3 Particle distribution of the Hwange coal and pinus sawdust.

moisture release through evaporation), oxygen adsorption stage, homogeneous and heterogeneous combustion stage, and burnout stage.

According to Figure 6.4, representing the degradation of 100HC, T_{init} represented the temperature at the start of the experiment; thus, the release of moisture begins accompanied by a slight decrease in mass. Subsequently, stage 2 is noted by the region with an almost constant or slight mass increase. This corresponded to the physical and chemical adsorption of the oxygen stage (region between T_1 and T_{ig}).

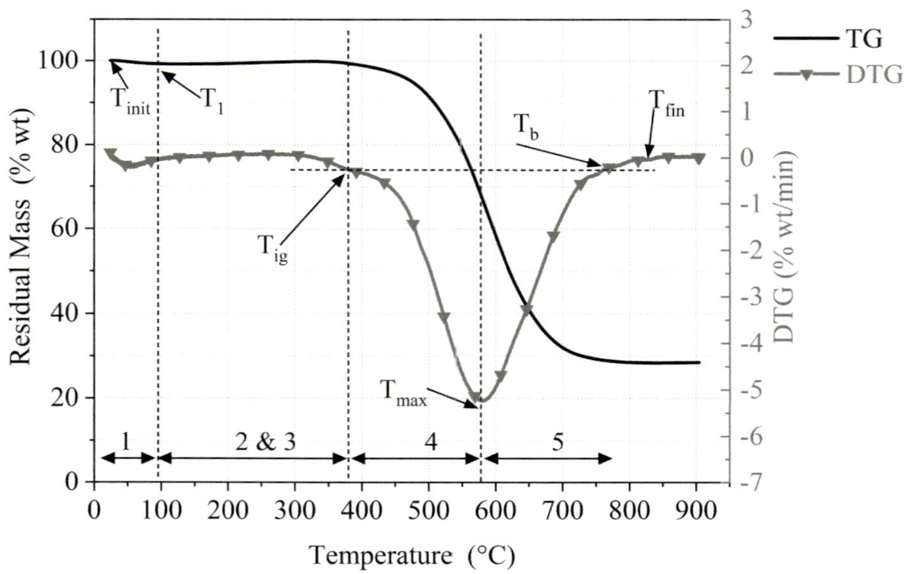

FIGURE 6.4 TG-DTG curve for 100HC at 12.5°C/min heating rate under an air atmosphere.

As highlighted by Wang et al. [35], the reason why the sample mass remains almost constant or increases at first during stages 2 and 3 is that, after oxygen adsorption, complexes containing oxygen are formed on the fuel particle surface. This process is also accompanied by the release of light volatiles (decrease in mass); thus, a contradicting phenomenon exists during stages 2 and 3. As highlighted in Figure 6.5, if the release of light volatiles is more significant than the mass gain due to oxygen adsorption, an overall mass loss will be experienced during stage 2 [36].

Of tremendous importance is the point at which ignition occurs symbolised by the region on the TG curves where rapid mass loss begins. Ignition is deduced to occur when the mass loss rate is at 1% weight/min as suggested by other researchers [37, 38]. Another method of locating the ignition point, which uses tangents that are drawn on the TG-DTG curves, can also be used. This method is, however, quite subjective to the user, especially when the effect of blending comes into play since some points of inflection become intertwined. The temperature at the ignition point is symbolised by T_{ig} and the time by t_{ig}. The fuel continues to lose weight until the burnout point where volatile and char combustion are judged to have been completed. The burnout point is also denoted by the point at which the mass loss rate is 1% weight/min after the fuel has lost most of its mass. T_b is used to denote the burnout temperature and t_b represents the time at burnout.

Figure 6.5 shows that the pro-homogeneous volatile and pro-heterogeneous combustion zones can be made apparent as distinct stages 3 and 4. During the pro-homogeneous volatile combustion zone, hydrocarbon active structures on the fuel surface increase exponentially, leading to an increase in oxygen adsorption and sample

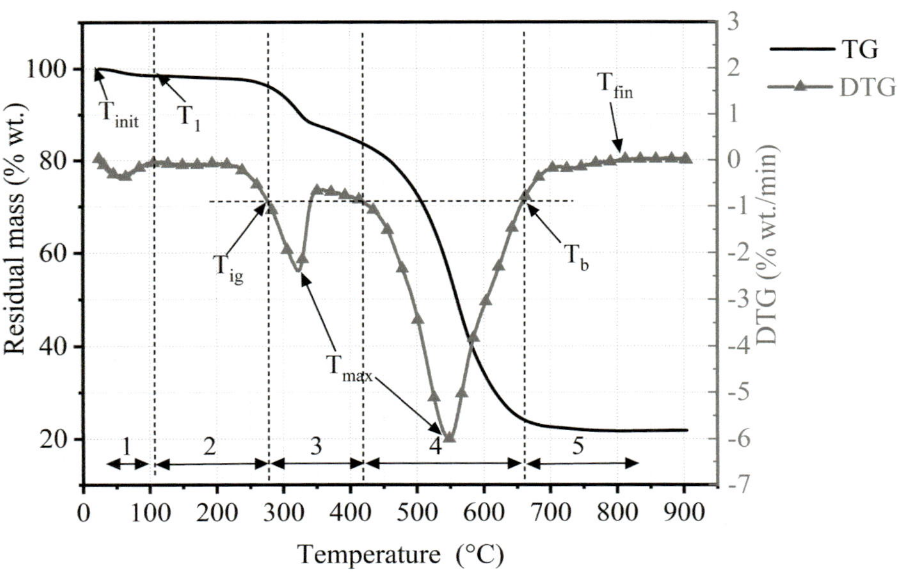

FIGURE 6.5 TG-DTG curve for 80HC20PS at 12.5°C/Min heating rate under an air atmosphere.

degradation rate. The products, tars, volatiles, CO, and CO_2 are then produced in large quantities, leading to an increased mass loss rate. Passing into the pro-heterogeneous combustion zone, volatiles continue to burn, whilst the fixed carbon or char is eventually ignited and combusted. As such, more CO and CO_2 gases are produced, leading to an increased mass loss rate [35]. T_{max} is used to show the temperature at which maximum mass loss rate occurs, visualised by a peak on the DTG curve, t_{max} the time at that point, and DTG_{max} the corresponding mass loss rate. Florentino-Madiedo et al. [39], during their tests with coal, sawdust, and paraffin, also obtained multiple peaks on their TG-DTG curves.

Having given perspective to the general allure of the TG-DTG curves for HC and PS blends, TG-DTG curve comparisons for the different fuel blends under different operating conditions can then be analysed. According to Figure 6.5, the TG curves show that an increase in the ratio of PS for the fuel blends shifts the high gradient zones to the left. This means most of the mass loss occurs earlier for PS blended samples. The ignition and burnout points are also shifted to the left slightly, which means blended samples are ignited at lower temperatures and the char is quickly combusted at lower temperatures as well.

Under a heating rate of 20°C/min and an air atmosphere, the ignition temperatures for 100HC, 90HC10PS, 80HC20PS, and 70HC30PS samples are 457.70, 299.81, 281.38, and 265.76°C, respectively. Similarly, the burnout temperature decreases by 14.64% from 820.05°C to 699.97°C, which is supported by the findings of Wu et al. [40] during their experiments with low-rank coal-cellulose blends.

Thermal degradation (TG/DTG) curves for 100HC, 90HC10PS, 80HC20PS, 70HC30PS, and 100PS fuel blends under an oxidative atmosphere at 5°C/min heating rate are represented in Figures 6.6 and 6.7, which show an apparent initial

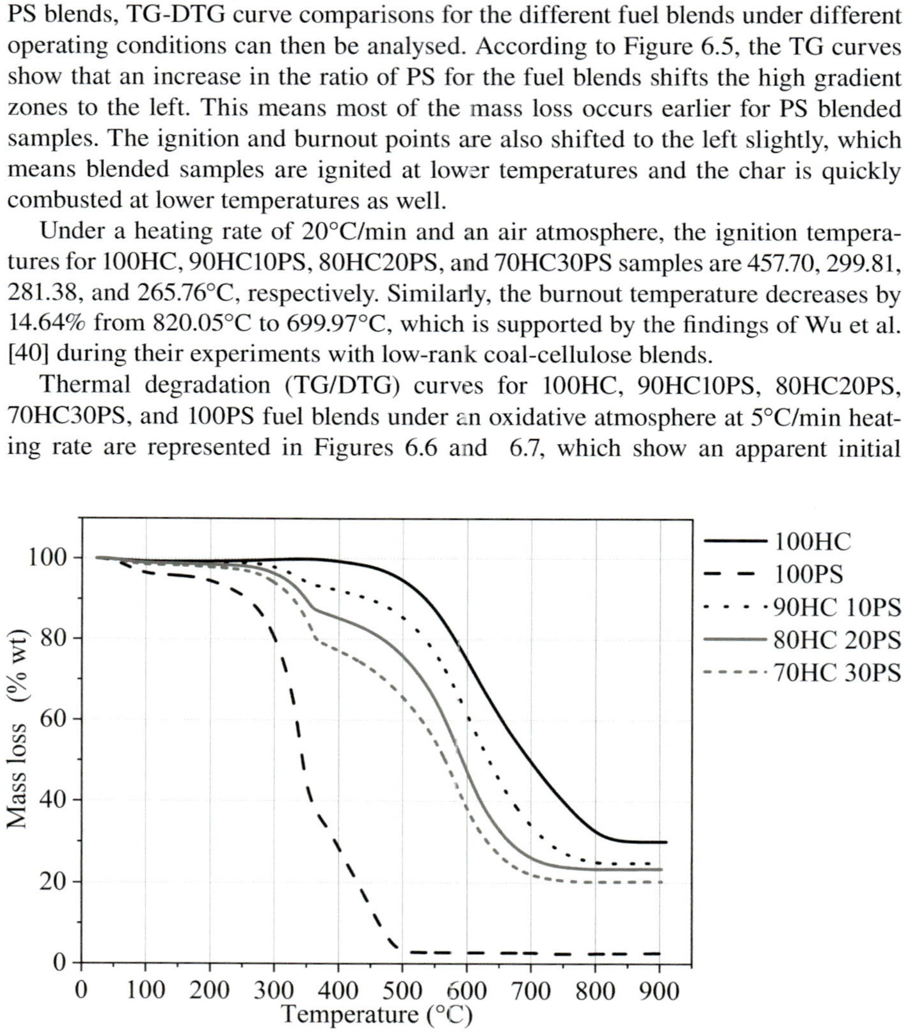

FIGURE 6.6 TG curves for all fuel samples at 5°C/Min heating rate under an air atmosphere.

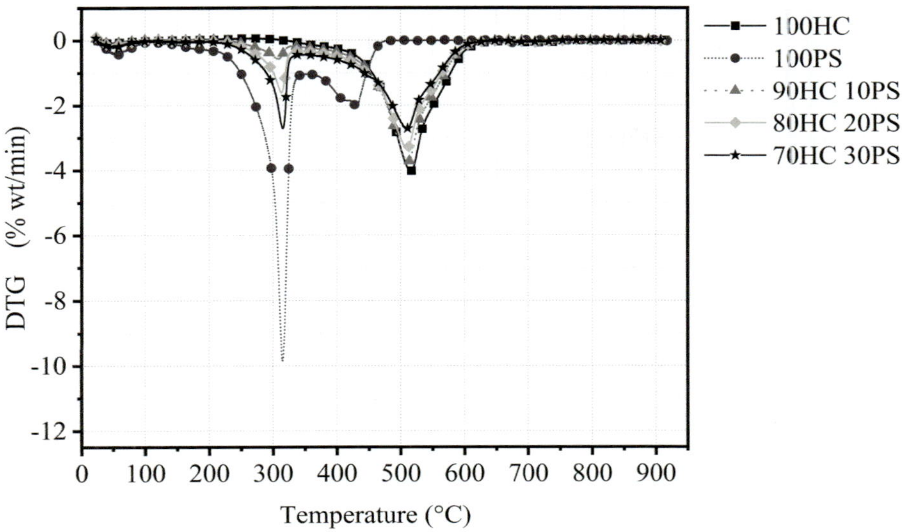

FIGURE 6.7 DTG for all fuel samples at 5°C/Min heating rate under an air atmosphere.

mass gain for the 100HC fuel blend from the initial temperature to around 350°C. Correlating with Figure 6.5, the first stage is highlighted as the initial mass loss phase due to inherent moisture loss. The first stage is demarcated by a temperature of around 115°C for all blending ratios, as all moisture is expected to be liberated. The 100PS sample loses about 4% of its mass during stage 1, whilst the 100HC loses only 1% of its mass. This meant the biomass sample contained more inherent moisture than coal [41]. The second stage displays a distinct difference in the effect of physical and chemical adsorption of oxygen towards the fuel sample. As the biomass proportion increases, the slight mass gain experienced during the initial phase of stage 2 becomes less pronounced.

This effect is linked to the coexistence of mass loss due to the release of light volatiles (alkyl side chains and oxygen-containing functional groups) occurring at the same time as weight gain due to oxygen adsorption [35]. The fuel blends that contain more coal are able to adsorb oxygen more than they can release light volatiles. During stage 2, the 100HC sample experiences a quasi-mass gain of around 1%, whilst the 90HC10PS sample does not experience any mass change, and the 100PS sample loses 8.4% of its initial weight. It is also important to note that in Figure 6.7, the DTG curves are close to the baseline during stages 1 and 2 due to the less pronounced mass loss.

Figures 6.6 and 6.7 show how synergistic effects of blending HC with PS are much more evident during Stage 3, which corresponds to the homogeneous and heterogeneous combustion stage. Stage 3 commences at around the ignition temperature and ends at the burnout temperature. Rapid release of volatiles, homogeneous volatile combustion, and heterogeneous combustion of the remaining char are pronounced during this stage [36]. Adiabatic flame temperature is attained during this stage as well since heat release reaches its peak.

Apart from the 100HC sample, a step is evident on the TG curve at around 315°C for fuel samples that contain Pinus sawdust, which corresponds to the formation of another peak on the DTG curve at that temperature. According to the ultimate analysis of the HC and PS as presented in Table 6.1, PS contains a higher volatile content than HC. As such, with volatiles having a lower flash point/ignition point than char, two peaks are formed on the DTG curve, with one signalling the pro-homogeneous volatile combustion zone, followed by one which highlights a pro-heterogeneous char combustion zone. Zhou et al. [24] highlighted why the first peak appears earlier than the second one, which is mainly due to volatile release when hemicellulose (220–315°C), cellulose (315–400°C), and part of lignin (137–667°C) decompose when biomass is subjected to heat.

The effect of high volatile content on the appearance of the DTG curve is also corroborated by Niu et al. [11] during their investigations with bituminous coal and municipal solid waste biomass. Figure 6.7 demonstrates how the pro-homogeneous volatile combustion DTG peak increased from 0.579 to 1.613, then 2.716 %wt./min for 90HC10PS, 80HC20PS, and 70HC30PS, respectively, signalling the increase in volatile composition. Homogeneous volatile combustion is associated with higher reaction rates than heterogeneous char combustion mainly because the former is based on volumetric combustion, whilst the latter is based on surface combustion [42].

The last stage corresponding to the burnout region occurs around the same temperature for the 5°C/min heating rate. Termination of char combustion takes place during this stage [42]. Since the particle diameter group 180 μm $< d_p \leq$ 250 μm for both HC and PS samples is used for the TGA, any influence of particle size towards char combustion was removed. Char combustion is a surface process, and char obtained from biomass has a higher porosity than char from coal. This tends to result in a higher surface area for biomass char which helps the overall burnout characteristics [20, 43]. Consequently, the 100PS sample attains burnout at a lower temperature of 445.90°C as compared to samples that contain HC that tend to burnout at around 565°C, corresponding to a 120°C difference. The final weight percentage on the TG curve shows a decreasing trend as PS increased which is directly linked to the ash content in the fuel samples, as shown in Table 6.1. By extension, the lesser the propensity to form ash towards the final stages of combustion, the more likely the furnace fouling is reduced [24].

In summary, Figure 6.6 shows how the TG curves shift from right to left, highlighting increased mass loss at lower temperatures. Figure 6.7 shows how the DTG curves make another peak apparent as the blending ratio of PS increases. The TG and DTG curves for 100HC display a trend associated with most coal species: a smooth, almost linear mass loss as well as a single peak on the DTG curve [44]. The TG and DTG curves for 100PS are mirrored by the findings of Bermejo et al. [45] as well during their investigations with quinoa plant residues. An interesting consistency for the DTG curves is the temperature at which maximum mass loss occurs. Looking at Figure 6.7, the left peak occurs at a temperature around 315.31°C for all the samples that contain PS blends. Similarly, the right peak occurs at around 510°C for all the samples that contain HC. This effect is attributed to the qualitative nature of volatiles and char for either PS or HC.

The ignition index (Φ_{ig}), which highlights the ignition reactivity of fuel samples, can be derived from experimental TGA analysis. A larger ignition index is representative of combustion occurring easily at lower temperatures. The ignition index is calculated from the maximum mass loss rate (DTG_{max}) as represented by Equation (6.1) [46].

$$\Phi_{ig} = \frac{\text{average}\left(DTG_{max}\right)}{t_{ig}\text{average}\left(t_{max}\right)} \tag{6.1}$$

with

Φ_{ig} representing the ignition index,

DTG_{max} representing the maximum mass loss rate,

t_{ig} representing the time at ignition, and

t_{max} representing the time at maximum mass loss rate.

Physically, a fuel sample can only be ignited once, though the DTG curve for a blended sample makes apparent multiple peaks. As such, in support of the findings by Jiang et al. [47] and Urban et al. [48], the two DTG_{max} values at the two local peaks that appear because of blending are averaged and used to evaluate the overall ignition index. This average (t_{max}) represents the average time for the two local peaks.

The burnout index (Φ_b) is used to highlight the ability of a fuel sample to combust completely [49, 50]. Using TGA experimental data, the burnout index is derived from the maximum mass loss rate (DTG_{max}) as represented by Equation (6.2) [46]:

$$\Phi_b = \frac{\text{average}\left(DTG_{max}\right)}{\text{average}\left(\Delta t_{1/2}\right)\cdot\text{average}\left(t_{max}\right)\cdot t_b} \tag{6.2}$$

with

Φ_b representing the burnout index,

DTG_{max} representing the maximum mass loss rate,

t_{max} representing the time at maximum mass loss rate,

t_b representing the time at the final burnout, and

$\Delta t_{1/2}$ representing the time range when $|DTG| > |DTG_{max}|/2$.

The average ($\Delta t_{1/2}$) represents the average of the time ranges when $|DTG| > |DTG_{max}|/2$. The stability index or first combustion index (Ψ_1) highlights the amount of combustion activity that a fuel sample possesses [48, 51]. The intensity index or second combustion index (Ψ_2) highlights the intensity and rate of the combustion process [52, 53]. Equations (6.3) and (6.4) represent the formulas used to derive the first combustion index and second combustion index:

$$\Psi_1 = \frac{\text{average}\left(DTG_{max}\right).DTG_{mean}}{\left(T_{ig}\right)^2.T_b} \tag{6.3}$$

$$\Psi_2 = \text{average}\left(T_{max}\right).\ln\left(\frac{\text{average}(\Delta T_{1/2})}{DTG_{mean}}\right) \tag{6.4}$$

with

Ψ_1 representing the first combustion index,

Ψ_2 representing the second combustion index,

T_{ig} representing the temperature at ignition,

T_{max} representing the temperature at maximum mass loss rate,

T_b representing the temperature at the final burnout,

$\Delta T_{1/2}$ representing the temperature range when $|DTG| > |DTG_{max}|/2$, and

DTG_{mean} representing the mean combustion rate.

Once the indices have been determined, the synergy caused by the blending of coal and biomass is determined using Equation (6.5), which seeks to determine the synergistic coefficient:

$$\omega = \frac{U_{experimental}}{\sum k_i U_i} - 1 \tag{6.5}$$

with

ω representing the synergistic effect coefficient,

$U_{experimental}$ representing the experimental combustion parameter of the fuel blend,

k_i representing the mass fraction of individual fuel within the blend, and

U_i representing the combustion parameter of individual fuel within the blend.

The same equation, though in different forms, was used by Wu et al. [40] when studying co-pyrolysis of coal and cellulose, Wang et al. [19] when studying co-combustion of coal and polyurethane, and Florentino-Madiedo et al. [39] when studying co-pyrolysis of torrefied sawdust, coal, and paraffin. From a statistical point of view, it is very important to understand that synergy can be determined from an additive, multiplicative, or empirical-based point of view [54]. The additive approach is the most common approach, which assumes parameters must behave in a linear manner when they are combined, thus the averaging done on the denominator in Equation (6.5). Hence, a zero coefficient means blending results in the expected parameter addition. A positive coefficient means the resulting parameter produces an extra effect beyond the addition of the two individual parameters.

Table 6.2 summarises the important points from the TG and DTG curves necessary to evaluate the combustion performance indexes for the various blends. The ignition temperature of the fuel blends containing PS is significantly lower when compared to that of 100HC mainly due to the presence of volatiles released from PS, as supported by Kongkaew et al. [53] during their investigations with blends containing rice straw. PS improves the ignition temperature, which is demonstrated by the decrease from around 450°C to less than 300°C for the 5°C/min heating rate. However, at a heating rate of 5°C/min, it is apparent that the ignition temperature does not improve significantly between the 100HC and 90HC10PS fuel samples.

As Table 6.2 highlights, a slower heating rate of 5°C/min shows a decreasing trend in maximum mass loss rate as blending increases, whilst the effect is reversed for a higher heating rate of 20°C/min. Riaza et al. [23] suggested that this trend can be explained by the existence of a volatile cloud around the fuel particle, which, in turn, hinders diffusion of oxygen to and from the surface of the fuel. These are two

TABLE 6.2

Important Points on TG and DTG Curves for Coal-Pinus Sawdust Fuel Blends in an Air Atmosphere

	T_{ig} (°C)	t_{ig} (min)	T_{max} (°C)	t_{max} (min)	DTG_{max} %wt/min	T_b (°C)	t_b (min)
Heating Rate @ β=5°C/min							
100HC	455.27	85.158	515.21	95.44	4.037	577.07	109.542
90HC 10PS	452.11	85.100	511.34	95.36	3.759	568.30	108.625
* 80HC 20PS	301.01	55.167	(312.62) 509.51	(57.18) 95.20	(1.608) 3.336	563.51	107.675
* 70HC 30PS	289.97	53.142	(315.92) 510.26	(57.37) 95.67	(2.716) 2.736	559.25	106.775
100PS	249.51	45.542	315.31	57.02	9.872	445.90	83.2
Heating Rate @ β=12.5°C/min							
	T_{ig}	t_{ig}	T_{max}	t_{max}	DTG_{max}	T_b	t_b
100HC	462.63	35.575	577.84	44.09	5.238	708.17	55.142
* 90HC 10PS	307.452	23.433	(323.773) 559.12	(24.625) 42.242	(1.323) 5.790	670.44	52.092
* 80HC 20PS	280.6285	21.475	(323.70) 548.43	(24.625) 41.658	(2.391) 6.001	655.34	51.092
* 70HC 30PS	266.366	20.408	(327.07) 540.36	(24.75) 41.108	(5.447) 5.366	634.50	49.417
Heating Rate @ β=20°C/min							
	T_{ig}	t_{ig}	T_{max}	t_{max}	DTG_{max}	T_b	t_b
100HC	457.70	22.267	605.66	29.12	5.741	820.05	39.725
* 90HC 10PS	299.81	14.892	(340.19) 599.30	(16.63) 28.86	(2.221) 6.673	768.58	37.625
80HC 20PS	281.38	14.008	573.17	27.125	7.271	717.25	35.133
70HC 30PS	265.76	13.342	350.68	16.458	10.748	699.96	34.300
100PS	197.91	10.333	341.82	16.32	34.403	498.81	24.158

* DTG = 1% wt./min at four separate points.

contradicting effects that exist at the same time, namely, char reaction vs. volatile cloud formation [55]. Because temperature has a direct relationship on diffusivity, higher heating rates tend to promote diffusion of oxygen towards the particle surface [42]. At higher heating rates, the effect of volatile cloud formation vs. char reaction seems to sway in favour of char reaction mainly because of the effect of temperature on diffusivity.

Figure 6.8 summarises the comparisons between the ignition temperature of the fuel samples at different heating rates. From a qualitative point of view, when considering a certain heating rate, the DTG peak for pro-homogeneous volatile combustion occurs around the same time regardless of the fuel blending ratio (e.g., ≈ 315°C and 57 min for a 5°C/min heating rate). The same phenomenon exists for the DTG peak corresponding to pro-heterogeneous combustion (e.g., ≈ 510°C and 95 mins for a 5°C/min heating rate). This aspect of DTG peak analysis is useful since each individual constituent has a respective temperature and time at which a DTG peak occurs. This means that as long as the blend contains PS, a DTG peak is expected around 315°C at 57 min if the heating rate is 5°C/min with the same reasoning applicable for other heating rates. Though not explained in detail, Zhou et al. [24], Jiang et al. [47], Li et al. [56], and Yan et al. [51] produced similar trends for blended coal and biomass samples.

Table 6.3 gives an overview of all the combustion parameters calculated using Equations (6.1) to (6.4) and data from Table 6.2. Figure 6.9 demonstrates the trends observed for the ignition index, burnout index, and second combustion index at 20°C/min heating rate under an air atmosphere. As shown in Figure 6.9, the ignition index

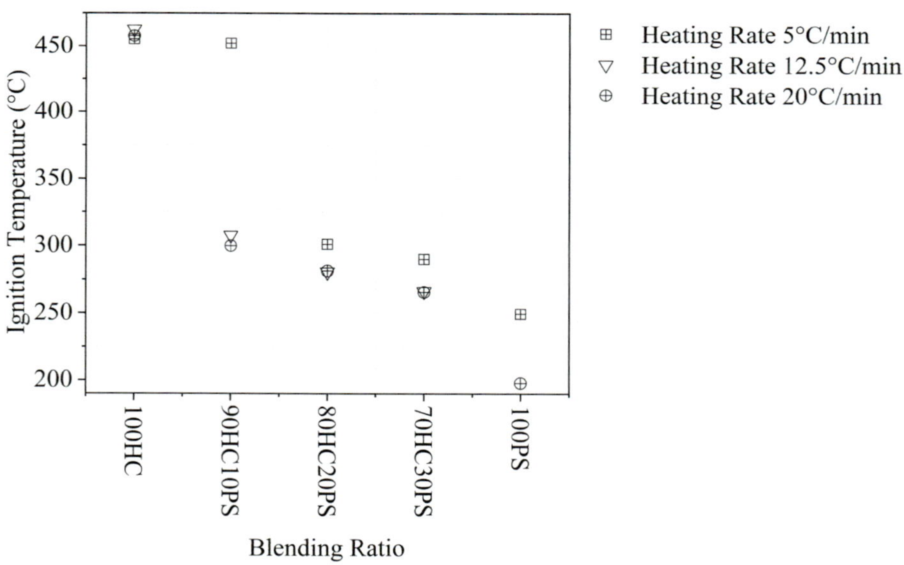

FIGURE 6.8 Ignition temperature variation for coal-Pinus sawdust fuel blends at 5°C/min and 20°C/min heating rate under an air atmosphere.

TABLE 6.3
Derived Combustion Parameters from TG and DTG Curves for Coal-Pinus Sawdust Fuel Blends in Air Atmosphere

	Heating Rate @ $\beta = 5°C/min$							
	DTG_{mean} %wt min	$\Delta T_{1/2}$ (°C)	$\Delta t_{1/2}$ (min)	$\Phi_{ig} \times 10^{-4}$ %wt min^3	$\Phi_b \times 10^{-5}$ %wt min^4	$D_i \times 10^{-4}$ %wt^2 min.°C^2	$\Psi_1 \times 10^{-8}$ %wt^2 min^2.°C^3	$\Psi_2 \times 10^3$ °C
100HC	0.4033	68.57	13.93	4.9670	2.7730	1.3072	1.3612	2.6461
90HC 10PS	0.4171	68.54	13.88	4.6321	2.6154	1.2411	1.3499	2.6087
80HC 20PS	0.4390	69.70	14.22	6.3520	2.2892	0.3140	2.8685	2.5819
70HC 30PS	0.4538	77.75	15.84	5.3815	1.6907	0.2434	2.6403	2.6246
	Heating Rate @ $\beta = 12.5°C/min$							
	DTG_{mean}	$\Delta T_{1/2}$	$\Delta t_{1/2}$	$\Phi_{ig} \times 10^{-4}$	$\Phi_b \times 10^{-5}$	$D_i \times 10^{-4}$	$\Psi_1 \times 10^{-8}$	$\Psi_2 \times 10^3$
100HC	1.01259	156.84	12.542	33.395	17.179	0.7868	3.4994	2.9139
90HC 10PS	1.0385	123.64	9.992	58.493	26.335	0.4115	9.4883	2.6724
80HC 20PS	1.1079	110.49	8.992	67.080	31.357	0.4086	12.8820	2.5241
70HC 30PS	1.1615	38.06	2.675	102.480	166.490	2.7435	14.0530	1.1413
	Heating Rate @ $\beta = 20°C/min$							
	DTG_{mean}	$\Delta T_{1/2}$	$\Delta t_{1/2}$	$\Phi_{ig} \times 10^{-4}$	$\Phi_b \times 10^{-5}$	$D_i \times 10^{-4}$	$\Psi_1 \times 10^{-8}$	$\Psi_2 \times 10^3$
	1.5717	260.70	12.58	89.00	39.44	0.6406	5.2525	3.0957
	1.6892	192.97	9.57	167.25	82.12	0.4152	17.0470	2.7135
	1.7222	152.93	7.70	191.36	99.09	0.4348	22.0510	2.5714
	1.7913	67.73	3.36	489.47	566.90	3.6092	38.9440	1.2739

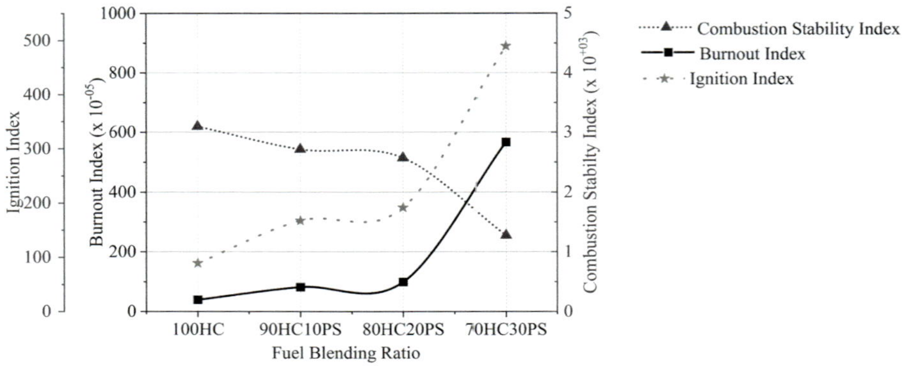

FIGURE 6.9 Ignition index, burnout index, and second combustion index at 20°C/min heating rate under an air atmosphere.

of the fuel blends increases with increasing blending ratio of PS. As highlighted by other authors, the higher the ignition index, the higher the propensity for combustion at lower temperatures or the occurrence of significant combustion at lower temperatures [49, 53]. The effect is mirrored by the first combustion index (Ψ_1), which is related to combustion activity. As blending increases, the actual combustion activity increases, supported by the volatiles released from biomass. According to Equation (6.3), the first combustion index is directly proportional to DTG_{mean}, which increases with blending and inversely proportional to burnout temperature, which decreases with blending. As such, the first combustion index tends to increase as blending increases. Niu et al. [11] supported the view and further highlighted the relationship between heating rate, ignition index, and first combustion index. Their results showed that better ignition and burnout indexes are directly proportional to heating rate.

The burnout index increases with the blending ratio of PS. Burnout is closely related to char combustion and, based on Equation (6.2), the index is inversely proportional to burnout time. Char from biomass is acknowledged to be more porous than that from coal, which then aids the surface combustion activity of biomass, thus increasing the burnout index. As Table 6.3 demonstrates, burnout time decreases with blending; hence, the calculated burnout index increases as well. As demonstrated by Zhou et al. [24] with their co-combustion experiments of coal and biomass (cornstalk or sawdust), blending increased the ignition properties, promoted complete burnout, and reduced emissions of toxic trace elements.

The second combustion index, which is related to the intensity and stability of combustion, shows a decreasing trend with the blending ratio. Equation (6.4) demonstrates that the second combustion index is heavily dependent on $\Delta T_{1/2}$, a parameter related to stability, and DTG_{mean} a parameter related to intensity. By deduction, the stability of combustion decreases with biomass blending.

Comparison of stability index at different heating rates shows that at a heating rate of 5°C/min, the second combustion index decreases by 1% as blending increases to 30%, whilst the index decreases by 60.8% and 58.8% at 12.5°C/min and 20°C/min, respectively. This means that in as much as the other combustion parameters increase with blending, a balance is necessary since combustion stability is decreased in turn [29]. In summary, blending did not result in a linear variation of combustion indexes. Optimisation is thus needed in such a way that the gains and losses are balanced with regard to combustion indexes.

Figure 6.10 shows the synergistic effect coefficient against temperature for the mass loss (TG) curves for the blends at 12.5°C/min heating rate. The 80HC20PS and 70HC30PS blends showed weak positive synergy around 348°C, which matched the region where the step in Figure 6.6 appeared.

This region is where the pro-homogeneous volatile combustion took place. However, negative synergy is experienced after about 492°C for all the fuel blends, which corresponds to the pro-heterogeneous char combustion region. The expected mass loss due to PS blending is not experienced during this region, thus the negative synergy. Florentino–Madiedo et al. [39] obtained almost similar results with regard to coal and biomass blending such that their synergy curves of mass loss were largely negative throughout the combustion process. Wu et al. [40] suggested that volatile generation results in a slightly positive synergistic coefficient towards mass loss as well.

Figure 6.11 shows the degree of synergy towards the rate of mass loss (DTG curves) as blending by PS increases. The 80HC20PS and 70HC30PS blends experience positive synergy until around 630°C, which corresponds to the burnout point. This means that the rate of mass loss for the 80HC20PS and 70HC30PS blended

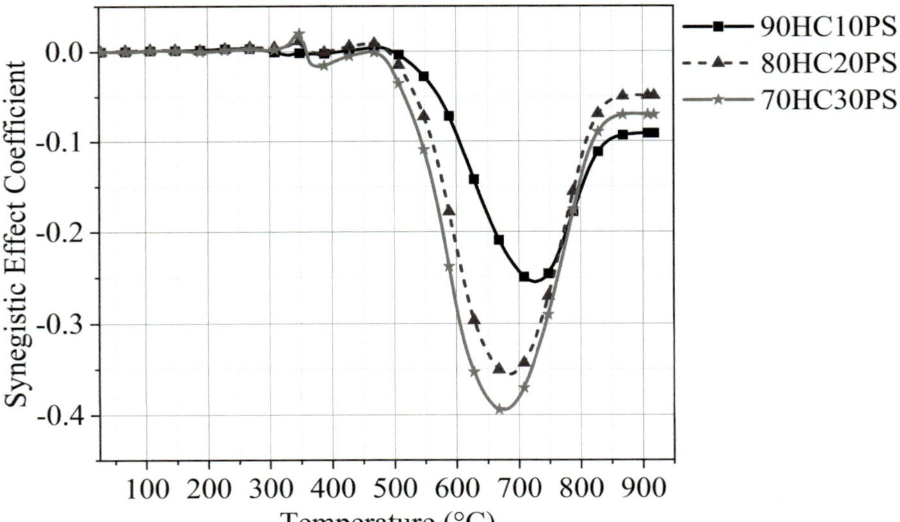

FIGURE 6.10 Degree of synergy towards TG curves for 90HC10PS, 80HC20PS, and 70HC30PS fuel samples at 12.5°C/min heating rate under an air atmosphere.

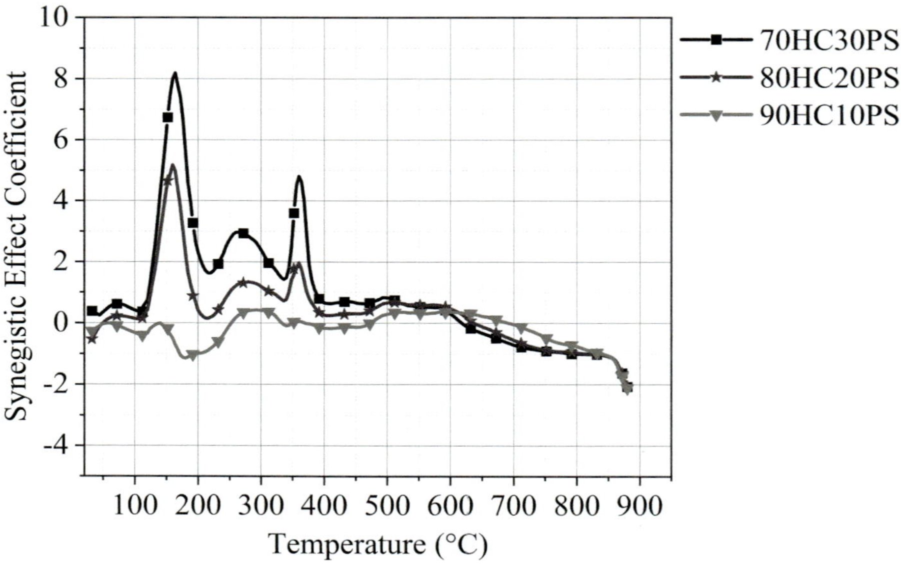

FIGURE 6.11 Degree of synergy towards DTG curves for 90HC10PS, 80HC20PS, and 70HC30PS fuel samples at 12.5°C/min heating rate under an air atmosphere.

samples increases greatly by PS throughout the combustion process unlike the 90HC10PS blend. As far as the 80HC20PS and 70HC30PS blends are concerned, Figure 6.11 shows that blending increases the first peak to 5.5 and 8.8, respectively, which corresponded to initial volatile release, whilst the second peak increases to 2.6 and 6.8, respectively, which corresponds to the pro-homogeneous volatile combustion region. Wu et al. [40] noted how a large number of H_2 containing free radicals and volatile intermediate products are formed more from biomass than coal. As such, the rate of volatile release (first peak) and rate of volatile combustion (second peak) experiences the most influence. In summary, a blending ratio of 20% or higher shows better synergy towards the rate of mass loss. This suggests that with regard to HC and PS, blending should be higher than 20% if the benefits towards combustion are to be experienced.

6.3 CO-FIRING

In the context of combustion, co-firing can be defined in various ways depending on the fuels being employed, the method of combustion, as well as the industrial application being considered. Tilliman [57] defined co-firing as "the practice of supplementing a base fuel with a dissimilar fuel", whilst Leckner [58] defined co-firing when applied to industrial boilers as "combustion of two or more fuels together in one boiler". Co-firing is an extension of fuel blending practices common to the fuel community. However, the use of biomass for co-firing in existing coal-fired power plants is not an easy task since issues to do with fouling and fuel handling problems become more apparent.

The proportion range of co-firing with biomass that is acceptable is introduced on a mass basis depending on the type of boiler. This is mainly because biomass is fibrous in nature and contains different elements compared to coal. As such, the balance between cost incurred due to increases in slagging and fouling of boiler material and the reduction of Sulphur Oxides (SOx) and Nitrogen Oxides (NOx) emissions is not constant [59]. Tillman suggested that for every tonne of biomass co-fired with coal, an equivalent amount of CO_2 is removed from the emissions atmosphere [60].

Co-firing for boilers can be found in two main forms, which are direct co-firing and indirect co-firing, as highlighted in Figures 6.12 and 6.13. Direct co-firing does not require any pre-treatment of biomass before entering the combustion zone. This might need two separate handling systems for coal and biomass or a single handling system. The limitations of this method are on the threshold of the substitution level which tends to be lower compared to the indirect methods [61]. For indirect co-firing, biomass is pre-treated by means of various chemical processes before being introduced into the boiler. The common pre-treatment processes are densification, pyrolysis, carbonisation, gasification, and liquefaction which tend to produce energy-dense fuels such as briquettes, biooils, biochar, and biogases. With advancement in

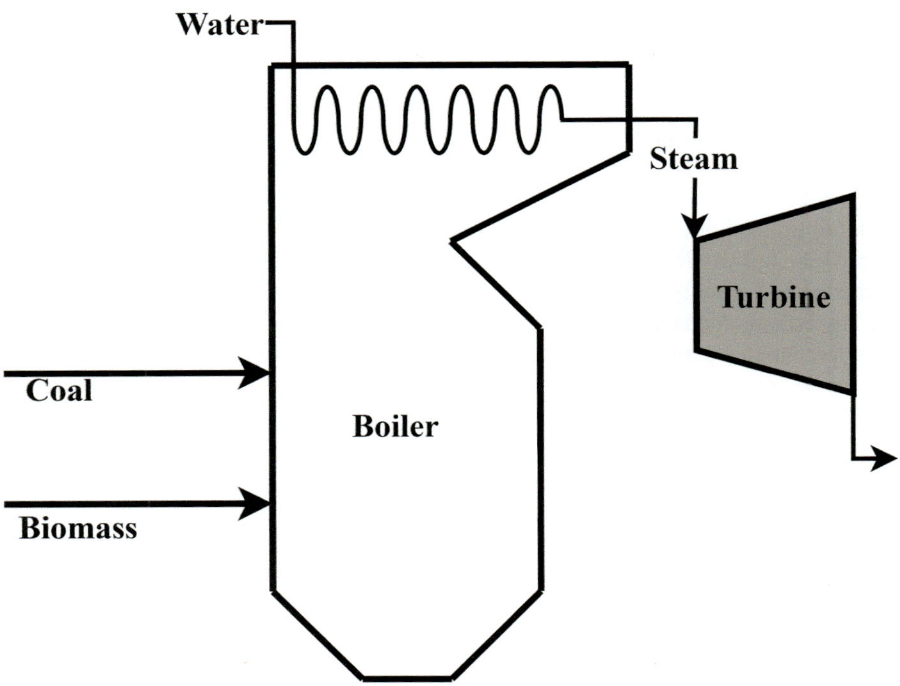

FIGURE 6.12 Direct co-firing.

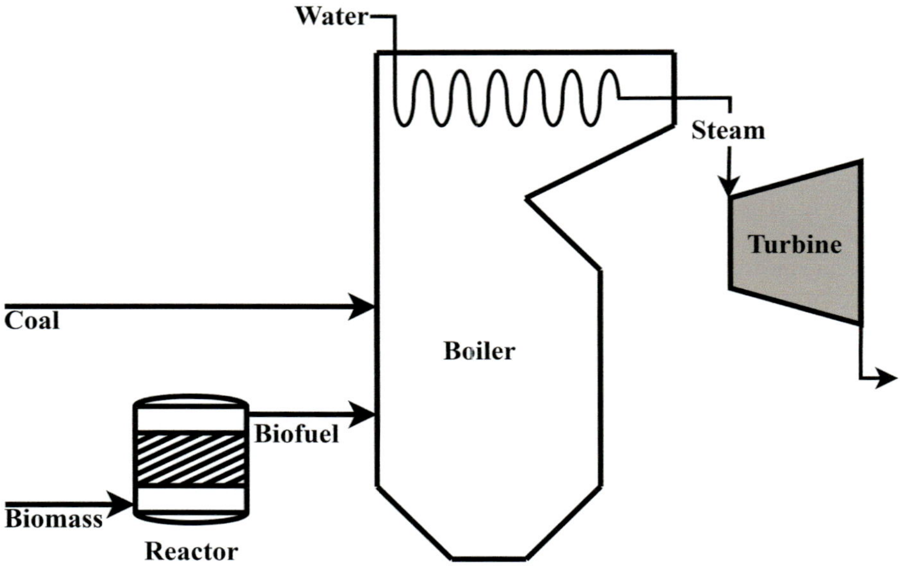

FIGURE 6.13 Indirect co-firing.

technologies, biofuel production will eventually become less capital-intensive for developing countries [62].

CFD provides an alternative that can help researchers understand the combustion behaviour of coal and biomass co-firing systems. Through CFD, detailed information on combustion ignition, burnout, stability, and intensity can be extracted from the models generated. The way coal and biomass fuels are introduced into a boiler or furnace during co-firing affects the heating profiles of the furnace. If the process is not optimised, it can result in localised hotspots, inefficient combustion, and relatively higher emissions [63–65]. Through CFD, the particle distribution within the furnace can be analysed, hence achieving optimised co-combustion. Biomass used for combustion tends to be larger and adopts a spread parameter according to Rosin Ramler, which is larger. The larger particles are difficult to ignite and, with regard to suspension combustion, difficult to entrain in the furnace. Through CFD, it becomes quite possible to track the burnout rate of each fuel species within the furnace [66].

Another problematic area with co-firing is the substitution ratio that can produce optimised parameters. Since most furnaces were originally designed to handle coal, retrofitting of biomass handling facilities and burners needs to be carried out in an optimised manner. Through CFD, different co-firing ratios can be studied in a rather quick and efficient manner provided the correct fuel properties are provided [67]. Since particulate matter results in the formation of fly ash, understanding its behaviour can assist with optimising furnace operating parameters. Amongst these is tube erosion which is probably one of the most common causes of power plant unplanned downtime [68]. Understanding the formation and movement of fly ash after co-firing by use of CFD has the potential to aid in plant optimisation. Some biomass species have a higher ash content than coal; hence, ash as a by-product of co-firing needs to be managed. CFD as a tool can also assist in understanding the locations within the furnace where ash fusion temperature is attained [69]. Emissions such as NOx, SOx, and particulate matter which are formed during the co-combustion process can be predicted with accuracy by use of CFD [70].

CFD modelling has been successfully employed to monitor and optimise various coal-biomass co-firing systems found in industry. Ma et al. [71] studied the combustion of coal and biomass in a 600-MW furnace primarily used to combust coal. The study objective was to maintain stable combustion through avoidance of high intensity zones, hotspots, within the furnace. A secondary deduction from the study was to monitor the O_2 and CO species as well as the heat flux due to the introduction of biomass. CFD accorded the researchers the ability to monitor the fuel mixing and injection patterns of the biomass at various positions within the furnace in detail. By extension, they managed to achieve optimum fuel mixing through intricate CFD studies of over-fire air burner configurations [72]. This strategy can be extended to other forms of co-firing systems such as (liquid-gas) and (gas-gas).

One of the most common parameters monitored during combustion within a furnace is temperature, as it is linked to most parameters such as flame temperature, flame speed, wall temperature, and pollutant formation. Wang et al. [73] adjusted the air to fuel ratio through air staging by use of CFD so as to achieve optimum temperature profiles during the combustion of straw and coal. Most biomass fuels have lower ash and sulphur content than coal which in turn leads to lower SOx and particulate matter in the combustion products [74]. CFD becomes a useful tool in determining the exact reduction in these emissions as it affects the design parameters of post-combustion emission abatement techniques.

Fixed bed combustion in the form of grate firing and fluidised bed combustion also stands to benefit from CFD studies as particle dynamics can be monitored in detail [75]. A cement kiln that wanted to monitor the effect of supplementing coal with biomass managed to utilise CFD to optimise the combustion process even though the biomass had moisture and ash content [76]. The good thing about fluidised beds is that they are well adapted for low-quality fuels; hence the proposed use of biomass waste.

Most fixed and fluidised bed combustion systems suffer from ununiform temperature distribution, which in turn results in incomplete combustion as well as slagging problems in most cases [77]. Teixeira et al. [78] monitored the slagging propensity of co-firing coal with biomass in a fluidised furnace with success. Of interest was the need to make sure specific combustion conditions such as furnace temperature, secondary air, and fusion index are monitored to have a successful co-combustion scenario. In summary, research by use of appropriate experimental techniques and CFD can yield useful results that can be used to optimise co-firing processes.

6.4 OXY-FUEL COMBUSTION

Oxyfuel combustion is achieved when an alternate oxidising atmosphere is utilised for the combustion process other than air (21% O_2 and 79% N_2 by volume). This entails various combinations such as 21% O_2 79% CO_2, 3% O_2 97% N_2, 10% O_2 50% CO_2 40% N_2 [5, 23]. As a technology, oxyfuel combustion has been used in specialised industries as a way of reducing CO_2 emissions and improving efficiency in power plants and cement kilns. With CO_2 emissions directly captured from the combustion process, oxyfuel systems provide a streamlined route for mitigating carbon emissions. Since most oxyfuel combustion systems require CO_2 as the other supplementary gas to O_2, this also supports the development of carbon capture systems [79, 80]. In extreme cases where O_2 is supplied in high proportions compared to the other gaseous constituents, high flame temperatures can be achieved as well as cleaner products [81]. Figure 6.14 shows a simplified flowchart of an oxyfuel combustion system. Most oxyfuel combustion systems utilise CO_2 as the secondary gas to the oxidiser O_2. As such, carbon capture and storage (CCS) is a very important aspect of most oxyfuel combustion systems.

Riaza et al. [82] used oxyfuel combustion in their studies on coal and olive waste biomass blends. Their CFD results showed increased combustion efficiency with respect to ignition and burnout characteristics, as shown in Figure 6.15.

CFD allows for detailed analysis of combustion processes whilst varying the atmosphere. The main challenge with oxyfuel combustion is the association with production and handling costs of O_2 and CO_2. Research and development efforts have focused on improving production efficiency of these gases as well as advancing of materials that can better withstand the extreme conditions of oxyfuel combustion [83]. In air-fuel combustion, N_2 is an inert bystander which doesn't participate in the chemical reactions, but it still absorbs heat, leading to a decrease in the overall temperature of the combustion process. With pure oxygen in oxyfuel combustion, the combustion is more efficient as there is no nitrogen to absorb heat; hence, higher temperatures. Without nitrogen, the products are primarily CO_2 and H_2O, which are easier to manage and capture, especially for carbon capture and storage (CCS) purposes.

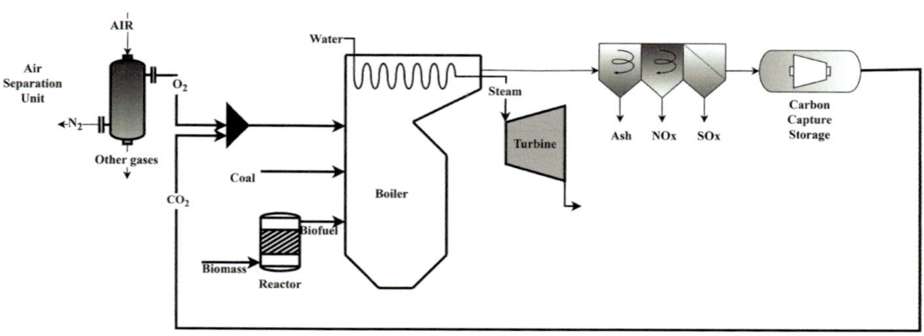

FIGURE 6.14 Power plant with oxyfuel combustion system.

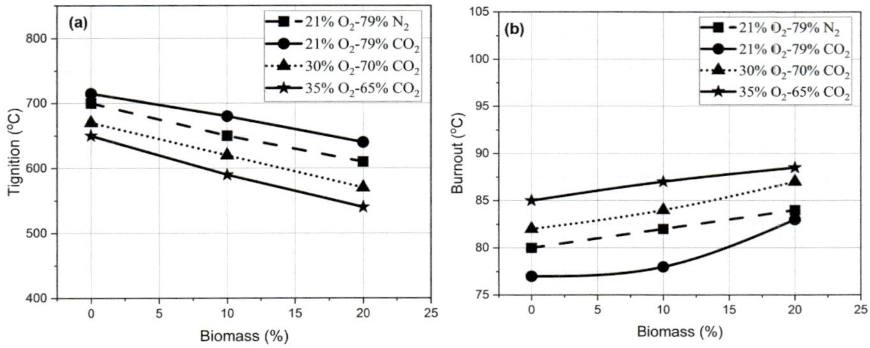

FIGURE 6.15 Oxyfuel combustion characteristics of coal and olive waste biomass blends: (a) ignition and (b) burnout.

CFD has also been used to successfully monitor the reduction in NOx during oxy-fuel combustion, with Perrone [84] demonstrating that due to the absence of oxygen in the oxidiser, only Fuel NOx influences the overall NOx emissions. Unfortunately, to optimise the oxyfuel combustion process, a lot of parameters need to be monitored in detail. CFD has allowed engineers to analyse flame structures, reaction kinetics, and heat flux. CFD allows for detailed analysis of combustion processes whilst varying the atmosphere [85]. The main challenge with oxyfuel combustion is associated with production and handling costs of O_2 and CO_2. Research and development efforts have focused on improving production efficiency of these gases as well as advancing of materials that can better withstand the extreme conditions of oxy-fuel combustion [83]. In air-fuel combustion, N_2 is an inert bystander as it doesn't participate in the chemical reactions but still absorbs heat, leading to a decrease in the overall temperature of the combustion process. By use of pure O_2 as experienced in oxy-fuel combustion, the combustion process becomes more efficient since there is no N_2 to act as a heat sink. Without N_2, the products are primarily CO_2 and H_2O, which are easier to manage and capture, especially for carbon capture and storage (CCS) purposes [86].

Apart from the problems associated with optimising oxyfuel combustion, there is substantial research going into the development of efficient air separation units which are generally energy-intensive during the cryogenic distillation and pressure swing adsorption. Producing 1 Nm^3 of O_2 through separation from air can consume up to about 0.45 kWh of energy [87]. Some researchers have managed to argue that the carbon footprint experienced during these stages is counteracted by the efficient combustion process experienced during oxyfuel combustion [88, 89]. Though not necessarily related to CFD, handling of O_2 in the presence of fuel requires high safety standards which also highlights the problems associated with oxyfuel combustion. Zheng et al. [90] managed to look into some of the technical challenges associated with setting up an oxyfuel combustion system. An important observation from their review was that, as the oxyfuel combustion system increases in size, more

TABLE 6.4
Applications of Oxyfuel Combustion

Oxyfuel Atmosphere Characteristics (Volume)	Fuel Characteristics	Industry Type	Reference
31% O_2 – 69% CO_2; 33% O_2 – 67% CO_2; 37% O_2 – 63% CO_2; 39% O_2 – 61% CO_2	Coal, wood, sludge, solid-recovered fuel (waste plastic, biomass, sewage sludge, meat, and bone meal, used tires, and used oils)	Cement kiln	Kroumian et al. [79]
25% O_2 – 75% flue gas; 30% O_2 – 70% flue gas; 35% O_2 – 65% flue gas	Coal	Power plant	Edge et al. [93]
0.35% O_2 – 66.6% H_2O – 33.2% CO_2; 1% O_2 – 66% H_2O – 33% CO_2; 2.5% O_2 – 65% H_2O – 32.5% CO_2	Methane	Steel plant	Kislinger et al. [94]
99.5 % O_2 – other gases	Natural gas and electric boosting	Glass fibre plant	Zhang et al. [98]
30 % O_2 – 70 % CO_2	Natural gas	Oil refinery	Escudero et al. [96]
100 % O_2; 30 % O_2 – 70 % CO_2; 50 % O_2 – 50 % CO_2; 70 % O_2 – 30 % CO_2	Methane	Hydrogen production	Sanusi and Mokheimer [97]

control systems are required per square metre of plant equipment. Real-time leak detection is one example of such a control system. Due to the high temperatures experienced during oxyfuel combustion, furnace materials such as burners, tubes, and linings tend to suffer higher thermal stress which means research into appropriate materials or optimum reaction positions needs to be taken into consideration as well. CFD is well positioned to carry out such research since it is physics based [91].

In summary, all this research into oxyfuel combustion has its main applications in power generation, cement industry, steel manufacturing, glass manufacturing, petrochemical industry, waste to energy systems, aerospace and rocket propulsion, and hydrogen production. Table 6.4 summarises applied research into oxyfuel combustion, with most being done through the use of CFD. To address SDG 7 and 13, the goal is to integrate oxyfuel combustion with renewable energy systems such as biomass waste combustion systems or wind and solar energy to supplement the energy required in the O_2 and CO_2 production [92].

6.5 CHEMICAL LOOPING COMBUSTION

In chemical looping combustion (CLC), rather than having the oxidiser supplied by blowers or fans in gaseous form, the oxidiser is bonded to a solid material which,

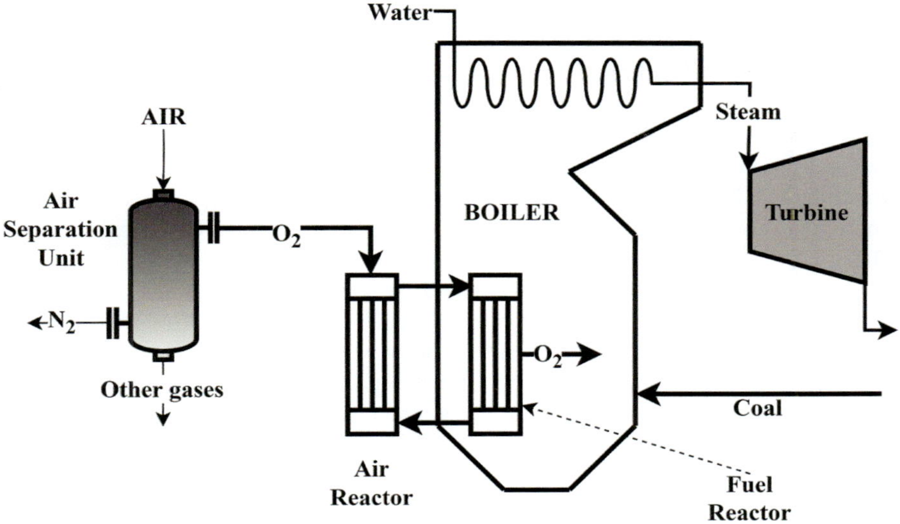

FIGURE 6.16 Chemical looping combustion with oxygen uncoupling schematic.

when inserted in the combustion chamber, releases the O_2 gradually so it can react with the fuel. After the O_2 in the solid material has depleted, the solid material is inserted in an air reactor where the O_2 is replenished; then the whole process repeats again [98]. This process is only specific for materials that have oxygen uncoupling capabilities. Since combustion is carried out in the presence of oxygen, chemical looping can be regarded as a form of oxyfuel combustion. Figure 6.16 gives a schematic of how chemical looping combustion takes place.

The fuel reactor is basically the combustion chamber. The oxygen carrier, which plays a huge part in the overall efficiency of the whole system, is typically a metal oxide such as iron oxide, copper oxide, or nickel oxide. The air reactor is responsible for replenishing the oxygen within the metal oxide. As with oxyfuel combustion, an air separation unit that produces the required oxygen offers tremendous advantages. Similarly, due to the oxidiser only having O_2 as the main constituent, the combustion exhaust tends to have a high concentration of CO_2 which can be easily cleaned and stored using carbon capture technology. Another major advantage of chemical looping combustion is its ability to control combustion reaction zones within the fuel reactor. This reduces the potential for incomplete combustion and high-temperature hotspots. Equation (6.6) demonstrates how the metal oxide is discharged to release oxygen in the fuel reactor (forward reaction) and charged with oxygen in the air reactor (reverse reaction) and discharged of oxygen when it enters the fuel reactor.

$$2M_xO_y \leftrightarrow 2M_xO_{y\text{-}1} + O_2 \quad (6.6)$$

These reactions occur in cycles between the two reactors, which require a metal oxide with high reactivity and good thermal stability without significant reduction in degradation. As such, a lot of research is being carried out to improve the degradation performance of metal oxides, especially making the process cost-effective. By combining fluidised bed and CLC technology, the bottleneck that had hindered the adoption of CLC on a large scale is slowly being shattered [99]. Since fluidised beds are well adapted to combust various types of fuels, including those with poor quality, CLC has been shown to be well adapted for various types of fuel inputs as well. Nonetheless, CLC is still facing challenges with commercial deployment, which is where CFD comes into play to assist with the optimisation of the CLC process [100, 101].

Banerjee and Agarwal [101] investigated the gas-solid interactions of a pseudo coal combustion system that uses iron metal oxide. Successful profiles of heat transfer, fluid dynamics, and gas-solid interactions were achieved. Efficient gas-solid interactions or heterogeneous reactions ensure high efficiency within the air and fuel reactors, as was demonstrated by the researchers. Gas concentration, temperature, and solid particle surface area have a large bearing on the overall efficiency of CLC. Adanez et al. [98] studied the effect of changing the metal oxides on the combustion efficiency, with manganese oxides having a better loading and unloading efficiency. Abad et al. [102] studied the residence time and temperature distribution in the fuel and air reactors, showing how the temperatures influence the reaction rates.

Designing of the optimum fuel and air reactors is also an active research area that has been tackled by various researchers. Adanez et al. [98] studied various reactor sizes and configurations to improve the optimal gas-solid heterogeneous reaction conditions. The bubbling fluidised bed was shown to effectively improve heat transfer for the fuel reactor without compromising on the metal oxide regeneration efficiency.

6.6 PULVERISED FUEL COMBUSTION

Pulverised fuel combustion is quite developed compared to other forms of solid fuel combustion, mainly due to its application in power and processing plants. CFD has been used to model almost every intricate process within the pulverised fuel combustion system, ranging from steam flow, combustion heat transfer, and solid particle movement. Figure 6.17 illustrates a schematic of a pulverised coal combustion system as used in thermal power plants.

The type of pulveriser depends on the application, with the most common being ball tube mills, impact mills, vertical roller mills, ring roll, and ball race mills. Ball tube mills consist of a rotating tube that is filled with balls, usually made of cast alloy material. When the tube is rotating, the balls tumble with the coal that would have been introduced, hence crushing the fuel to the intended size [103]. Impact mills have hinged hammers continuously hitting the coal against fixed wear-resistant plates. This action results in the crushing of the coal to the intended size [104]. Vertical roller mills generally have two rollers arranged concentrically. Coal is pulverised as it passes through the annulus of these rollers [105]. The ring roll and ball race mills have quite a unique and modern configuration that allows good grinding with minimum downtime. The setup resembles a ball bearing where balls will

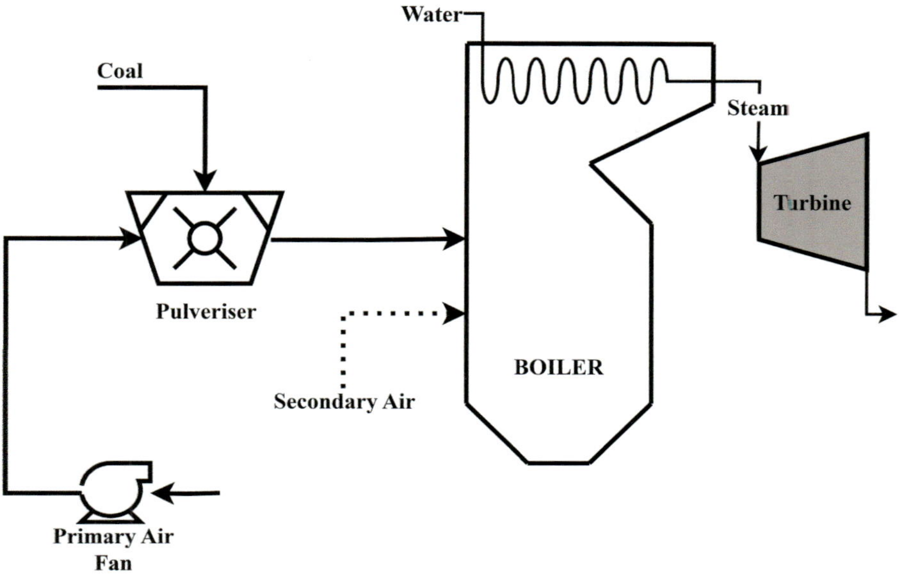

FIGURE 6.17 Schematic of pulverised fuel combustion with emphasis on the pulveriser.

constantly rotate against the outer ring. Coal is ground as it passes through the space between the ball and the ring. With ring roll mills, instead of having a ball, a wheel-like device that is attached to a cylindrical shaft is used. This means the wheel or roll will be able to rotate against the ring, hence allowing the grinding to take place [106].

Table 6.5 summarises various applications where CFD has been used to analyse pulverised combustion systems. Thakre and Goldin [107] used CFD to monitor intricate parameters related to species mass fraction, peak temperature and NOx emissions within a 2.4 MW swirl stabilised furnace. Their main objective was to validate their CFD model, which they did with reasonable accuracy, especially for the peak temperatures along the centreline. Abdul-Satel [108] used CFD to monitor the combustion of various ratios of syngas in a 5 MW furnace. Unfortunately, the researchers did not validate their results, which is one of the most common issues associated with CFD modelling. Torresi et al. [109] used CFD to study the operation of a newly designed coal burner on a 50 MW furnace. Velocity profiles of the combustion gases, species mass fractions and temperature distributions were all used to justify the new design.

Since most industrial furnaces are quite difficult to modify once installed due to the costs involved, the only plausible option available becomes modification of operating parameters. Zhang et al. [111] investigated the effect of varying the O_2 concentration during the pre-calcination process, a stage which requires combustion heat during cement manufacturing. As expected, an increase in the O_2 resulted in a higher furnace temperature. This resulted in better decomposition of $CaCO_3$

TABLE 6.5
Industrial Applications of Pulverised Fuel Combustion

Furnace Type and Size	Fuel Characteristics	Code	Parameters of Interest	Reference
2.4-MW swirl stabilised furnace	Coal	Simcenter STAR CCM+	Peak temperature, flame structure, species mass fraction, NOx emission	Thakre and Goldin [107]
5-MW furnace	Syngas	Aspen Plus	Temperature, pressure, density, velocity	Abdul-Sater et al. [108]
50-MW front-fired furnace	Coal	Fluent	Velocity, temperature, species mass fraction	Torresi et al. [109]
Fluidised bed	Coal and biomass	Gambit and Saturne	Temperature, SOx, NO, CO_2,	Wladyslaw [110]
Triple sprayed furnace	Coal	Commercial code	CaO, CO_2, temperature, $CaCO_3$, NO	Zhang et al. [111]
600-MW wall-fired utility boiler	Coal	Fluent	Burnout of char, CO	Liu et al. [112]
75 tonnes per hour circulating fluidised boiler	Municipal solid waste coal	Barracuda VR	Temperature. Wall erosion, O_2, CO_2, SO_2, NO	Xie et al. [113]
0.58-MW pilot furnace	Coal and NH_3	Fluent	Ignition, temperature, burnout, NOx	Zhang et al. [114]
4-MW burner	Coal and NH_3	Fluent	CO, NO, Volatile, NH_3, reaction rates	Wang et al. [115]
350-MW down-fired boiler	Coal	Fluent	Slagging, temperature, species mass fraction	Bian et al. [116]
150-MW tangential fired boiler	Coal and bagasse	Simcenter STAR CCM+	Heat flux, species mass fraction, particle residence time, particle diameter, velocity	Samantaray et al. [117]
660-MW wall-fired boiler	Coal and biomass derived fuel	Fluent	Temperature, NOx, SOx, CO_2	Xia et al. [118]

which meant the researchers had to determine an optimum point at which there was maximum decomposition at reasonable heat supply. Another benefit from their study was the reduction in NO in the product, which reduced the ammonia injection requirements. Liu et al. [112] used CFD to monitor the flow field around the burner when the oxidiser is changed from normal air to oxyfuel combustion. Through various iterations, controlling the over fire and secondary nozzles was deduced to be the best method of controlling flame stability without serious retrofitting. The optimum O_2 volume fraction in oxyfuel combustion that could produce similar properties comparable to air-fired combustion was found to be 30%, which is shared by other researchers.

Xie et al. [113] investigated the combustion of coal in a circulating fluidised bed (CFB) of coal by use of CFD. CFD allowed the researchers to analyse solid-gas interactions in detail for the various components of the CFB. Since these components are contained within the combustion zone, it becomes plausible to have insights about them. CFD was utilised by Zhang et al. [114] to investigate moderate or intense low oxygen dilution (MILD) combustion with success for various blends of pulverised coal and NH_3. As with most applications of CFD, this technology is deemed to be at the research stage requiring further validation, especially its potential to reduce CO_2 and NOx emissions. A blending ratio of 50% NH_3 has the potential to decrease furnace temperatures, decrease ignition time by 52%, decrease ignition temperature by 33%, and increase particle burnout by 52%. Detailed investigations about the combustion reactions that exist during coal-NH_3 combustion were performed by Wang et al. [115]. Results obtained through CFD profiles of species concentrations of intermediate reactions showed that NH_3 reactions are more competitive with combustion and NO formation reactions. The burner that was modelled showed rapid NH_3 decomposition in the primary combustion zone which can be linked to the excess air in the primary air. With CFD, the researchers deduced that improving the turbulence in the secondary combustion zone is very important as it reduces the tendency of converting NH_3 into H_2, which in turn affects the coal burnout performance. Xia et al. [118] employed CFD to study the effect of co-firing coal with biomass-derived fuels on a 660 MW furnace. Blending with biomass showed a significant increase in CO_2, lower NOx by 31%, decrease in outlet temperature by 70 K, and decrease in SOx by 37% at the furnace outlet. This study was able to verify the optimum positions for co-firing biomass fuel into the furnace.

6.7 LOW FUEL PLASMA IGNITION

Low fuel plasma ignition (LFPI) is a technology that utilises plasma discharges from ionised gas to ignite a fuel of interest. The conventional process of igniting coal requires first igniting a gaseous fuel which then ignites a liquid fuel that then ignites the coal of interest. Low fuel plasma ignition allows for direct ignition of coal without the need for other intermediate stages, hence the term low fuel. The plasma arc is created when a gas passes through two electrodes which are at a high voltage. One of the major advantages of low plasma ignition is its ability to ignite inherently difficult-to-ignite fuels such as low calorific value fuels, high moisture fuels, and high ash content fuels. Table 6.6 summarises various industrial applications of low fuel plasma ignition [119, 120].

Butakov et al. [121] through their studies of plasma ignition of pulverised coal showed experimentally how ignition of coal at 13 m/s velocity can be achieved in

TABLE 6.6
Industrial Applications of Low Fuel Plasma Ignition

Plasma Characteristics	Fuel Characteristics	Industrial Application	Reference
10 kW AC plasmatron	Pulverised coal	220 tonnes of steam per hour boiler	Butakov et al. [121]
40–100 kW plasma torch	Pulverised coal	300 MW boiler	Messerle et al. [122]
DC plasma torch	Pulverised lignite	500 kW pilot furnace	Baig et al. [123]
4–7 kW DC arc plasmatron	Lignite dust	400 kW pilot furnace	Youssefi et al. [124]

under 200 seconds. On an industrial scale, coal with a volatile content of 30% was ignited successfully. Messerle et al. [122] applied numerical modelling on a 300 MW boiler equipped with 12 front and back burners. Integration of 12 plasma ignitors showed a 7% decrease in combustion products temperature, 40% reduction in NOx, 6% reduction in O_2, and 43% reduction in unburnt carbon at the furnace outlet. A detailed study of the variation of plasma torch power and positions along the furnace was carried out by Baig et al. [123]. Their results demonstrated why it is essential to have sufficient plasma power for fuels with low ignitability, as high temperatures allow for sufficient ignition and devolatilisation. There is a limiting point at which any increase in plasma power does not result in an increase of carbon oxidation. Their studies also showed that at low thermal loads, the position of the plasma torch does not affect the efficiency of ignition, but rather an increase in fuel flow rate. Low calorific value fuels such as lignite were investigated with success by Youssefi et al. [124], showing the potential of using plasma ignition to improve the Wobble index of burners.

Challenges associated with LFPI include energy consumption of the plasma generation due to the high voltage and current needed. The silver lining is that compared to conventional methods of ignition, plasma ignition results in a high level of instantaneous energy whilst conventional methods result in prolonged consumption of energy. Another shortcoming is the need to replace electrodes which suffer high wear rates and corrosion due to their exposure to high temperatures and reactive species, which results in a high cost of implementation [119, 120]. Retrofitting LFPI to existing systems requires major modifications of existing systems, which increases capital costs. CFD allows researchers to investigate the complex plasma interactions coupled with combustion dynamics. In addition to the usual Navier-Stokes equations, electromagnetic models are used to solve the electric fields, magnetic fields, and flow of ionised gases. With CFD, the researchers can optimise the discharge location of the plasma, so sufficient heat flux is delivered to the fuel for ignition [125].

Hybrid systems combining plasma ignition with renewable fuels (such as hydrogen or biofuels) are emerging as solutions for reducing carbon emissions [126]. CFD modelling can be used to optimise the interaction between plasma and these clean fuels, ensuring stable ignition and efficient combustion. Future applications of plasma-assisted hybrid combustion systems require CFD to simulate fuel blending, plasma ignition, and emissions control. Industrial waste often includes difficult-to-burn materials that require high-energy inputs for ignition. Plasma ignition can enhance the efficiency of waste-to-energy plants by ensuring that low-energy waste materials are ignited reliably. CFD simulations will help optimise the plasma-assisted ignition system in these applications to improve waste conversion efficiency and minimise environmental impacts [127–129].

6.8 MACHINE LEARNING IN COMBUSTION MODELLING

Since combustion includes principles related to fluid mechanics, thermodynamics, and mass transfer, machine learning is well positioned to assist in optimising combustion due to the vast amount of data. Machine learning (ML) can be used to predict thermal properties of combustion such as temperature, pressure, heat, and chemical properties of combustion such as chemical reaction mechanisms and species composition. Understanding these parameters is essential with regard to understanding combustion, hence the motivation behind combustion machine learning. Basic theory of reaction mechanisms gives insights on how combustion is an exothermic reaction which involves multiple intermediate reactions. The optimisation of combustion processes requires accurate prediction of these multiple intermediate reactions as loosely represented by the Arrhenius equation. Traditional reduction techniques rely on analytical methods, but ML-based reduction strategies provide greater flexibility and adaptability to specific combustion problems. ML can automatically identify key species and reactions that contribute significantly to combustion dynamics, allowing for more efficient simulation models. Reaction rate parameters in combustion models are often uncertain due to limited experimental data. ML approaches, such as Bayesian optimisation and genetic algorithms, have been used to fine-tune reaction rate coefficients, improving the accuracy of combustion models. By analysing large datasets, ML algorithms can discover underlying chemical principles, predict reaction pathways, and identify previously unknown reaction mechanisms. This data-driven approach aids in refining kinetic models beyond what is possible through traditional theoretical and experimental techniques.

When fluid mechanics and thermodynamics are applied to combustion, it means mass and energy conservation parameters need to be satisfied as well. These parameters, which are also referred to as combustion manifolds, are temperature, pressure, specific volume, air to fuel ratio, and their related derived parameters. Trying to design experimental work that helps study the relationships between these parameters generally leads to expensive and labour-intensive scenarios. As with dimensional analysis, parametrisation or reduction of dimensionality plays a huge role in simplifying the relationships between combustion parameters; hence, the applications of combustion machine learning. Since flames behave in a fluid-like manner and with the regime of most flames being turbulent, application of Navier-Stokes equations is inevitable. Solving these complex Navier-Stokes equations with respect to reactions and transport processes advents closure problems, which can also be ameliorated by machine learning.

6.8.1 Species Mass Fractions and Product Yield

Prediction of species mass fractions and intermediate species experienced during the combustion process has been an active research area for a long time. This is mainly because chemical reaction mechanisms are difficult to formulate due to the low residence time of the process. Machine learning has also been applied successfully in the field of biomass pretreatment. The main focus of most of these studies is related to the bio-oil yield and its corresponding properties, as summarised in Table 6.7.

The random forest model was used to create an optimising tool that can predict the bio-oil production (yield, calorific value, H/C and O/C ratios) by use of lignin, cellulose, hemicellulose, C, O, H, N, and proximate analysis as inputs. Feature analysis demonstrated that yield prediction depends on the lignin, cellulose, and hemicellulose parameters hence producing a coefficient of determination of 0.92 [130]. The influence of lignin, cellulose, and hemicellulose on yield was similarly experienced by Cheng et al. [131] though lipid content was given the highest coefficient of influence on bio-oil yield. Dimensional reduction is a very important output of machine learning as various parameters such as reaction conditions, temperature, time, heating rate, extraction solvent, reactor type, and reactor size have a large influence on bio-oil yield [132–135].

6.8.2 Combustion Kinetic Parameters

Of interest to most solid fuel combustion applications are combustion kinetic parameters which are sometimes extended to determine combustion indices. These applications are summarised in Table 6.8. Ji et al. [136] studied the possibility of using neural networks to predict the kinetic properties of biomass pyrolysis which is normally done through the interpretation of TGA analytical data. The approach was able to integrate chemical reaction neural networks with fundamental physics laws in predicting the residual mass with success. Wang et al. [137] with their investigation of various biomass samples, determined that random forest algorithms can be used to predict the activation energy of these samples with 99.64% accuracy. The model inputs that were employed were proximate data, elementary data, and expected kinetic modelling methods namely Kissinger Akahira Sunose (KAS), Ozawa Flynn Wall (OFW), and Friedman. Li et al. [138] used a feedforward ANN to conduct sensitivity analysis on the kinetic mechanisms necessary to represent the decomposition of ethanol and the combustion of hydrogen with oxygen. A shallow neural network with one hidden layer was able to approximate complex reaction behaviours with less computational cost than CFD.

Combustion indexes such as ignition index, burnout index, devolatilisation index, first combustion index which highlights combustion activity a fuel possesses, and second combustion index which highlights the intensity and rate of combustion are derived from TGA data [1]. Sezer et al. [139] used an ANN to

TABLE 6.7
Application of Machine Learning in Combustion Species Mass Fractions and Biomass Thermochemical Treatment Properties

Application Area	Input Properties	Output	Predictive Performance		Reference
			Algorithms	R^2	
Combustion reactivity of biomass, syngas	C, O, H, and TGA heating rate and temperature	Hydrogen composition	Artificial neural network	0.9991	Sezer et Özveren [144]
Biomass densification through torrefaction	Proximate analysis, torrefaction temperature, and time	Elementary composition (C, H, O) and HHV	Feedforward backpropagation algorithm-based artificial neural network model and adaptive neuro-fuzzy inference system	0.82–0.94	Kartal and Özveren [145]
Bio oil characterisation	Spectropy data	C, H, O, LHV, effective hydrocarbon efficiency, unsaturated concentration	Support vector regression	0.92–0.995	Chen et al. [146]
Bio oil characterisation	Chemical composition (lignin cellulose hemicellulose) C, O, H, N proximate analysis	Bio oil yield, viscosity, calorific value, H/C, O/C	Random forests	0.92	Zhang et al. [130]
Bio oil characterisation	Feedstock type, biochemical composition, reaction conditions,	Bio crude yield	Various		Cheng et al. [131]
Biochar properties	Ultimate and proximate analysis	HHV	Gradient boosting regression, random forests, support vector machine	0.90–0.98	Chen et al. [147]

TABLE 6.8
Machine Learning Applications in Combustion Kinetic Parameter Prediction

Application Area	Input Properties	Output	Predictive Performance		Reference
			Algorithms	R^2	
Biomass pyrolysis	Physics-based laws, TGA data	Kinetic properties of biomass pyrolysis	Chemical reaction neural network		Ji et al. [136]
Biomass characterisation	Biomass type, proximate and ultimate analysis, kinetic chemical model	Activation energy	Random forest	0.9964	Wang et al. [137]
Decomposition of ethanol and the combustion of hydrogen with oxygen		Kinetic mechanisms	Feedforward artificial neural network		Li et al. [138]
Biomass combustion	C,O,H, and TGA heating rate and temperature	1st and 2nd combustion indexes	Artificial neural network	0.9800	Sezer et al. [139]
Biomass densification through torrefaction	Proximate analysis, torrefaction temperature, and time	Activation energy	Feedforward backpropagation algorithm-based artificial neural network (ANN) model, adaptive neuro-fuzzy inference system (ANFIS)	0.9600	Kartal and Özveren [140]

TABLE 6.9
Machine Learning Applications in Combustion Thermal Property Determination

Application Area	Input Properties	Output	Predictive Performance		Reference
			Algorithms	R^2	
		Enthalpy of combustion	Artificial neural network	0.9630	Jameel et al. [141]
Biomass and syngas combustion	C, O, H, and TGA heating rate and temperature	Chemical exergy	Artificial neural network		Sezer et al. [142]
Biomass densification through torrefaction	Proximate analysis, torrefaction temperature, and time	Chemical exergy	Feedforward backpropagation algorithm-based artificial neural network (ANN), adaptive neuro-fuzzy inference system (ANFIS)	0.92	Kartal and Özveren [143]

predict the first and second combustion indexes of various lignocellulosic biomass with success. The input variables into the model were simply the elementary composition and TGA parameters namely the heating rate and temperature progression.

6.8.3 Fuel Thermal Properties

Engineers tend to gravitate towards thermal properties of fuels as they are interested in the heat produced when a fuel combusts. Table 6.9 summarises the various cases where ML was employed to predict various thermal properties of fuels of interest. Jameel et al. [141] also showed that an ANN with 14 input layers, 26 hidden layers, and 1 output layer can be employed to predict the enthalpy of combustion of various solid fuels with an overall precision of approximately 96.3%. Kartal and Özveren [140, 143, 145] used feedforward backpropagation algorithm based on an artificial neural network model and adaptive neuro-fuzzy inference system (ANFIS) to predict elementary composition (C, H, O), HHV, chemical exergy, and activation energy of various biomass samples. The prediction capabilities were commendable as they managed to break down the analysis such that the R^2 for each predicted output was determined. summarises the various cases where ML was employed to predict various thermal properties of fuels of interest.

Parameters important to combustion such as chemical exergy and hydrogen composition within syngas from various biomass fuels were also predicted successfully by Sezer et al. [142, 144]. The fact that the model was able to perform dimensional reduction considering the relatively large number of input variables is quite commendable. Dimensional reduction was also employed by Chen et al. [146] whilst using spectral data from FTIR analysis to predict bio-oil composition using the support vector regression (SVR) algorithm.

6.8.4 Flame Properties

Ihme et al. [148] investigated the machine learning algorithms that can be applied to predict the ignition behaviour of n-dodecane/air mixture as summarised in Table 6.10. The experimental conditions are shown in Figure 6.18, whilst a comparison of the different algorithms that can be used is shown in Figure 6.19. The input parameters were initial temperature (y axis) and n-dodecane mole fraction (x axis), whilst the output parameters that needed classification were high temperature ignition (adiabatic temperature exceeding 1050 K), low temperature ignition (adiabatic temperature below 1050 K), and no ignition. The feedforward neural network that used 60% of the data as training data was deemed to produce the best prediction.

Sahoo et al. [149] employed machine learning models (regression model, support vector machine (SVM), and artificial neural network (ANN)) to predict emissions and combustion performance characteristics of a spark ignition engine successfully. Sahoo et al. [149] employed machine learning models (regression model, support vector machine (SVM), and artificial neural network (ANN)) to predict emissions

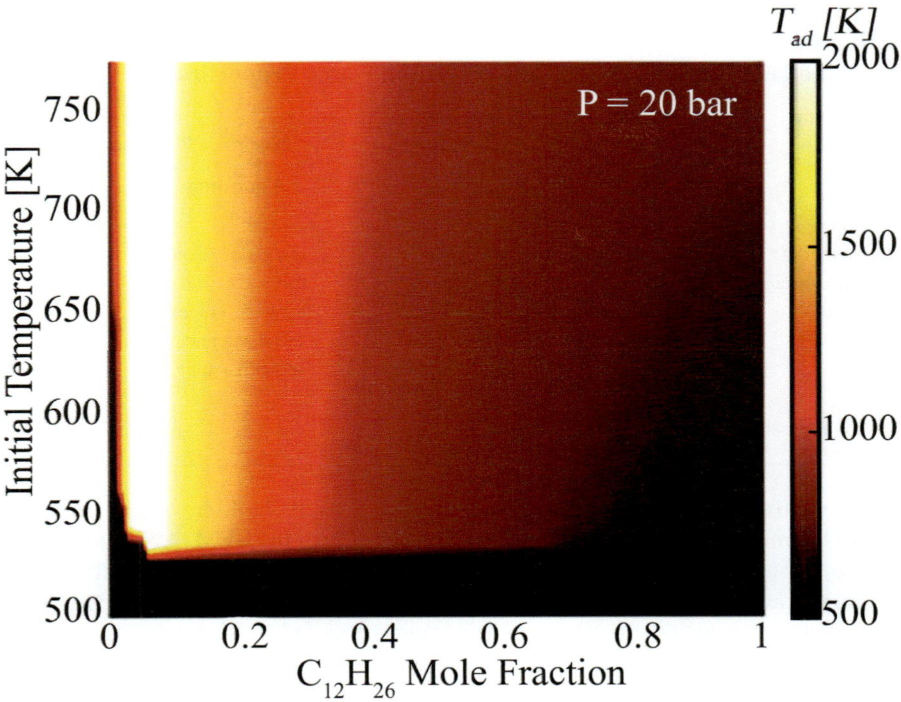

FIGURE 6.18 Flame ignition experimental properties of n-dodecane oxygen mixtures.

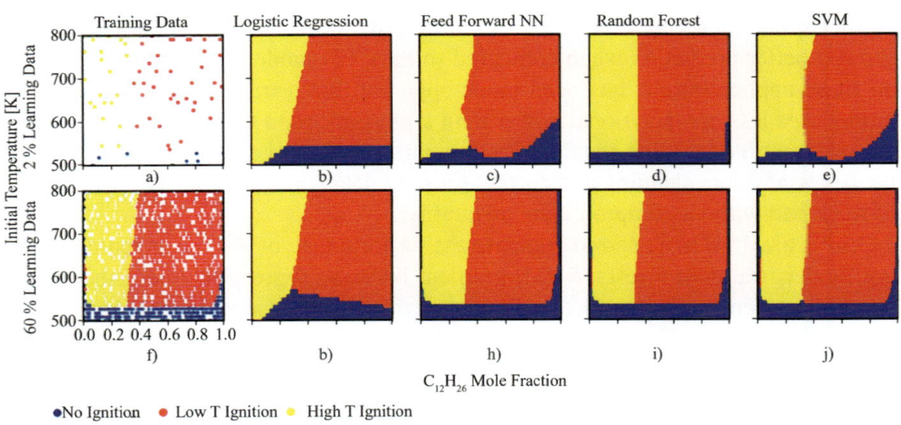

FIGURE 6.19 Flame ignition experimental properties of n-dodecane oxygen mixtures.

TABLE 6.10
Machine Learning as Applied to Combustion Flame Properties

Application Area	Input Properties	Output	Predictive Performance	
			Algorithms	R^2
Ignition behaviour of n-dodecane/air mixture	Initial temperature, n-dodecane mole fraction, adiabatic flame temperature	Classification (high temperature ignition/low temperature ignition/no ignition)	Logistic regression, feed forward neural network, random forest, support vector machine	

Source: Ihme et al. [148].

and combustion performance characteristics of a spark ignition engine successfully. A remarkable 98% R^2 error level was attained with the SVM. ANNs were also employed successfully by Fu et al. [150] to predict combustion efficiency (fuel consumption) and emissions (NOx, unburnt hydrocarbon, and carbon monoxide) from an internal combustion engine by using engine speed and load as the model inputs. The input data that was used for compiling the testing data was generated by high-resolution cameras. Wan et al. [151] also used a convolutional neural network (CNN) to predict the combustion regimes of CH_4/air jet flames, giving insights into the distribution of species mass fractions during the combustion process.

Hwang et al. [152] trained a linear regression-based machine learning model to predict spray angles for combustion in an internal combustion engine. The results showed better prediction when compared to the CFD model that was employed. In all these applications, it has been made apparent that a trained machine learning model uses less computational power than a physics-based CFD model [153]. Even though machine learning requires vast amounts of data to train the models, use of the necessary data cleaning and correct training data generally aids in the reduction of workload without compromising on quality.

Liu et al. [154] managed to compare the performance of various machine learning models (random forests, SVM, gradient boosting regression trees, and ANN) using equivalence ratio, spark timing, and engine speed as model inputs and engine exhaust temperature as model outputs [153]. Even though machine learning requires vast amounts of data to train the models, use of the necessary data cleaning and correct training data generally aids in the reduction of workload without compromising on quality.

6.9 OPTIMISATION OF THE CO-COMBUSTION PROCESSES

Various tools can be used to optimise the co-combustion process of coal and biomass depending on the parameter that needs to be improved [66]. These tools basically employ a specific design of experiments and corresponding surface response theory to adjust their parameters until they arrive at an optimum condition [155]. Optimetrics is the general term used to encompass all the methods that can be employed to optimise a process through design variations or input parameters, thus identifying the best performing configuration. Input parameters can include geometry dimensions of the furnace, geometry dimensions of the burner, thermodynamic-related parameters, fluid mechanic-related parameters, as well as chemical reaction-related parameters.

Under optimetrics there is parametric, sensitivity, optimisation, and tuning analysis. Parametric analysis allows the researcher to vary one or more input parameters across a defined range. Sensitivity analysis identifies which input parameters have the most influence on the output parameter. Sensitivity analysis is key as it allows CFD to focus on the parameters that need to be varied rather than wasting computational resources. Optimisation works by defining a goal or output parameter such as maximum wall temperature then playing around with the input parameters to arrive at that goal. Recent advances have allowed researchers to perform tuning analysis which is an interactive adjustment of input parameters that shows real-time performance of output parameters [71, 156]. At the heart of any optimisation problem is the objective function which is the value that needs to be minimised or maximised. To achieve this objective function, design variables are programmed into CFD with their respective gradients. When two design variables are traced on the x and y axes with the objective function traced on the z axis, it becomes possible to extract a response surface which guides the optimisation process.

6.10 ENVIRONMENTAL BENEFITS OF CO-COMBUSTION

Emission reduction has received a lot of backing from the United Nations and its partners over the years, as shown by the postulation of Sustainable Development Goals (SDGs) 3, 7 and 13 which keep on pushing for pollution reduction, clean energy, and action on climate change by 2040 [157]. Financial backing from international institutions is on an upwards trend for projects that are considered to be green. Combustion is still the predominant mode of heat generation in industry through the large share of thermal plants, processing, and chemical industries [158]. As such, research into cleaner modes of combustion is very much an active research area.

Studies have demonstrated that optimising the combustion of coal and biomass blends results in increased combustion efficiency and reduced pollutants in the form of SOx, NOx, and particulate matter [159]. It is important to understand that experimental analysis of combustion is expensive, requiring multiple measuring instruments such as gas analysers, instruments that can analyse flame behaviour, and heat and temperature analysers. This makes CFD an important alternative capable of optimising combustion parameters in a cost-efficient way [160]. Basically, when a solid fuel undergoes combustion, it heats up and, in the process, loses its moisture, followed by the release of gaseous volatiles, leaving a porous solid char material and mineral elements behind. Both the volatiles and char undergo chemical reactions with oxygen, releasing heat in the process. The mineral elements react with each other, forming ash in the process [42]. All these processes should be modelled with as much precision as possible to attain an acceptable combustion model.

6.11 SUMMARY

This chapter examined coal biomass co-combustion with a focus on ignition, burnout, and flame stability, supported by a case study on coal and Pinus sawdust blends. Broader strategies including co-firing, oxy-fuel, chemical looping, pulverised fuel, and plasma ignition were reviewed for their potential to enhance efficiency and reduce emissions. The role of machine learning was highlighted in predicting species behaviour, fuel properties, and flame dynamics, offering complementary tools to traditional models. Optimisation methods and environmental benefits were also considered, showing consistent reductions in CO_2, SOx, and NOx. Overall, the chapter demonstrates that integrating conventional modelling with advanced combustion strategies and data-driven approaches strengthens both the technical performance and sustainability of co-combustion systems.

REFERENCES

1. Marangwanda, G.T.; Madyira, D.M.; Ndungu, P.G.; Chihobo, C.H. Combustion Characterisation of Bituminous Coal and Pinus Sawdust Blends by Use of Thermo-Gravimetric Analysis. *Energies (Basel)* **2021**, *14*, 1–20, doi:10.3390/en14227547.
2. Haykiri-Acma, H.; Yaman, S. Synergistic Investigation for Co-Combustion of Biochars and Lignite—Thermogravimetric Analysis Approach. *Journal of Thermal Science and Engineering Application* **2019**, *11*, doi:10.1115/1.4040992.
3. Pallarés, J.; Gil, A.; Cortés, C.; Herce, C. Numerical Study of Co-Firing Coal and Cynara Cardunculus in a 350 MWe Utility Boiler. *Fuel Processing Technology* **2009**, *90*, 1207–1213, doi:10.1016/j.fuproc.2009.05.025.
4. Williams, A.; Pourkashanian, M.; Jones, J.M. Combustion of Pulverised Coal and Biomass. *Progress in Energy and Combustion Science* **2001**, *27*, 587–610, doi:10.1016/S0360-1285(01)00004-1.
5. Bhuiyan, A.A.; Naser, J. CFD Modelling of Co-Firing of Biomass with Coal under Oxy-Fuel Combustion in a Large Scale Power Plant. *Fuel* **2015**, *159*, 150–168, doi:10.1016/j.fuel.2015.06.058.
6. Mohammed Redha, A.; Lau, A.; Holuszko, M.; Vakil, A.; Sokhansanj, S. CFD Investigation of Pelletization Effect on Co-Firing Coal with Wheat Straw. *Canadian Journal of Chemical Engineering* **2022**, *100*, 237–253, doi:10.1002/cjce.24106.
7. Sami, M.; Annamalai, K.; Wooldridge, M. Co-Firing of Coal and Biomass Fuel Blends. *Progress in Energy and Combustion Science* **2001**, *27*, 171–214, doi:10.1016/S0360-1285(00)00020-4.
8. Gubba, S.R.; Ingham, D.B.; Larsen, K.J.; Ma, L.; Pourkashanian, M.; Tan, H.Z.; Williams, A.; Zhou, H. Numerical Modelling of the Co-Firing of Pulverised Coal and Straw in a 300MWe Tangentially Fired Boiler. *Fuel Processing Technology* **2012**, *104*, 181–188, doi:10.1016/j.fuproc.2012.05.011.
9. Yin, C.; Rosendahl, L.A.; Kær, S.K. Towards a Better Understanding of Biomass Suspension Co-Firing Impacts via Investigating a Coal Flame and a Biomass Flame in a Swirl-Stabilized Burner Flow Reactor under Same Conditions. *Fuel Processing Technology* **2012**, *98*, 65–73, doi:10.1016/j.fuproc.2012.01.024.

10. Ma, L.; Gharebaghi, M.; Porter, R.; Pourkashanian, M.; Jones, J.M.; Williams, A. Modelling Methods for Co-Fired Pulverised Fuel Furnaces. *Fuel* **2009**, *88*, 2448–2454, doi:10.1016/j.fuel.2009.02.030.
11. Niu, S.; Chen, M.; Li, Y.; Song, J. Co-Combustion Characteristics of Municipal Sewage Sludge and Bituminous Coal. *Journal of Thermal Analysis and Calorimetry* **2018**, *131*, 1821–1834, doi:10.1007/s10973-017-6716-3.
12. Al-Qayim, K.; Nimmo, W.; Hughes, K.; Pourkashanian, M. Kinetic Parameters of the Intrinsic Reactivity of Woody Biomass and Coal Chars via Thermogravimetric Analysis. *Fuel* **2017**, *210*, 811–825, doi:10.1016/j.fuel.2017.09.010.
13. Hurt, R.H.; Calo, J.M. Semi-Global Intrinsic Kinetics for Char Combustion Modeling. *Combustion and Flame* **2001**, *125*, 1138–1149, doi:10.1016/S0010-2180(01)00234-6.
14. Zhang, J.; Prationo, W.; Zhang, L.; Zhang, Z. Computational Fluid Dynamics Modeling on the Air-Firing and Oxy-Fuel Combustion of Dried Victorian Brown Coal. *Energy & Fuels* **2013**, *27*, 4258–4269, doi:10.1021/ef400032t.
15. Chen, X.; Liu, L.; Zhang, L.; Zhao, Y.; Zhang, Z.; Xie, X.; Qiu, P.; Chen, G.; Pei, J. Thermogravimetric Analysis and Kinetics of the Co-Pyrolysis of Coal Blends with Corn Stalks. *Thermochim Acta* **2018**, *659*, 59–65, doi:10.1016/j.tca.2017.11.005.
16. Boumanchar, I.; Chhiti, Y.; M'hamdi Alaoui, F.E.; Elkhouakhi, M.; Sahibed-dine, A.; Bentiss, F.; Jama, C.; Bensitel, M. Investigation of (Co)-Combustion Kinetics of Biomass, Coal and Municipal Solid Wastes. *Waste Management* **2019**, *97*, 10–18, doi:10.1016/j.wasman.2019.07.033.
17. Xing, X.; Wang, S.; Zhang, Q. Thermogravimetric Analysis and Kinetics of Mixed Combustion of Waste Plastics and Semicoke. *Journal of Chemistry* **2019**, *2019*, 10, doi:doi.org/10.1155/2019/8675986.
18. Haykiri-Acma, H.; Yaman, S. Synergistic Investigation for Co-Combustion of Biochars and Lignite—Thermogravimetric Analysis Approach. *Journal of Thermal Science and Engineering Application* **2019**, *11*, doi:10.1115/1.4040992.
19. Wang, H.; Tian, Y.; Chen, X.; Li, J. Kinetic and Synergetic Effect Analysis of the Co-Combustion of Coal Blended with Polyurethane Materials. *ACS Omega* **2020**, *5*, 26005–26014, doi:10.1021/acsomega.0c03449.
20. Chen, G.B.; Chatelier, S.; Lin, H.T.; Wu, F.H.; Lin, T.H. A Study of Sewage Sludge Co-Combustion with Australian Black Coal and Shiitake Substrate. *Energies (Basel)* **2018**, *11*, doi:10.3390/en11123436.
21. Álvarez, L.; Yin, C.; Riaza, J.; Pevida, C.; Pis, J.J.J.; Rubiera, F. Biomass Co-Firing under Oxy-Fuel Conditions: A Computational Fluid Dynamics Modelling Study and Experimental Validation. *Fuel Processing Technology* **2014**, *120*, 22–33, doi:10.1016/j.fuproc.2013.12.005.
22. Pérez-Jeldres, R.; Cornejo, P.; Flores, M.; Gordon, A.; García, X. A Modeling Approach to Co-Firing Biomass/Coal Blends in Pulverized Coal Utility Boilers: Synergistic Effects and Emissions Profiles. *Energy* **2017**, *120*, 663–674, doi:10.1016/j.energy.2016.11.116.
23. Riaza, J.; Álvarez, L.; Gil, M.V.; Pevida, C.; Pis, J.J.; Rubiera, F. Effect of Oxy-Fuel Combustion with Steam Addition on Coal Ignition and Burnout in an Entrained Flow Reactor. *Energy* **2011**, *36*, 5314–5319, doi:10.1016/j.energy.2011.06.039.
24. Zhou, C.; Liu, G.; Wang, X.; Qi, C. Co-Combustion of Bituminous Coal and Biomass Fuel Blends: Thermochemical Characterization, Potential Utilization and Environmental Advantage. *Bioresource Technology* **2016**, *218*, 418–427, doi:10.1016/j.biortech.2016.06.134.
25. Bhuiyan, A.A.; Blicblau, A.S.; Naser, J. Co-Firing of Biomass and Slagging in Industrial Furnace: A Review on Modelling Approach. *Journal of the Energy Institute* **2017**, *90*, 838–854, doi:10.1016/j.joei.2016.08.010.

26. Priyanto, D.E.; Matsunaga, Y.; Ueno, S.; Kasai, H.; Tanoue, T.; Mae, K.; Fukushima, H. Co-Firing High Ratio of Woody Biomass with Coal in a 150-MW Class Pulverized Coal Boiler: Properties of the Initial Deposits and Their Effect on Tube Corrosion. *Fuel* **2017**, *208*, 714–721, doi:10.1016/j.fuel.2017.07.053.
27. Yin, C.; Rosendahl, L.A.; Condra, T.J.; Kær, S.K. Use of Numerical Modeling in Design for Co-Firing Biomass in Wall-Fired Burners. *Chemical Engineering Science* **2004**, *59*, 3281–3292, doi:10.1016/j.ces.2004.04.036.
28. Al-Abbas, A.H.; Naser, J. Effect of Chemical Reaction Mechanisms and NOx Modeling on Air-Fired and Oxy-Fuel Combustion of Lignite in a 100-KW Furnace. *Energy & Fuels* **2012**, *26*, 3329–3348, doi:10.1021/ef300403a.
29. Tan, P.; Ma, L.; Xia, J.; Fang, Q.; Zhang, C.; Chen, G. Co-Firing Sludge in a Pulverized Coal-Fired Utility Boiler: Combustion Characteristics and Economic Impacts. *Energy* **2017**, *119*, 392–399, doi:10.1016/j.energy.2016.12.084.
30. Chang, J.; Wang, X.; Zhou, Z.; Chen, H.; Niu, Y. CFD Modeling of Hydrodynamics, Combustion and NOx Emission in a Tangentially Fired Pulverized-Coal Boiler at Low Load Operating Conditions. *Advanced Powder Technology* **2021**, *32*, 290–303, doi:10.1016/j.apt.2020.12.008.
31. Chen, L.; Yong, S.Z.; Ghoniem, A.F. Oxy-Fuel Combustion of Pulverized Coal: Characterization, Fundamentals, Stabilization and CFD Modeling. *Progress in Energy and Combustion Science* **2012**, *38*, 156–214, doi:10.1016/j.pecs.2011.09.003.
32. Mohanna, H. Combustion of Pulverized Biomass: Impact of Fuel Preparation and Flow Conditions, Normandie Université, 2021.
33. Xinjie, L.; Shihong, Z.; Xincheng, W.; Jinai, S.; Xiong, Z.; Xianhua, W.; Haiping, Y.; Hanping, C. Co-Combustion of Wheat Straw and Camphor Wood with Coal Slime: Thermal Behaviour, Kinetics, and Gaseous Pollutant Emission Characteristics. *Energy* **2021**, *234*, 121292, doi:10.1016/j.energy.2021.121292.
34. Bhuiyan, A.A.; Blicblau, A.S.; Islam, A.K.M.S.; Naser, J. A Review on Thermo-Chemical Characteristics of Coal/Biomass Co-Firing in Industrial Furnace. *Journal of the Energy Institute* **2018**, *91*, 1–18, doi:10.1016/j.joei.2016.10.006.
35. Wang, H.; Tian, Y.; Li, J.; Chen, X. Experimental Study on Thermal Effect and Gas Release Laws of Coal-Polyurethane Cooperative Spontaneous Combustion. *Scientific Reports* **2021**, *11*, 1–13, doi:10.1038/s41598-021-81537-5.
36. Lü, H.F.; Deng, J.; Li, D.J.; Xu, F.; Xiao, Y.; Shu, C.M. Effect of Oxidation Temperature and Oxygen Concentration on Macro Characteristics of Pre-Oxidised Coal Spontaneous Combustion Process. *Energy* **2021**, *227*, 120431, doi:10.1016/j.energy.2021.120431.
37. Wang, C.; Liu, Y.; Zhang, X.; Che, D. A Study on Coal Properties and Combustion Characteristics of Blended Coals in Northwestern China. *Energy & Fuels* **2011**, *25*, 3634–3645, doi:10.1021/ef200686d.
38. Wang, C.; Zhang, Y.; Wang, P.; Zhang, J.; Du, Y.; Che, D. Effects of Silicoaluminate Oxide and Coal Blending on Combustion Behaviors and Kinetics of Zhundong Coal under Oxy-Fuel Condition. *Journal of Thermal Analysis and Calorimetry* **2018**, *134*, 1975–1986, doi:10.1007/s10973-018-7627-7.
39. Florentino-Madiedo, L.; Vega, M.F.; Barriocanal, C. Evaluation of Synergy during Co-Pyrolysis of Torrefied Sawdust, Coal and Paraffin. A Kinetic and Thermodynamic Study. *Fuel* **2021**, *292*, 120305, doi:10.1016/j.fuel.2021.120305.
40. Wu, Z.; Zhang, J.; Zhang, B.; Guo, W.; Yang, G.; Yang, B. Synergistic Effects from Co-Pyrolysis of Lignocellulosic Biomass Main Component with Low-Rank Coal: Online and Offline Analysis on Products Distribution and Kinetic Characteristics. *Applied Energy* **2020**, *276*, 115461, doi:10.1016/j.apenergy.2020.115461.
41. Roberts, D.L. The Inherent Moisture Content of South African Permian Coals. *International Journal of Coal Geology* **1991**, *17*, 297–311, doi:10.1016/0166-5162(91)90036-I.

42. Turns, S. *An Introduction to Combustion: Concepts and Applications*; McGraw-Hill series in mechanical engineering; 2nd ed.; McGraw-Hill: New York, 1996.
43. Zhou, Z.; Chen, L.; Guo, L.; Qian, B.; Wang, Z.; Cen, K. Computational Modeling of Oxy-Coal Combustion with Intrinsic Heterogeneous Char Reaction Models. *Fuel Processing Technology* **2017**, *161*, 169–181, doi:10.1016/j.fuproc.2017.03.009.
44. Chen, P.; Zhang, L.; Huang, K. Kinetic Modeling of Coal Thermal Decomposition under Air Atmosphere. *Energy & Fuels* **2016**, *30*, 5158–5166, doi:10.1021/acs.energyfuels.6b00902.
45. Bermejo, S.P.; Prado-guerra, A.; Isabel, A.; García, P.; Fernando, L.; Prieto, C. Study of Quinoa Plant Residues as a Way to Produce Energy through Thermogravimetric Analysis and Indexes Estimation. *Renewable Energy* **2020**, *146*, 2224–2233, doi:10.1016/j.renene.2019.08.056.
46. Korotkikh, A.G.; Slyusarskiy, K.V.; Larionov, K.B.; Osipov, V.I. Comparison of Coal Reactivity during Conversion into Different Oxidizing Medium. *Journal of Physics: Conference Series* **2016**, *754*, doi:10.1088/1742-6596/754/5/052005.
47. Jiang, X.; Wang, C.; Wang, X.; Liu, Y.; Song, Z.; Lin, Q. Co-Combustion of Coal and Carbonaceous Wastes: Thermogravimetric Analysis. *Energy Sources, Part A: Recovery, Utilization, and Environmental Effects* **2020**, 1–15, doi:10.1080/15567036.2019.1704314.
48. Urban, B.; Shirazi, Y.; Maddi, B.; Viamajala, S.; Varanasi, S. Flash Pyrolysis of Oleaginous Biomass in a Fluidized-Bed Reactor. *Energy & Fuels* **2017**, *31*, 8326–8334, doi:10.1021/acs.energyfuels.7b01306.
49. Xu, T.; Ning, X.; Wang, G.; Liang, W.; Zhang, J.; Li, Y.; Wang, H.; Jiang, C. Combustion Characteristics and Kinetic Analysis of Co-Combustion between Bag Dust and Pulverized Coal. *International Journal of Minerals, Metallurgy, and Materials* **2018**, *25*, 1412–1422, doi:10.1007/s12613-018-1695-3.
50. Mureddu, M.; Dessì, F.; Orsini, A.; Ferrara, F.; Pettinau, A. Air-and Oxygen-Blown Characterization of Coal and Biomass by Thermogravimetric Analysis. *Fuel* **2018**, *212*, 626–637, doi:10.1016/j.fuel.2017.10.005.
51. Yan, Y.F.; Zhang, Z.E.; Zhang, L.; Zhang, L. Influence of Coal Properties on the Co-Combustion Characteristics of Low-Grade Coal and City Mud. *Global NEST Journal* **2014**, *16*, 329–338, doi:10.30955/gnj.001315.
52. Wnorowska, J.; Ciukaj, S.; Kalisz, S. Thermogravimetric Analysis of Solid Biofuels with Additive under Air Atmosphere. *Energies (Basel)* **2021**, *14*, 2257, doi:10.3390/en14082257.
53. Kongkaew, N.; Pruksakit, W.; Patumsawad, S. Thermogravimetric Kinetic Analysis of the Pyrolysis of Rice Straw. *Energy Procedia* **2015**, *79*, 663–670, doi:10.1016/j.egypro.2015.11.552.
54. Yuan, S.; Chen, H. Mathematical Rules for Synergistic, Additive, and Antagonistic Effects of Multi-Drug Combinations and Their Application in Research and Development of Combinatorial Drugs and Special Medical Food Combinations. *Food Science and Human Wellness* **2019**, *8*, 136–141, doi:10.1016/j.fshw.2019.01.003.
55. Sarroza, A.C.; Bennet, T.D.; Eastwick, C.; Liu, H. Characterising Pulverised Fuel Ignition in a Visual Drop Tube Furnace by Use of a High-Speed Imaging Technique. *Fuel Processing Technology* **2017**, *157*, 1–11, doi:10.1016/j.fuproc.2016.11.002.
56. Li, H.; Li, Y.; Jin, Y. Co-Combustion Analyses of Coal and Sewage Sludge with High Moisture Content. *Energy Sources, Part A: Recovery, Utilization, and Environmental Effects* **2015**, *37*, 1896–1903, doi:10.1080/15567036.2011.648306.
57. Tillman, D.A. Biomass Cofiring: The Technology, the Experience, the Combustion Consequences. *Biomass Bioenergy* **2000**, *19*, 365–384, doi:10.1016/S0961-9534(00)00049-0.

58. Leckner, B. Possibilities and Limitations of Co-Firing of Biomass. In Proceedings of the 1st Project Conference AGS; 1st Project Conference AGS: Stockholm, 2006.
59. Fernando, R. *Fuels for Biomass Cofiring*; IEA Coal Research: London, 2005.
60. Tillman, D.A. Cofiring Benefits for Coal and Biomass. *Biomass Bioenergy* **2000**, *19*, 363–364, doi:10.1016/S0961-9534(00)00048-9.
61. Savolainen, K. Co-Firing of Biomass in Coal-Fired Utility Boilers. *Applied Energy* **2003**, *74*, 369–381, doi:10.1016/S0306-2619(02)00193-9.
62. Stöcker, M. Biofuels and Biomass-To-Liquid Fuels in the Biorefinery: Catalytic Conversion of Lignocellulosic Biomass Using Porous Materials. *Angewandte Chemie International Edition* **2008**, *47*, 9200–9211, doi:10.1002/anie.200801476.
63. Chen, C.; Zhao, L.; Wu, X.; Wang, J. Numerical and Experimental Study on Oxy-Fuel Coal and Biomass Co-Firing in a Bubbling Fluidized Bed. *Energy & Fuels* **2019**, *33*, 5829–5839, doi:10.1021/acs.energyfuels.8b04163.
64. Tan, P.; Ma, L.; Fang, Q.; Zhang, C.; Chen, G. Application of Different Combustion Models for Simulating the Co-Combustion of Sludge with Coal in a 100 MW Tangentially Coal-Fired Utility Boiler. *Energy & Fuels* **2016**, *30*, 1685–1692, doi:10.1021/acs.energyfuels.5b02236.
65. Wang, P.; Wang, C.; Du, Y.; Feng, Q.; Wang, Z.; Yao, W.; Liu, J.; Zhang, J.; Che, D. Experiments and Simulation on Co-Combustion of Semicoke and Coal in a Full-Scale Tangentially Fired Utility Boiler. *Energy & Fuels* **2019**, *33*, 3012–3027. doi:10.1021/acs.energyfuels.8b04482.
66. Marangwanda, G.T.; Madyira, D.M. Evaluating Combustion Ignition, Burnout, Stability, and Intensity of Coal–Biomass Blends Within a Drop Tube Furnace Through Modelling. *Energies* **2025**, *18*, 1322, doi:10.3390/en18061322.
67. Marangwanda, G.T.; Madyira, D.M.; Chihobo, H.C.; Babarinde, T.O. Modelling Co-Combustion of Bituminous Coal and Pine Sawdust: Thermal Behaviour. *Fuel Communications* **2021**, *9*, 100035, doi:10.1016/j.jfueco.2021.100035.
68. Kalu, I.E.; Inglis, H.M.; Kok, S. Failure Assessment Methodology for Boiler Tubes with Localized External Erosion Defects. *International Journal of Pressure Vessels and Piping* **2020**, *188*, 104190, doi:10.1016/j.ijpvp.2020.104190.
69. Liu, H.; Wang, Y.; Xue, J.; Zhang, Y.; Yu, P.; Che, D. Experimental Study on Combustion, Ash Fusibility and Slagging Propensity during Co-Combustion of Organic Solid Waste and Lignite. *Journal of the Energy Institute* **2023**, *106*, 101145, doi:10.1016/j.joei.2022.11.005.
70. Bian, C.; Huang, J.; Sun, R. Numerical Optimization of Combustion and NOX Emission in a Retrofitted 600MWe Tangentially-Fired Boiler Using Lignite. *Applied Thermal Engineering* **2023**, *226*, 120228, doi:10.1016/j.applthermaleng.2023.120228.
71. Ma, L.; Li, Y.; Fang, Q.; Liang, J.; Zhang, Z.; Tang, L.; Shen, G.; Chen, S. Combustion Improvement and Burnout Enhancement by Optimizing Biomass Nozzle Position in a 600 MW Opposed Wall-Fired Boiler with Coal Co-Firing Biomass. *Applied Thermal Engineering* **2025**, *276*, 126946, doi:10.1016/j.applthermaleng.2025.126946.
72. Ma, L.; Li, Y.; Fang, Q.; Li, K.; Liang, J.; Chen, S.; Chen, C.; Zhang, S. Numerical Study and Industrial Application of Coal Co-Firing Biomass in a 600 MW Opposed Wall-Fired Boiler with Feeding Biomass through Inner OFA Nozzles. *Applied Thermal Engineering* **2025**, *279*, 127684, doi:10.1016/j.applthermaleng.2025.127684.
73. Wang, X.; Hu, Z.; Deng, S.; Xiong, Y.; Tan, H. Effect of Biomass/Coal Co-Firing and Air Staging on NOx Emission and Combustion Efficiency in a Drop Tube Furnace. *Energy Procedia* **2014**, *61*, 2331–2334, doi:10.1016/j.egypro.2014.11.1196.
74. Barnes, D.I. Understanding Pulverised Coal, Biomass and Waste Combustion – A Brief Overview. *Applied Thermal Engineering* **2015**, *74*, 89–95, doi:10.1016/j.applthermaleng.2014.01.057.

75. Venier, C.M.; Torres, E.; Fouga, G.G.; Rodriguez, R.A.; Mazza, G.; Reyes Urrutia, A. Computational Modeling of Biomass Fast Pyrolysis in Fluidized Beds with Eulerian Multifluid Approach. *Fluids* **2024**, *9*, 301, doi:10.3390/fluids9120301.
76. Ngadi, Z.; Lahlaouti, M.L. CFD Modeling of Petcoke Co-Combustion in a Real Cement Kiln: The Effect of the Turbulence-Chemistry Interaction Model Applied with K-ϵ Variations. *International Review of Applied Sciences and Engineering* **2022**, *13*, 148–163, doi:10.1556/1848.2021.00326.
77. Milacic, E.; Nunez Manzano, M.; Madanikashani, S.; Heynderickx, G.J.; van Geem, K.M.; van de Greef, A.A.M.; Richter, A.; Kriebitzsch, S.H.L.; Buist, K.A.; Baltussen, M.W.; et al. Experimental Study on the Temperature Distribution in Fluidised Beds. *Chemical Engineering Science* **2022**, *248*, 117062, doi:10.1016/j.ces.2021.117062.
78. Teixeira, P.; Lopes, H.; Gulyurtlu, I.; Lapa, N.; Abelha, P. Evaluation of Slagging and Fouling Tendency during Biomass Co-Firing with Coal in a Fluidized Bed. *Biomass Bioenergy* **2012**, *39*, 192–203, doi:10.1016/j.biombioe.2012.01.010.
79. Kroumian, C.; Maier, J.; Peloriadi, K.; Scheffknecht, G.; Grammelis, P. Evaluation of 100% Alternative Fuel Combustion under Oxyfuel Conditions in a Pilot-Scale Burner for Application in Retrofit Oxyfuel Cement Kiln. *Fuel* **2025**, *381*, 133697, doi:10.1016/j.fuel.2024.133697.
80. Liu, Q.; Shi, Y.; Zhong, W.; Yu, A. Co-Firing of Coal and Biomass in Oxy-Fuel Fluidized Bed for CO2 Capture: A Review of Recent Advances. *Chinese Journal of Chemical Engineering* **2019**, *27*, 2261–2272, doi:10.1016/j.cjche.2019.07.013.
81. Cheng, F.; Zhang, Y.; Zhang, G.; Zhang, K.; Wu, J.; Zhang, D. Eliminating Environmental Impact of Coal Mining Wastes and Coal Processing By-Products by High Temperature Oxy-Fuel CFB Combustion for Clean Power Generation: A Review. *Fuel* **2024**, *373*, 132341, doi:10.1016/j.fuel.2024.132341.
82. Riaza, J.; Gil, M. V.; Álvarez, L.; Pevida, C.; Pis, J.J.; Rubiera, F. Oxy-Fuel Combustion of Coal and Biomass Blends. *Energy* **2012**, *41*, 429–435, doi:10.1016/j.energy.2012.02.057.
83. Ji, L.; Wu, X.; Gao, Y.; Zhang, Y.; Yan, Z.; Pu, L.; Liu, X. Coal Blending with Peanut Shells: Thermal Behavior, Kinetics, Carbon Reduction and Pollution Reduction Analyses. *Water, Air, & Soil Pollution* **2024**, *235*, 634, doi:10.1007/s11270-024-07425-6.
84. Perrone, D. A Study of an Oxy-Coal Combustion with Wet Recycle Using CFD Modelling. *Energy Procedia* **2015**, *82*, 900–907, doi:10.1016/j.egypro.2015.11.837.
85. Abdul Gani, Z.F.; Arun Pandian, P. Modelling of Coal Combustion in a Drop Tube Furnace in Air and Oxy Fuel Environment. *Materials Today: Proceedings* **2021**, *47*, 4431–4437, doi:10.1016/j.matpr.2021.05.269.
86. Toporov, D.D. *Combustion of Pulverised Coal in a Mixture of Oxygen and Recycled Flue Gas*; Elsevier: Waltham, 2014.
87. Huo, C.; Sun, J.; Song, P. Energy, Exergy and Economic Analyses of an Optimal Use of Cryogenic Liquid Turbine Expander in Air Separation Units. *Chemical Engineering Research and Design* **2023**, *189*, 194–209, doi:10.1016/j.cherd.2022.11.030.
88. Castillo, R. Thermodynamic Analysis of a Hard Coal Oxyfuel Power Plant with High Temperature Three-End Membrane for Air Separation. *Applied Energy* **2011**, *88*, 1480–1493, doi:10.1016/j.apenergy.2010.10.044.
89. Stanger, R.; Wall, T.; Spörl, R.; Paneru, M.; Grathwohl, S.; Weidmann, M.; Scheffknecht, G.; McDonald, D.; Myöhänen, K.; Ritvanen, J.; et al. Oxyfuel Combustion for CO2 Capture in Power Plants. *International Journal of Greenhouse Gas Control* **2015**, *40*, 55–125, doi:10.1016/j.ijggc.2015.06.010.
90. Zheng, C.; Liu, Z.; Xiang, J.; Zhang, L.; Zhang, S.; Luo, C.; Zhao, Y. Fundamental and Technical Challenges for a Compatible Design Scheme of Oxyfuel Combustion Technology. *Engineering* **2015**, *1*, 139–149, doi:10.15302/j-eng-2015008.

91. Montgomery, M.; Bjurman, M.; Hjørnhede, A.; Rombrecht, H.B.; Lisk, A.; Krautz, H.J. High Temperature Corrosion Investigation in an Oxyfuel Combustion Test Rig. *Materials and Corrosion* **2015**, *66*, 257–269, doi:10.1002/maco.201307245.
92. Ghorbani, B.; Rahnavard, Z.; Ahmadi, M.H.; Jouybari, A.K. An Innovative Hybrid Structure of Solar PV-Driven Air Separation Unit, Molten Carbonate Fuel Cell, and Absorption–Compression Refrigeration System (Process Development and Exergy Analysis). *Energy Reports* **2021**, *7*, 8960–8972, doi:10.1016/j.egyr.2021.10.108.
93. Edge, P.J.; Heggs, P.J.; Pourkashanian, M.; Stephenson, P.L. Integrated Fluid Dynamics-Process Modelling of a Coal-Fired Power Plant with Carbon Capture. *Applied Thermal Engineering* **2013**, *60*, 242–250, doi:10.1016/j.applthermaleng.2013.06.024.
94. Kislinger, C.; Radünz, C.; Daurer, G.; Schwarz, S.; Plank, B.; Demuth, M.; Gaber, C.; Hochenauer, C. Experimental and Numerical Analysis of the Time-Dependent Scale Formation of 1.4307 Stainless Steel in Various Air–Fuel and Oxy–Fuel Combustion Atmospheres. *Materials Today Communications* **2025**, *45*, 112284, doi:10.1016/j.mtcomm.2025.112284.
95. Zhang, O.; Yuan, L.; Teng, X.; Zhao, F.; Fan, X.; Tao, C.; Liu, R. CFD-Based Investigation of Flow Field in a Glass Furnace Reactor under an Oxy-Fuel Electric Boosting Mode. *Scientific Reports* **2025**, *15*, 1–15, doi:10.1038/S41598-025-87261-8.
96. Escudero, A.I.; Espatolero, S.; Romeo, L.M. Oxy-Combustion Power Plant Integration in an Oil Refinery to Reduce CO2 Emissions. *International Journal of Greenhouse Gas Control* **2016**, *45*, 118–129, doi:10.1016/j.ijggc.2015.12.018.
97. Sanusi, Y.S.; Mokheimer, E.M.A. Numerical Modeling and Simulation of Oxy-Combustion Exhaust Gas Recycling for Fuel Reforming. *Energy and Fuels* **2017**, *31*, 5385–5394, doi: 10.1021/acs.energyfuels.7b00772.
98. Adánez, J.; Abad, A.; Mendiara, T.; Gayán, P.; de Diego, L.F.; García-Labiano, F. Chemical Looping Combustion of Solid Fuels. *Progress in Energy and Combustion Science* **2018**, *65*, 6–66, doi:10.1016/j.pecs.2017.07.005.
99. Schnellmann, M.A.; Donat, F.; Scott, S.A.; Williams, G.; Dennis, J.S. The Effect of Different Particle Residence Time Distributions on the Chemical Looping Combustion Process. *Applied Energy* **2018**, *216*, 358–366, doi:10.1016/j.apenergy.2018.02.046.
100. Deng, Z.; Xiao, R.; Jin, B.; Song, Q.; Huang, H. Multiphase CFD Modeling for a Chemical Looping Combustion Process (Fuel Reactor). *Chemical Engineering & Technology* **2008**, *31*, 1754–1766, doi:10.1002/ceat.200800341.
101. Banerjee, S.; Agarwal, R.K. Computational Fluid Dynamics Modeling and Simulations of Fluidized Beds for Chemical Looping Combustion. *Handbook of Chemical Looping Technology* **2018**, 303–332, doi:10.1002/9783527809332.CH10.
102. Abad, A.; Adánez-Rubio, I.; Gayán, P.; García-Labiano, F.; de Diego, L.F.; Adánez, J. Demonstration of Chemical-Looping with Oxygen Uncoupling (CLOU) Process in a 1.5 KWth Continuously Operating Unit Using a Cu-Based Oxygen-Carrier. *International Journal of Greenhouse Gas Control* **2012**, *6*, 189–200, doi:10.1016/j.ijggc.2011.10.016.
103. Gupta, A.; Yan, D. Tubular Ball Mills. In *Mineral Processing Design and Operations*; Gupta, A., Yan, D., Eds.; Elsevier: Amsterdam, 2016; pp. 189–240.
104. Li, H.; Jiang, S.; Zeng, R.; Geng, J.; Niu, Z. Numerical Simulation and Analysis of the Airflow Field in the Crushing Chamber of the Hammer Mill. *ACS Omega* **2024**, doi:10.1021/acsomega.4c02187.
105. Ali, M.; López, A.; Pasha, M.; Ghadiri, M. Optimisation of the Performance of a New Vertical Roller Mill by Computational Fluid Dynamics Simulations. *Powder Technology* **2024**, *433*, 119282, doi:10.1016/J.POWTEC.2023.119282.

106. Sato, K.; Meguri, N.; Shoji, K.; Kanemoto, H.; Hasegawa, T.; Maruyama, T. Breakage of Coals in Ring-Roller Mills Part I. The Breakage Properties of Various Coals and Simulation Model to Predict Steady-State Mill Performance. *Powder Technology* **1996**, *86*, 275–283, doi:10.1016/0032-5910(95)03061-1.
107. Thakre, P.; Goldin, G. CFD Modeling of Pulverized Coal Combustion Using Relax to Chemical Equilibrium Model With Turbulence-Chemistry Interaction. *Proceedings of the ASME Turbo Expo* **2021**, *3*, doi:10.1115/gt2020-14190.
108. Abdul-Sater, H.; Lenertz, J.; Bonilha, C.; Lu, X.; Fetvedt, J. A CFD Simulation of Coal Syngas Oxy-Combustion in a High-Pressure Supercritical CO2 Environment. In Proceedings of the Volume 4A: Combustion, Fuels and Emissions; American Society of Mechanical Engineers, June 26 2017.
109. Torresi, M.; Fortunato, B.; Camporeale, S.M.; Saponaro, A. CFD Modeling of Pulverized Coal Combustion in an Industrial Burner. *Proceedings of the ASME Turbo Expo* **2013**, *1*, 657–666, doi:10.1115/gt2012-69506.
110. Wladyslaw, M. Co-Combustion of Pulverized Coal and Biomass in Fluidized Bed of Furnace. *Journal of Energy Resources Technology, Transactions of the ASME* **2017**, *139*, doi:10.1115/1.4036958/373363.
111. Zhang, L.; Wei, X.; Zhao, J.; Sen, L. Numerical Simulation of Oxy-Fuel Combustion with Different O_2/CO_2 Fractions in a Large Cement Precalciner. *Energy & Fuels* **2020**, *34*, 4949–4957, doi:10.1021/acs.energyfuels.9b04237.
112. Liu, X.; Wang, C.; Lv, Q.; Zhu, T.; Li, D.; Du, Y.; Che, D. Effects of O_2 Feeding Strategy and Over-Fire Air Configuration on Oxy-Fuel Combustion Characteristics in an Opposed Wall-Fired Utility Boiler. *Energy & Fuels* **2018**, *32*, 2479–2489, doi:10.1021/acs.energyfuels.7b03745.
113. Xie, J.; Zhong, W.; Shao, Y.; Liu, Q.; Liu, L.; Liu, G. Simulation of Combustion of Municipal Solid Waste and Coal in an Industrial-Scale Circulating Fluidized Bed Boiler. *Energy & Fuels* **2017**, *31*, 14248–14261, doi:10.1021/acs.energyfuels.7b02693.
114. Zhang, Z.; Zhao, Z.; Zha, X.; Li, X.; Luo, C.; Wu, F.; Zhang, L. Computational Study on Cocombustion Characteristics of Pulverized Coal Blending with NH_3 under Moderate or Intense Low-Oxygen Dilution Combustion Mode. *Energy & Fuels* **2022**, *36*, 12750–12762, doi:10.1021/acs.energyfuels.2c02774.
115. Wang, G.; Zhang, B.; Zhao, J.; Chen, J.; Han, X.; Zhang, H.; Fan, W.; Xu, J. Numerical Investigation of Coal/Ammonia Co-Firing in a Prototype Burner with Air-Staging and Various Ammonia Injection Modes. *Combustion Science and Technology* **2025**, 1–24, doi:10.1080/00102202.2025.2540323.
116. Bian, L.; Li, Z.; Lu, Y.; Chen, Z.; Li, Z.; Sun, D. Numerical Simulation on Combustion and Slagging Characteristics of a Down-Fired Boiler Under Different Blending Ratios of Anthracite/Bituminous Coal. *Combustion Science and Technology* **2025**, 1–29, doi: 10.1080/00102202.2025.2541851.
117. Samantaray, U.; Nemalipuri, P.; Mailapalli, V.K.; Vitankar, V.; Das, H.C.; Pradhan, M.K. Computational Study on the Blended Fuel Combustion in an Industrial Tangentially Fired Boiler. *International Journal of Coal Preparation and Utilization* **2025**, 1–27, doi:10.1080/19392699.2025.2532744.
118. Xia, G.; Wu, X.; Huang, C.; Yin, X.; Liu, H.; Guo, X.; Cheng, J.; Guo, X. Numerical Simulation of a 660 MWe Wall-Fired Boiler Using Biomass-Derived Fuel Gas Co-Fired with Coal. *Energy Sources, Part A: Recovery, Utilization, and Environmental Effects* **2025**, *47*, 11351–11370, doi:10.1080/15567036.2025.2502159.
119. Messerle, V.E.; Karpenko, E.I.; Ustimenko, A.B. Plasma Assisted Power Coal Combustion in the Furnace of Utility Boiler: Numerical Modeling and Full-Scale Test. *Fuel* **2014**, *126*, 294–300, doi:10.1016/j.fuel.2014.02.047.

120. Messerle, V.E.; Ustimenko, A.B. Thermodynamic and Kinetic Modeling and Experiment on Plasma Ignition of Pulverized High-Ash Coal. *Applications in Energy and Combustion Science* **2024**, *17*, 100248, doi:10.1016/j.jaecs.2024.100248.
121. Butakov, E.B.; Burdukov, A.P.; Alekseenko, S.V.; Yaganov, E.N. Plasma Ignition System to Start Up Pulverized Coal Boilers: Experimental Simulation and Full-Scale Test. *Journal of Engineering Thermophysics* **2022**, *31*, 375–383, doi:10.1134/S1810232822030018.
122. Messerle, V.E.; Ustimenko, A.B.; Umbetkaliev, K.A. Plasma Ignition of Solid Fuels at Thermal Power Plants. Part 2. 3D Modeling of the Furnace of a Pulverized Coal Boiler. *Thermophysics and Aeromechanics* **2023**, *29*, 981–992, doi:10.1134/S0869864322060191.
123. Baig, M.H.; Azher, K.; Youssefi, R. Optimizing the Effect of Ignition and Combustion on Carbon Oxidation Using Pulverized Fuel Burner. *Arabian Journal for Science and Engineering* **2023**, *48*, 3323–3337, doi:10.1007/s13369-022-07150-4.
124. Youssefi, R.; Maier, J.; Scheffknecht, G. Pilot-Scale Experiences on a Plasma Ignition System for Pulverized Fuels. *Energies* **2021**, *14*, 4726, doi:10.3390/en14164726.
125. Gorokhovski, M.A.; Jankoski, Z.; Lockwood, F.C.; Karpenko, E.I.; Messerle, V.E.; Ustimenko, A.B. Enhancement of Pulverized Coal Combustion by Plasma Technology. *Combustion Science and Technology* **2007**, *179*, 2065–2090, doi:10.1080/00102200701386115.
126. Pettinau, A.; Kurt, E.; Manuel Lopez-Guede, J.; Messerle, V.E.; Ustimenko, A.B.; Lavrichshev, O.A.; Nugman, M.K. The Gasification and Pyrolysis of Biomass Using a Plasma System. *Energies* **2024**, *17*, 5594, doi:10.3390/en17225594.
127. Kumar, U.; Paul, M.C. CFD Modelling of Biomass Gasification with a Volatile Break-up Approach. *Chemical Engineering Science* **2019**, *195*, 413–422, doi:10.1016/j.ces.2018.09.038.
128. Sedej, O.; Mbonimpa, E. CFD Modeling of a Lab-Scale Microwave Plasma Reactor for Waste-to-Energy Applications: A Review. *Gases* **2021**, *1*, 133–147, doi:10.3390/gases1030011.
129. Sakhraji, M.; Ramos, A.; Monteiro, E.; Bouziane, K.; Rouboa, A. Plasma Gasification Process Using Computational Fluid Dynamics Modeling. *Energy Reports* **2022**, *8*, 1541–1549, doi:10.1016/j.egyr.2022.08.069.
130. Zhang, T.; Cao, D.; Feng, X.; Zhu, J.; Lu, X.; Mu, L.; Qian, H. Machine Learning Prediction of Bio-Oil Characteristics Quantitatively Relating to Biomass Compositions and Pyrolysis Conditions. *Fuel* **2022**, *312*, 122812, doi:10.1016/j.fuel.2021.122812.
131. Cheng, F.; Belden, E.R.; Li, W.; Shahabuddin, M.; Paffenroth, R.C.; Timko, M.T. Accuracy of Predictions Made by Machine Learned Models for Biocrude Yields Obtained from Hydrothermal Liquefaction of Organic Wastes. *Chemical Engineering Journal* **2022**, *442*, 136013, doi:10.1016/j.cej.2022.136013.
132. Cheng, F.; Luo, H.; Colosi, L.M. Slow Pyrolysis as a Platform for Negative Emissions Technology: An Integration of Machine Learning Models, Life Cycle Assessment, and Economic Analysis. *Energy Conversion and Management* **2020**, *223*, 113258, doi:10.1016/j.enconman.2020.113258.
133. Chen, X.; Zhang, H.; Song, Y.; Xiao, R. Prediction of Product Distribution and Bio-Oil Heating Value of Biomass Fast Pyrolysis. *Chemical Engineering and Processing – Process Intensification* **2018**, *130*, 36–42, doi:10.1016/j.cep.2018.05.018.
134. Chen, W.-H.; Lo, H.-J.; Aniza, R.; Lin, B.-J.; Park, Y.-K.; Kwon, E.E.; Sheen, H.-K.; Grafilo, L.A.D.R. Forecast of Glucose Production from Biomass Wet Torrefaction Using Statistical Approach along with Multivariate Adaptive Regression Splines, Neural Network and Decision Tree. *Applied Energy* **2022**, *324*, 119775, doi:10.1016/j.apenergy.2022.119775.

135. Zhang, W.; Li, J.; Liu, T.; Leng, S.; Yang, L.; Peng, H.; Jiang, S.; Zhou, W.; Leng, L.; Li, H. Machine Learning Prediction and Optimization of Bio-Oil Production from Hydrothermal Liquefaction of Algae. *Bioresource Technology* **2021**, *342*, 126011, doi:10.1016/j.biortech.2021.126011.
136. Ji, W.; Richter, F.; Gollner, M.J.; Deng, S. Autonomous Kinetic Modeling of Biomass Pyrolysis Using Chemical Reaction Neural Networks. *Combustion and Flame* **2022**, *240*, 111992, doi:10.1016/j.combustflame.2022.111992.
137. Wang, S.; Shi, Z.; Jin, Y.; Zaini, I.N.; Li, Y.; Tang, C.; Mu, W.; Wen, Y.; Jiang, J.; Jönsson, P.G.; et al. A Machine Learning Model to Predict the Pyrolytic Kinetics of Different Types of Feedstocks. *Energy Conversion and Management* **2022**, *260*, 115613, doi:10.1016/j.enconman.2022.115613.
138. Li, S.; Yang, B.; Qi, F. Accelerate Global Sensitivity Analysis Using Artificial Neural Network Algorithm: Case Studies for Combustion Kinetic Model. *Combustion and Flame* **2016**, *168*, 53–64, doi:10.1016/j.combustflame.2016.03.028.
139. Sezer, S.; Kartal, F.; Özveren, U. Prediction of Combustion Reactivity for Lignocellulosic Fuels by Means of Machine Learning. *Journal of Thermal Analysis and Calorimetry* **2022**, *147*, 9793–9809, doi:10.1007/s10973-022-11208-8.
140. Kartal, F.; Özveren, U. Prediction of Activation Energy for Combustion and Pyrolysis by Means of Machine Learning. *Thermal Science and Engineering Progress* **2022**, *33*, 101346, doi:10.1016/j.tsep.2022.101346.
141. Abdul Jameel, A.G.; Al-Muslem, A.; Ahmad, N.; Alquaity, A.B.S.; Zahid, U.; Ahmed, U. Predicting Enthalpy of Combustion Using Machine Learning. *Processes* **2022**, *10*, 2384, doi:10.3390/pr10112384.
142. Sezer, S.; Kartal, F.; Özveren, U. Prediction of Chemical Exergy of Syngas from Downdraft Gasifier by Means of Machine Learning. *Thermal Science and Engineering Progress* **2021**, *26*, 101031, doi:10.1016/j.tsep.2021.101031.
143. Kartal, F.; Özveren, U. Investigation of the Chemical Exergy of Torrefied Biomass from Raw Biomass by Means of Machine Learning. *Biomass Bioenergy* **2022**, *159*, 106383, doi:10.1016/j.biombioe.2022.106383.
144. Sezer, S.; Özveren, U. Investigation of Syngas Exergy Value and Hydrogen Concentration in Syngas from Biomass Gasification in a Bubbling Fluidized Bed Gasifier by Using Machine Learning. *International Journal of Hydrogen Energy* **2021**, *46*, 20377–20396, doi:10.1016/j.ijhydene.2021.03.184.
145. Kartal, F.; Özveren, U. Prediction of Torrefied Biomass Properties from Raw Biomass. *Renewable Energy* **2022**, *182*, 578–591, doi:10.1016/j.renene.2021.10.042.
146. Chen, C.; Liang, R.; Ge, Y.; Li, J.; Yan, B.; Cheng, Z.; Tao, J.; Wang, Z.; Li, M.; Chen, G. Fast Characterization of Biomass Pyrolysis Oil via Combination of ATR-FTIR and Machine Learning Models. *Renewable Energy* **2022**, *194*, 220–231, doi:10.1016/j.renene.2022.05.097.
147. Chen, J.; Ding, L.; Wang, P.; Zhang, W.; Li, J.; A. Mohamed, B.; Chen, J.; Leng, S.; Liu, T.; Leng, L.; et al. The Estimation of the Higher Heating Value of Biochar by Data-Driven Modeling. *Journal of Renewable Materials* **2022**, *10*, 1555–1574, doi:10.32604/jrm.2022.018625.
148. Ihme, M.; Chung, W.T.; Mishra, A.A. Combustion Machine Learning: Principles, Progress and Prospects. *Progress in Energy and Combustion Science* **2022**, *91*, 101010, doi:10.1016/j.pecs.2022.101010.
149. Sahoo, S.; Kumar, V.N.S.P.; Srivastava, D.K. Quantitative Analysis of Engine Parameters of a Variable Compression Ratio CNG Engine Using Machine Learning. *Fuel* **2022**, *311*, 122587, doi:10.1016/j.fuel.2021.122587.

150. Fu, J.; Yang, R.; Li, X.; Sun, X.; Li, Y.; Liu, Z.; Zhang, Y.; Sunden, B. Application of Artificial Neural Network to Forecast Engine Performance and Emissions of a Spark Ignition Engine. *Applied Thermal Engineering* **2022**, *201*, 117749, doi:10.1016/j.applthermaleng.2021.117749.
151. Wan, K.; Hartl, S.; Vervisch, L.; Domingo, P.; Barlow, R.S.; Hasse, C. Combustion Regime Identification from Machine Learning Trained by Raman/Rayleigh Line Measurements. *Combustion and Flame* **2020**, *219*, 268–274, doi:10.1016/j.combustflame.2020.05.024.
152. Hwang, J.; Lee, P.; Mun, S.; Karathanassis, I.K.; Koukouvinis, P.; Pickett, L.M.; Gavaises, M. Machine-Learning Enabled Prediction of 3D Spray under Engine Combustion Network Spray G Conditions. *Fuel* **2021**, *293*, 120444, doi:10.1016/j.fuel.2021.120444.
153. Owoyele, O.; Pal, P. A Novel Machine Learning-Based Optimization Algorithm (ActivO) for Accelerating Simulation-Driven Engine Design. *Applied Energy* **2021**, *285*, 116455, doi:10.1016/j.apenergy.2021.116455.
154. Liu, J.; Huang, Q.; Ulishney, C.; Dumitrescu, C.E. Machine Learning Assisted Prediction of Exhaust Gas Temperature of a Heavy-Duty Natural Gas Spark Ignition Engine. *Applied Energy* **2021**, *300*, 117413, doi:10.1016/j.apenergy.2021.117413.
155. Joshi, V.V.; Swaminathan, G.; Prabhakaran, S.S. Life Cycle Assessment of the Co-Combustion System of Single-Use Plastic Waste and Lignite Coal to Promote Circular Economy. *Journal of Cleaner Production* **2021**, *329*, 129579, doi:10.1016/j.jclepro.2021.129579.
156. Wang, S.; Luo, K.; Fan, J. CFD-DEM Coupled with Thermochemical Sub-Models for Biomass Gasification: Validation and Sensitivity Analysis. *Chemical Engineering Science* **2020**, *217*, 115550, doi:10.1016/j.ces.2020.115550.
157. International Energy Agency. *World Energy Outlook 2017*; Paris, 2017.
158. Bloomberg. *The Clean Energy Country Competitiveness Index*; London, 2017.
159. Akhtar, J.; Yaseen, A.; Munir, S. Effect of Rice Husk Co-Combustion with Coal on Gaseous Emissions and Combustion Efficiency. *Energy Sources, Part A: Recovery, Utilization, and Environmental Effects* **2018**, *40*, 1010–1018, doi:10.1080/15567036.2018.1468513.
160. Du, Y.; Wang, C.; Lv, Q.; Li, D.; Liu, H.; Che, D. CFD Investigation on Combustion and NO Emission Characteristics in a 600 MW Wall-Fired Boiler under High Temperature and Strong Reducing Atmosphere. *Applied Thermal Engineering* **2017**, *126*, 407–418, doi:10.1016/j.applthermaleng.2017.07.147.

7 Co-Combustion Case Study
Co-Combustion of Coal and Pinus Sawdust

7.1 INTRODUCTION

In an effort to lessen the carbon impact of combusting fossil fuels, engineers can address several sustainable development goals by co-firing coal with Pinus sawdust in combustion systems. As noted by Ji et al. [1], biomass contains more volatiles than coal; hence, better ignition and intensity properties are exhibited when we combust their blends with an increasing proportion of biomass. Unfortunately, because of the wetness and fibrous nature of biomass, using Pinus sawdust in combustion systems leads to fuel handling problems [2]. Inasmuch as great progress has been made in modelling coal combustion, biomass still poses a problem mainly due to the vast library of species available [3]. As such, optimum blending conditions must be investigated and implemented so we can benefit from the use of biomass in combustion systems [4].

This case study will:

- demonstrate the methodological workflow for developing and validating a co-combustion computational fluid dynamics (CFD) model using commercially available solvers,
- evaluate the predictive accuracy of the developed model by comparing simulation outputs with experimental data on ignition, burnout, temperature, and emission characteristics, and
- assess the broader implications of coal biomass co-firing for cleaner energy production, emission control, and policy frameworks supporting sustainable energy transitions.

Through these objectives, a full modelling implementation will be explained in this chapter. Starting with a review of the commercial platforms available on the market, the background to why the case study proceeded in such a manner is explained. The experimental setup that was used to validate the co-combustion model will be explained in this chapter. A thorough co-combustion model setup is then presented before various validation and sensitivity analysis techniques are explained. The evaluation of results from the case study focusing on combustion-relevant parameters is then presented before giving a general overview of the industrial benefits of such a case study.

DOI: 10.1201/9781003647942-7

7.2 COMMERCIAL CFD SOFTWARE

The CFD software market anchored on computational ability is projected to continue growing as AI-enabled workflows are infused into these platforms [5]. From a researcher's perspective, ANSYS FLUENT and CFX are the most common platforms used to perform CFD analysis. Their ability to analyse complex scenarios with relatively much ease as well as customise applications to match the users' needs makes them attractive to both researchers and engineers. These applications range from general internal or external flows, turbomachinery, reacting flows, electronics cooling, and various multiphysics applications. Simcenter STAR-CCM+ is another platform that has a strong presence in industry, especially with its new modules that allow particle modelling through discrete element modelling. This software has been used to analyse external flow, turbomachinery, multiphase flow, particle hydrodynamics, and robust fluid–structure interaction. COMSOL Multiphysics is another platform that has been preferred by non-expert CFD users due to its versatility and, as its name suggests, the software can handle strongly coupled physics.

With combusting fuels coming in the form of solids, liquids, and gases, there exist domain-specific CFD software that are well adapted for these specific applications. CHEMKIN is one platform that is quite common for gaseous reacting processes, especially those that have multiple reaction mechanisms governing them. AVL FIRE was developed specifically for rocket engines; hence its strengths with combustion of hydrogen and alternative fuels. CONVERGE, in as much as it is a domain-specific platform, has developed into commercial software due to its strong performance with internal combustion engines as well as hydrogen combustion. An innovative approach employed by CONVERGE is the autonomous meshing of complex geometries, which reduces simulation setup time significantly. CFD Turbo, though a module that can be coupled into various platforms, has proved its capabilities when analysing various turbomachinery.

As with all computer software, there exist open source CFD platforms. These include OpenFoam, SU2, Code Saturne, MFIX, and FDS, only to mention just a few. Open source software tends to have deployable toolboxes that are well adapted for process optimisation. SU2 is strong with aerospace topology optimisation, discrete phase simulations, as well as turbomachinery analysis. Code Saturne is suited for nuclear energy, HVAC analysis, thermal, and hydraulic analysis. MFIX, short for multiphase flow with interphase exchanges, is a platform mainly used to analyse multiphase reactors that contain a lot of gas–solid interactions. As such, most users are found in the chemical processing industry. FDS, short for fire dynamics simulator, is used to perform CFD analysis of fire-driven flows; hence, it is good with building smoke management.

Using the search string ("CFD" AND "ANSYS" AND "COMBUSTION") AND ("Ansys" OR "Star ccm+" OR "Comsol" OR "Openfoam"), a general Scopus and Web of Science database search for article keywords focused on CFD analysis and combustion reveals that ANSYS has 54 documents, STAR CCM+ has 9 documents, Comsol has 4 documents, and OpenFoam has 137 documents. These statistics give a crude background to the use of these software in research.

7.3 CASE STUDY: EXPERIMENTAL SETUP OF DROP TUBE FURNACE FOR COAL AND PINUS SAWDUST COMBUSTION

To illustrate the co-combustion modelling of coal and pinus sawdust, the experimental side is presented first. The physical and chemical properties of the fuel samples as obtained from the experiments are summarised in Table 7.1.

The specifications of the drop tube furnace (DTF) are given in Table 7.2. Figure 7.1 shows a flow diagram of how the DTF operates, whilst Figure 7.2 illustrates the furnace components in detail. The alumina tube within the DTF is modelled as the furnace, with a total height of 2032 mm and an internal diameter of 70 mm. The outer furnace diameter has a larger diameter due to the refractory lining made of alumina silicate and heating tubes arranged concentrically around the alumina tube.

The fuel blends were introduced using a 2-mm cooled injector at a temperature of 373 K and collected using a 6-mm cooled sampling probe at the reactor exit. Within the fuel feeding mechanism, fuel blends weighing 15 g were placed in a test tube connected to the primary carrier gas at an average flow rate of 2.5 NL/min ($1.5 <$ flow rate < 3.5). The test tube was placed on a mechanism with an electromechanical vibrator. As such, the incoming primary carrier gas fluidised the fuel and transported it into the outgoing pipe leading up to the furnace. Blockages in the injector were anticipated to be minimal because the maximum diameter of the fuel particles used in the DTF was 0.25 mm, which was significantly less than the threshold of 0.333 mm. To avoid obstructions, the injector diameter should be at least eight times the maximum particle size [6]. The water-cooled sampling probe was maintained at a suction flow rate of between 6 NL/min and 15 NL/min. The secondary gas was preheated to 1,273 K for all the cases, and the flow rate was set between 10 NL/min and 20 NL/min depending on the furnace residence time that was required.

The experimental procedure was split into two stages: the first was carried out under a nitrogen atmosphere, whilst the second was under an oxygen enriched atmosphere. As such, the product of the first stage that could be measured was the solid char, since the released volatiles were entrained in the exhaust gases that were not sufficiently monitored. After collecting and testing the char, the second stage then required putting the char within the DTF again under an oxygen enriched atmosphere (3% O_2, 97% N_2), thus producing ash as the final product. Limitations around the specified DTF operating temperatures necessitated this procedure of using two stages.

The collecting probe was set at various distances from the furnace injection point which were 520 mm, 920 mm, and 1,320 mm along the furnace centreline depending on the experimental procedure. The sampling probe was equipped with a bag filter that collected the char or the ash which was analysed accordingly. The first stage did not require large residence times; thus, a 520-mm probing position was sufficient. Three set furnace temperatures were investigated which were 1,273 K, 1,473 K, and 1,673 K. The injector was assumed to supply a normal to inlet boundary flow direction without any swirl.

TABLE 7.1
Chemical Properties of Hwange Coal, Pinus Sawdust, and Fuel Blends

	Proximate Analysis * (weight %)				Ultimate Analysis * (weight%)				
	** Fixed Carbon	Volatile Matter	Ash	HHV (MJ/kg)	C	H	** O	N	S
100HC	53.97	23.10	22.93	22.51	58.671	2.946	13.245	1.613	0.593
90HC 10PS	48.21	29.91	21.88	21.54	58.489	3.367	14.246	1.448	0.570
80HC 20PS	46.35	31.82	21.83	21.18	56.954	3.734	15.632	1.296	0.555
70HC 30PS	46.02	33.74	20.24	21.37	56.885	4.093	17.100	1.143	0.535
100PS	15.62	80.68	3.70	18.07	49.504	6.035	40.404	0.358	0

* On a dry basis, ** By difference.

TABLE 7.2
DTF Summarised Geometry Specifications and Operating Conditions

Parameter	Description
Reactor inner diameter	70 mm
Reactor height, total length	2,032 mm
Furnace wall temperature settings	1,273 K 1,473 K 1,673 K
Carrier gas	Stage 1 (100% N_2) Stage 2 (3% O_2 97% N_2) 0.25 < Primary flow rate (NL/min) < 0.35 10 < Secondary flow rate (NL/min) < 20 Secondary gas preheated to 1,273 K Primary gas at 373 K
Fuel blends	1.5 < flow rate (g/min) < 3.6 373 K

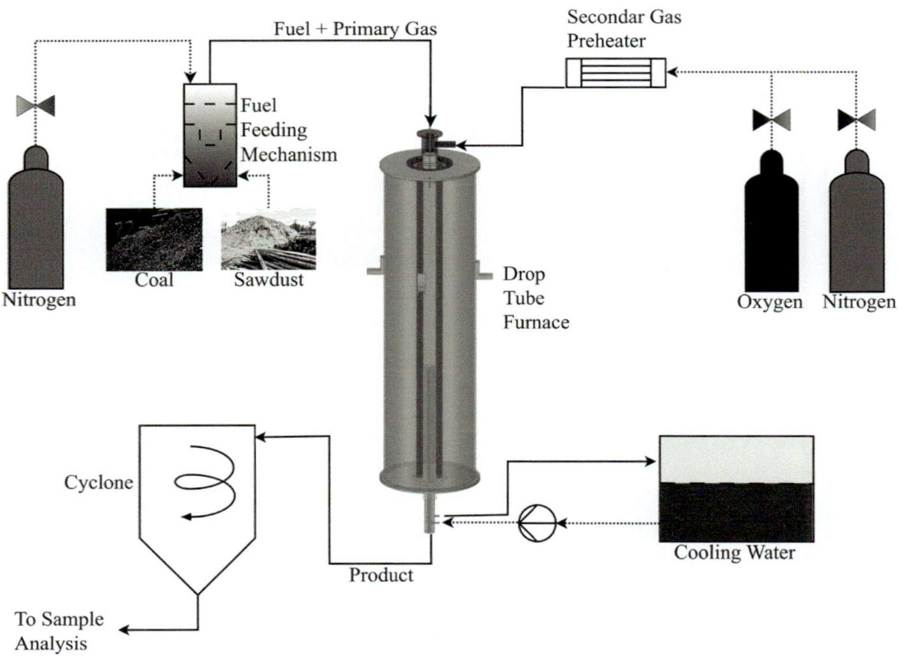

FIGURE 7.1 Flow diagram of DTF experimental facility.

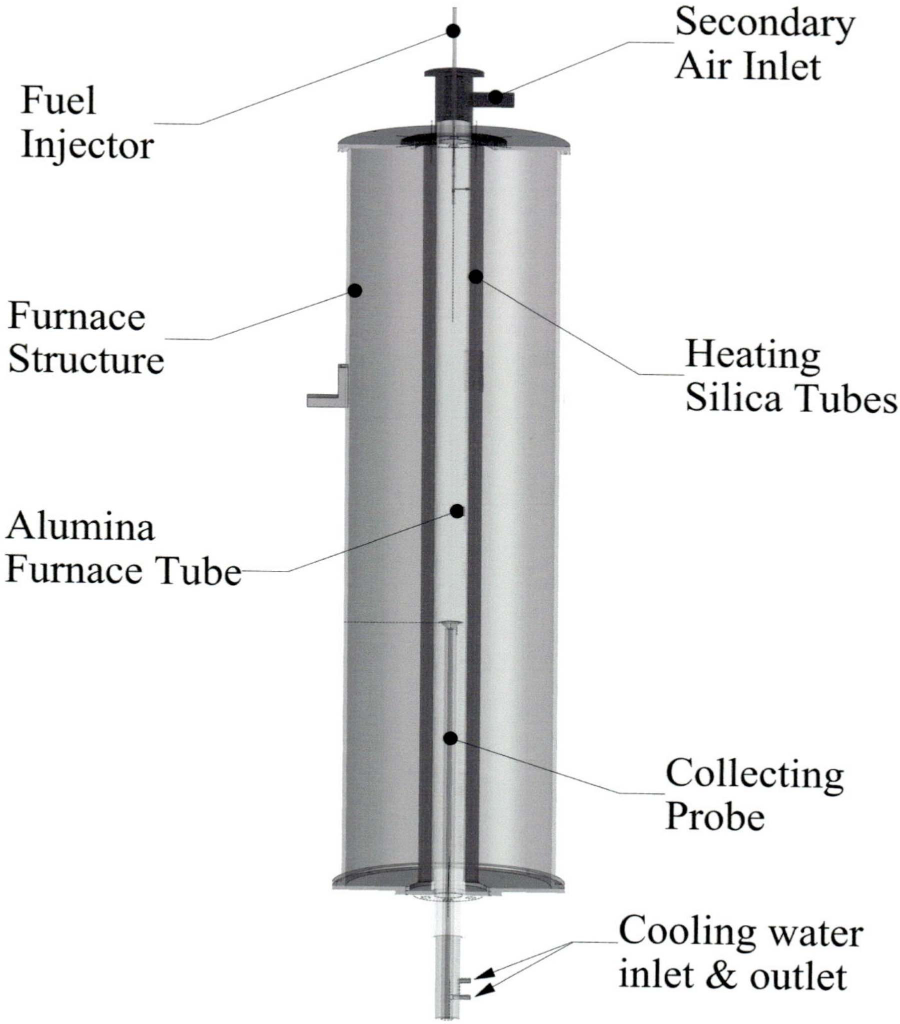

FIGURE 7.2 CAD drawing of drop tube furnace.

7.3.1 DTF Analysis Methods

Within a DTF, the fuel sample is subjected to an isothermal heating rate within the range of 10^4 and 10^5 K/s, quite high when compared to a TGA analysis. During the first stage of the experiment which uses a nitrogen atmosphere, volatiles evolve and are entrained in the carrier gas, whilst the solid char gets collected at the bottom. During the second stage when using an oxygen-enriched atmosphere, the char combusts leaving ash to be collected at the collection probe. Particle properties were determined indirectly based on experimental measurements of various parameters such as fluid velocity and particle size distribution.

7.3.2 DTF Particle Residence Time

To simplify particle trajectory analysis, a one-dimensional flow was assumed along the furnace axis (z-direction thus velocity is represented with a v). The particle residence time at each position was calculated as given in Equation (7.1).

$$t = \sum_{i=0}^{i=1\text{-}n} \frac{L_i}{v_g + v_t} \tag{7.1}$$

with

t representing the particle residence time (s),

L_i representing the length of slice (m) (whole length of DTF was sliced into sections),

v_g representing the velocity of carrier gas (m/s), and

v_t representing the terminal velocity of particle (m/s).

The terminal velocity of the particle was determined using Stokes' Law as given by Equation (7.2) as applied to laminar flows.

$$v_t = g.d_p^2.\frac{(\rho_p - \rho_g)}{18.\mu_g} \tag{7.2}$$

with

g representing the gravitational acceleration (m/s^2),

d_p representing the particle diameter (m),

ρ_p representing the particle density (kg/m^3),

ρ_g representing the carrier gas density (kg/m^3), and

μ_g representing the carrier gas viscosity (Pa.s).

The dynamic viscosity of carrier gas at different temperatures was evaluated through the Sutherland [7] expression as described by Equation (7.3).

$$\mu_g = \mu_0 \cdot \left(\frac{T_g}{T_0}\right) \cdot \left(\frac{T_0 + C_{\mu s}}{T_g + C_{\mu s}}\right) \tag{7.3}$$

with

μ_0 representing the reference gas dynamic viscosity (Pa.s) [N_2 gas:1.663 × 10^{-05} Pa.s],

T_g representing the gas temperature (K),

T_0 representing the reference gas temperature (K) [N_2 gas: 273 K], and

$C\mu_s$ representing the Sutherland constant for viscosity [N_2 gas:10^7].

The velocity of the carrier gas within the furnace was determined as a function of reactor volume per unit length and average gas volume flow rate. The instantaneous gas volume flow rate was dependent on temperature, pressure, and volume such that it is the average gas volume flow rate as determined according to Equation (7.4).

$$Q_v = v_g \cdot \left(\frac{V_{reactor}}{L_{reactor}} \right) = Q_{v,0} \cdot \left(\frac{P_0}{P_g} \right) \left(\frac{T_g}{T_0} \right) \tag{7.4}$$

with

Q_v representing the corrected gas volume flow rate (m^3/s),
$Q_{v,0}$ representing the inlet gas volume flow rate (m^3/s at STP),
$V_{reactor}$ representing the volume of reactor (m^3),
$L_{reactor}$ representing the length of reactor (m),
P_0 representing the inlet carrier gas pressure at STP,
T_0 representing the inlet carrier gas temperature at STP,
P_g representing the combustion zone gas pressure, and
T_g representing the combustion zone gas temperature.

To verify the flow regime within the DTF by use of Reynolds number, Equation (7.5) was called upon.

$$Re = \frac{\rho_g v_g D_{reactor}}{\mu_g} \tag{7.5}$$

The maximum Reynolds number that could be attained in the experiments was 214, which is within the laminar range. The highest carrier gas density obtainable at 1,273 K was 0.2778 kg/m^3, gas velocity was 5.3 m/s, and the lowest dynamic viscosity was 47.88×10^{-5} Pa.s.

7.3.3 DTF Volatile Yield

During both stages of the experiment, a couple of assumptions had to be considered due to lack of 100% collection efficiency by the probe. This was caused by the propensity of solid char material to stick to the furnace walls and collection probe, coupled with some volatile condensation as well. As such, ash measurement was used as a precursor in monitoring the mass loss. This is because ash constituents such as Al_2O_3, Fe_2O_3 and SiO_2 require temperatures above 2,100 K to volatilise. Thus, the ash tracing method was sufficient to determine the percentage of evolved volatiles on a dry basis as represented by Equation (7.6).

$$VM_{DTF} = \frac{1}{Ash_1} \left[Ash_1 . \left(1 - Ash_0\right) - Ash_0 . \left(1 - Ash_1\right) \right] \tag{7.6}$$

with

VM_{DTF} representing the volatile matter from DTF analysis on a dry basis,
Ash_0 representing the mass of ash in initial particle on a dry basis, and
Ash_1 representing the mass of ash in sampled residue particle on a dry basis (collected char).

7.3.4 DTF Particle Temperature during Devolatilization

To determine the particle temperature, it was imperative to acknowledge the conservation of energy and mass as the fuel particle travelled through the furnace. This meant considering all modes of heat transfer as given by Equation (7.7). However, flow within the DTF is assumed to be one-dimensional and the fuel particles are supplied on a dry basis (no vapourisation energy required). As such, a more useful energy balance is formed as represented by Equation (7.8).

$$\frac{d\left(m_p.c_{p,p}.T_p\right)}{dt} = H_{rad} + H_{conv} - Q_{vap} \tag{7.7}$$

$$\frac{h_{reac}.\Delta m_p}{\Delta t} = H_{rad} + H_{conv} \tag{7.8}$$

with

m_p representing the particle mass (kg),
$C_{p,p}$ representing the particle specific heat capacity (J/kg·K),
T_p representing the particle temperature (K),
H_{rad} representing the radiation heat transfer (W),
H_{conv} representing the convection heat transfer (W),
Q_{vap} representing the heat of vaporisation (W),
Δh representing the specific enthalpy of reaction (J/kg), and
Δt representing the residence time (s).

The radiative heat exchange between the DTF walls and the particle surface was then given by Equation (7.9).

$$H_{rad} = \varepsilon_p.\sigma.A_P.\left(T_p^4 - T_w^4\right) \tag{7.9}$$

with

ε_p representing the particle surface emissivity,
σ representing the Stefan Boltzmann constant,
A_p representing the particle surface area (m^2), and
T_w representing the DTF wall temperature (K).

The convective heat exchange between the hot carrier gas within the furnace and the particle was also given by Equation (7.10):

$$H_{conv} = h_{conv}.A_P.\left(T_p - T_g\right) \tag{7.10}$$

with

h_{conv} representing the convective heat transfer coefficient ($W/m^2.K$) and
T_g representing the combustion zone gas temperature (K).

To deduce the convective heat transfer coefficient, it was logical to acknowledge the existence of the boundary layer through inclusion of Prandtl number principles for a spherical particle. Since the flow within the furnace was kept within the laminar flow regime as verified by Equation (7.5), necessary calculations were performed as given by Equation (7.11).

$$h_{conv} = \frac{Nu.\lambda_g}{d_p} \tag{7.11}$$

with

Nu representing the Nusselt number and

λ_g representing the thermal conductivity of carrier gas (W/m.K).

The thermal conductivity of carrier gas at different temperatures was evaluated through the Sutherland [7] expression as well as described by Equation (7.12).

$$\lambda_g = \lambda_0 . \left(\frac{T_g}{T_0}\right) . \left(\frac{T_0 + C_{\lambda s}}{T_g + C_{\lambda s}}\right) \tag{7.12}$$

with

λ_g representing the thermal conductivity (W/m.K),

λ_0 representing the reference thermal conductivity (W/m·K) [N_2 gas: 0.0242 W/m·K], and

$C\lambda_s$ representing the Sutherland constant for thermal conductivity [N_2 gas:150].

The Nusselt number is thus determined through the relation coined by Whitaker [8,9] as given by Equation (7.13). All the values were evaluated at properties of the carrier gas free stream except the viscosity at the particle surface, which was evaluated using the particle surface properties.

$$Nu = 2 + \left(0{,}4.Re^{0.5} + 0{,}06.Re^{\frac{2}{3}}\right).Pr^{0.4}.\left(\frac{\mu_g}{\mu_p}\right)^{0.25} \tag{7.13}$$

with

Re representing the carrier gas Reynolds number,

Pr representing the carrier gas Prandtl number,

μ_g representing the carrier gas dynamic viscosity (Pa.s), and

μ_p representing the carrier gas dynamic viscosity at particle surface (Pa.s).

In summary, the left hand of Equation (7.8) reduces to zero when re-written with assumptions specific to the devolatilisation process. This is because during devolatilisation, the change in combustible mass within the particle tend towards zero for short residence times.

7.3.5 DTF Particle Temperature during Char Combustion

Similar equations were called upon to determine the particle temperature during char combustion. The energy balance in Equation (7.8) was written to account for the surface reactions taking place on the char surface to become Equation (7.14).

$$\frac{h_{reac}.\left[m_{p,initial} - m_{p,final}\right]}{\Delta t} = \varepsilon.\sigma.A_p.\left(T_p^4 - T_w^4\right) + h.A_p.\left(T_p - T_g\right) \tag{7.14}$$

with

h_{reac} representing the enthalpy of reaction (J/kg) (Char + $O_2 \rightarrow CO_2$),

$[m_{p,initial} - m_{p,final}]$ representing the change in mass of combustible matter in char particle (kg), and

Δt representing the residence time (s).

The ash tracing method was sufficient to determine the extent of combustion on a dry basis as represented by Equation (7.15). This parameter is directly linked to the measurement of the unburnt carbon carried out in the co-combustion model.

$$CE = \left(\frac{Ash_{0,char}}{1 - Ash_{0,char}}\right) \cdot \left(\frac{1 - Ash_{1,char}}{Ash_{1,char}}\right) \tag{7.15}$$

with

CE representing the combustion extent on a dry basis,

$Ash_{0,char}$ representing the mass of ash in initial char particle on a dry basis, and

$Ash_{1,char}$ representing the mass of ash in sampled residue particle on a dry basis (collected ash).

7.4 CASE STUDY: DEVELOPMENT OF A CO-COMBUSTION MODEL

The geometry is set up as 3D by using both ANSYS SPACECLAIM 2021 R1 [10] and ANSYS ICEM CFD 2021 R1 [11]. The model setup is compiled in ANSYS FLUENT 2021 R1 [12]. User defined functions (UDFs) are employed in the model setup to customise mainly the furnace freeboard temperature profile and other relevant parameters. UDFs allow researchers to include specific sub-models relevant to the needs of a problem. The modelling process should be split into two distinct stages to emulate the DTF experimental setup. The first step involves passing the fuel samples through an inert atmosphere containing N_2 only, whilst the second step allows a 3% O_2 97% N_2 atmosphere to resemble combustion.

In summary, the co-combustion model will investigate the combustion of South African (SA) coal in a 100% N_2 atmosphere, Hwange coal (HC) and Pinus sawdust (PS) blends in a 100% N_2 atmosphere, and HC and PS blends in a 97% N_2 and 3% O_2 atmosphere by volume.

7.4.1 Furnace Geometry and Meshing

Figure 7.3 shows the computer-aided design (CAD) of the drop tube furnace (DTF). Figure 7.4 shows how the DTF furnace is modelled as a 30° section by employing the periodic symmetric boundary about the axis as it allows us to capture the axial and radial variation of parameters with more precision as compared to a 2D analysis. Figures 7.5 and 7.6 illustrate detailed specifications of the modelled DTF section.

Hex meshing is carried out since a structured mesh is used for the furnace meshing. The shortest edge has a distance of 1.109 mm which is used to guide the initial seed element size for the coarse mesh. Refinement is then carried out depending on the required emphasis. Table 7.3 summarises the three mesh densities that are considered when carrying out the grid sensitivity test for the modelling study.

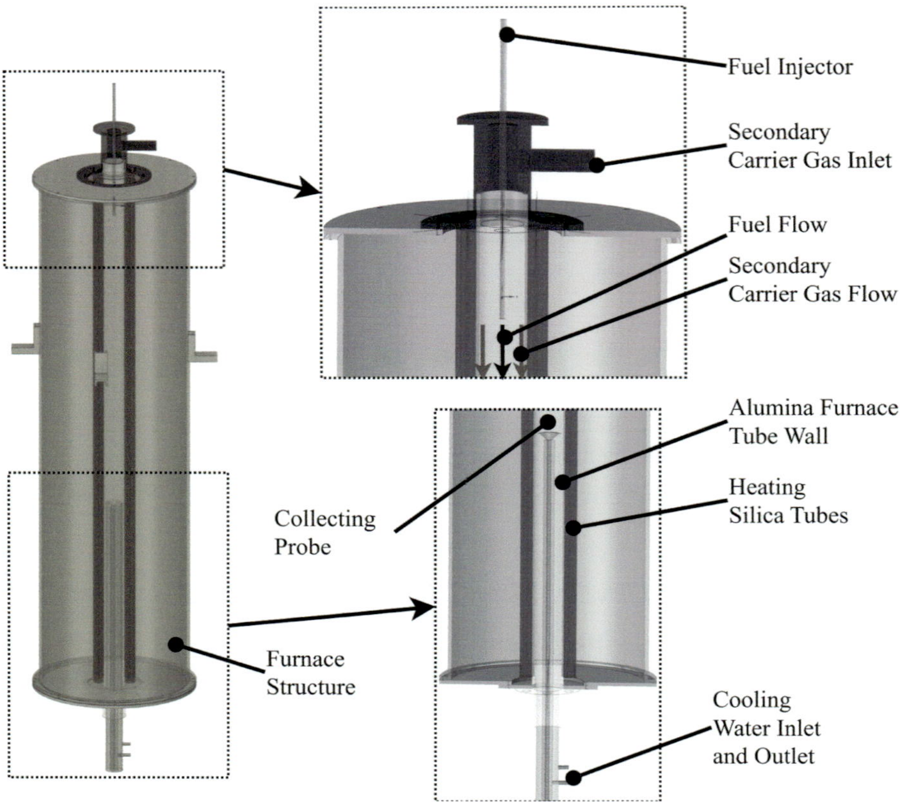

FIGURE 7.3 Close up of drop tube furnace CAD drawing.

An illustration of the coarse mesh employed within the study is given in Figures 7.7 and 7.8. Figure 7.7 shows a close up of the meshing at furnace inlets (fuel, primary, and secondary carrier gas inlets), whilst Figure 7.8 shows a close up of meshing at the furnace outlet.

7.4.2 Boundary Conditions

A pressure-based solver is used for all the runs. The discrete phase is represented by the solid fuel particles, solid combustion intermediates and solid combustion products whilst the continuous phase is represented by the gaseous carrier gas, gaseous combustion intermediates and products. The model is set up to have forty continuous phase iterations per discrete phase model (DPM) iteration which is a modest value that balances the time taken tracking discrete phase particles vs flow of gaseous constituents. A surface injection originating from the fuel inlet, Rosin-Rammler diameter distribution, and fuel velocity inlet of 1.85 m/s models the fuel inlet parameters. Similarly, since the fuel blending was on a mass basis, the total fuel mass flow rate

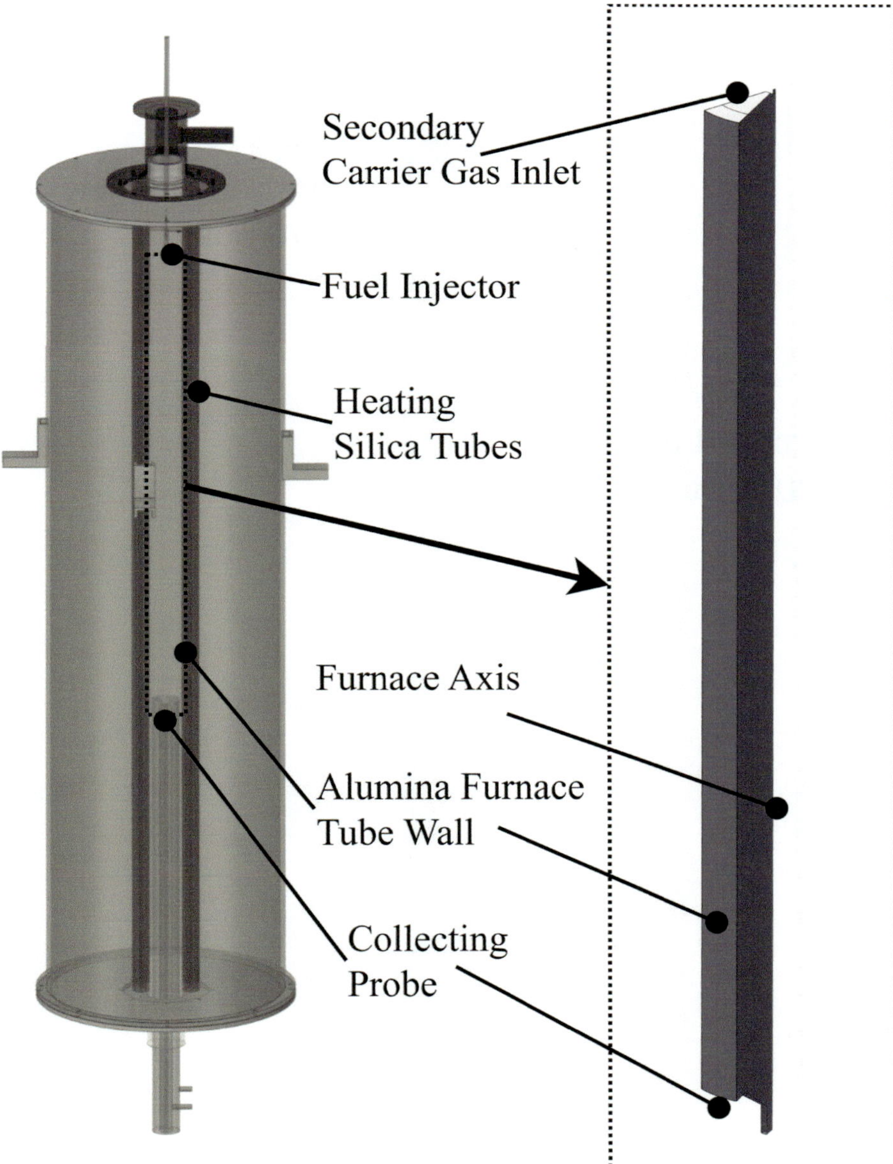

FIGURE 7.4 CAD drawing showing how drop tube furnace is modelled as a 30° section.

is kept as 1.5 g/min which translates to 2.08×10^{-06} kg/s for a 30° modelled section. As such the modelled primary carrier gas mass flow rate for a 30° section is 4.85×10^{-07} kg/s whilst the secondary carrier gas is 2.43×10^{-05} kg/s. The sawdust particles are treated to a shape factor of 0.83 since a cylindrical shape is assumed due to the

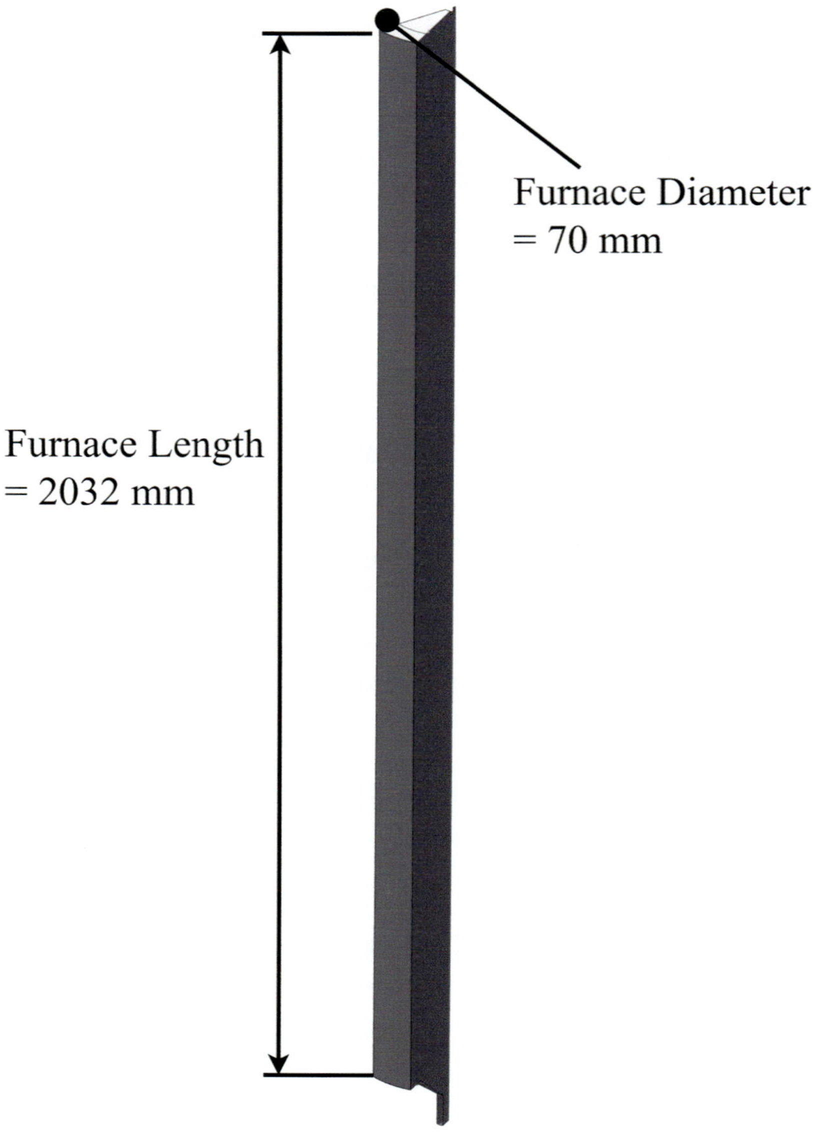

FIGURE 7.5 Modelled DTF 30° section specifications.

fibrous nature of biomass. The coal particles are assumed to be spherical because when small coal particles heat up leading to devolatilisation, they tend to soften up and adopt a spherical shape [13]. The discrete random walk model is preferred for the stochastic tracking of particles after they interact with the continuous phase turbulence. To help with the balance between accuracy and computational power

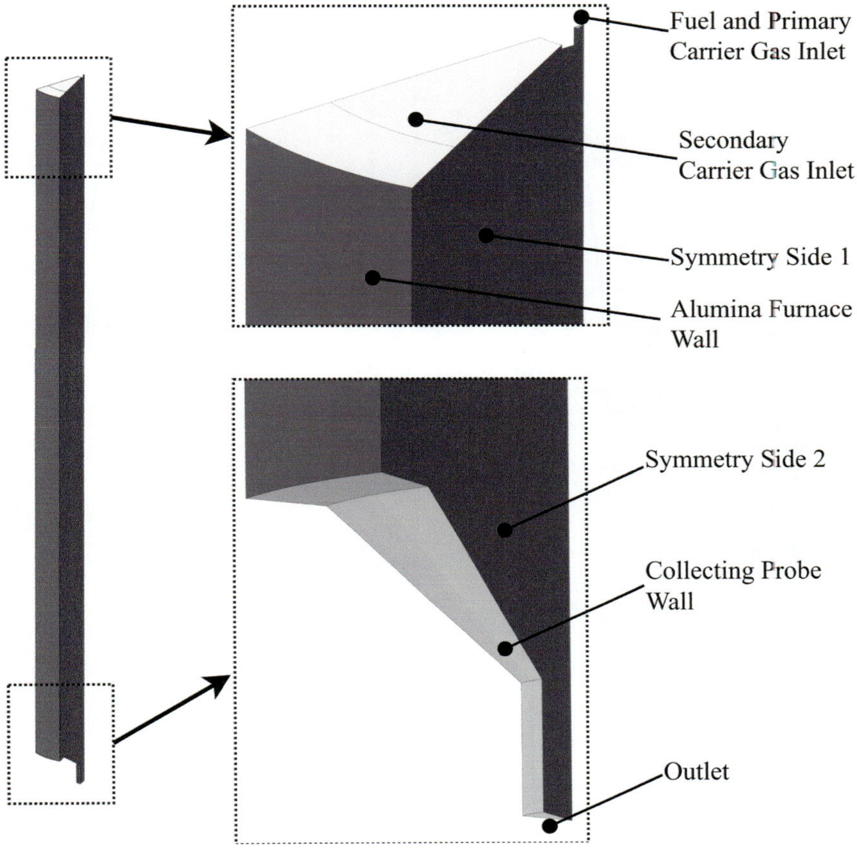

FIGURE 7.6 Modelled DTF 30° section showing detailed inlets, outlets, symmetry, and wall boundaries.

TABLE 7.3
Grid Analysis Mesh Properties

Mesh Elements	Inflation parameters: minimum cell size, max cell size, and boundary layer growth rate	Warpage (<10 acceptable)	Angle (>18° acceptable)	2x2x2 Determinant (>0.2 acceptable)
159,755	Min 0.0001 Max 0.001095 1.13628	Min 0 Max 8.2	Min 26.55 Max 88.11	Min 0.562 Max 0.997
399,200	Min 0.0001 Max 0.0009103 1.10851	Min 0 Max 9.15	Min 26.55 Max 88.11	Min 0.569 Max 0.996
60,208	Min 0.0001 Max 0.00114547 1.16126	Min 0 Max 4.29	Min 25.92 Max 89.19	Min 0.659 Max 0.997

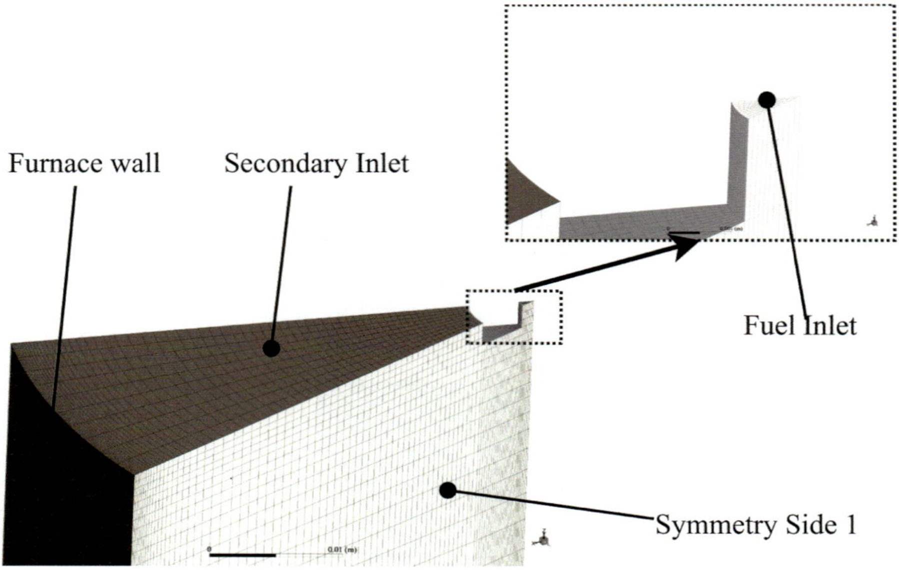

FIGURE 7.7 DTF mesh showing inlets.

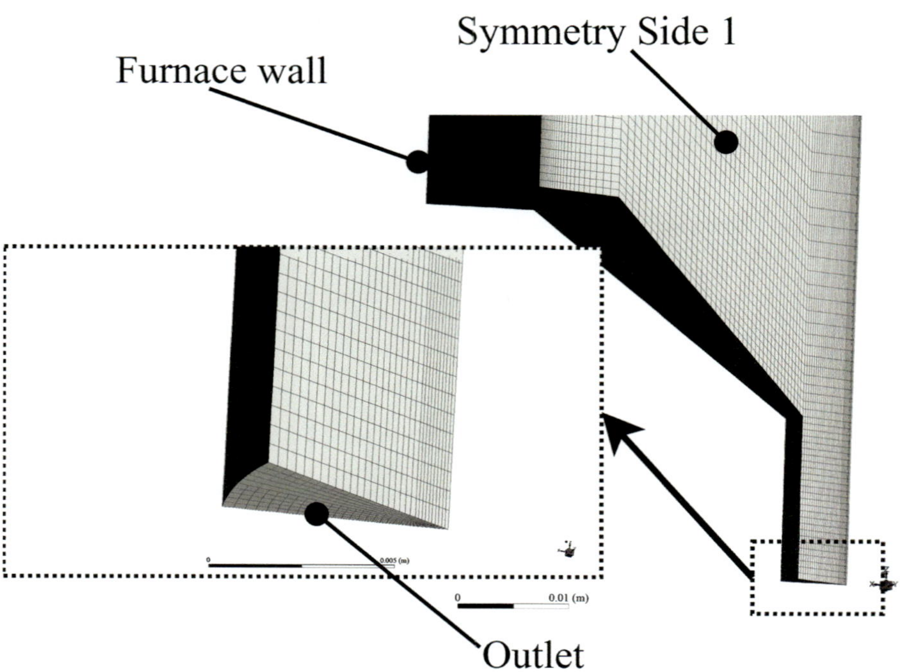

FIGURE 7.8 DTF mesh showing outlet.

TABLE 7.4
Hwange Coal and Pine Sawdust Devolatilisation Parameters

Fuel	$A(s^{-1})$ [x 10^{+14}]	E_a(kJ/mol) [x 10^{+06}]
Devolatilisation		
100 HC	4.20	87.27
90HC10PS	3.82	100.01
80HC20PS	3.44	105.55
70HC30PS	3.06	109.72

required, a value of 10 is used which is the number of times turbulent effects are considered per iteration.

The first stage is carried out under a nitrogen atmosphere which resembles devolatilisation as there was no presence of oxygen. The main chemical reactions that are expected are related to the release of volatiles. Table 7.4 summarises the parameters used for the kinetic doublet of activation energy (E_a) and pre-exponential factor (A) as obtained from TGA experiments.

The second stage is carried out under a 97% nitrogen atmosphere and 3% oxygen atmosphere, which resembled combustion of the char fuel. This is justifiable since all the volatiles are carried along with the carrier gas in the first stage and thus could not initiate the homogeneous combustion process in this case. As such, only char combustion submodels were required to model the chemical interactions within stage 2. The heterogeneous surface reaction mechanisms are also known as multiple surface reaction mechanisms because reactions are assumed to take place within the boundary layer and bulk flow simultaneously. Table 7.5 summarises the mechanisms employed for the multiple surface reaction mechanism of coal and sawdust.

The kinetic parameters are determined by use of TGA experiments. The gasification reactions of char with CO_2 or H_2O are overlooked since the reaction temperatures expected in the reactor are not high enough to warrant their influence. Furthermore, the concentrations of CO_2 and H_2O are low compared to N_2 [14].

TABLE 7.5
Hwange Coal and Pine Sawdust Heterogeneous Combustion Parameters

	Rate Exponent	$A(s^{-1})$	B	E_a (J/kg.mol)
$C(s) + 0.5O_2(g) \rightarrow CO(g)$				
100HC	$[O_2]^1$	$1.16 \times 10^{+03}$	0	$52.9 \times 10^{+06}$
90HC10PS	$[O_2]^1$	$2.78 \times 10^{+04}$	0	$63.99 \times 10^{+06}$
80HC20PS	$[O_2]^1$	$9.53 \times 10^{+05}$	0	$90.52 \times 10^{+06}$
70HC30PS	$[O_2]^1$	$6.118 \times 10^{+07}$	0	$103.85 \times 10^{+06}$

Modelling of NOx emissions includes the two most important mechanisms responsible for NOx formation, which include fuel NOx and thermal NOx. Prompt NOx is disregarded since the air-to-fuel ratio is lean enough to avoid fuel-rich pockets within the combustion zone. Calculations are performed to determine the fuel nitrogen modelling parameters based on the volatile matter and fixed carbon content from characterisation experiments. These calculations entail the use of Equations (7.16) and (7.17), which are used to determine the nitrogen in char (N_{char}) and nitrogen in volatile (N_{vol}) parameters [12,15].

$$N_{char} = \frac{N_{split} \times N_{daf}}{FC_{daf}} \tag{7.16}$$

$$N_{vol} = \frac{\left(1 - N_{split}\right) \times N_{daf}}{VM_{daf}} \tag{7.17}$$

with

N_{char} representing the nitrogen in char mass fraction,
N_{vol} representing the nitrogen in volatile mass fraction,
N_{split} representing the split ratio between N_{char} and $N_{vol,}$
N_{daf} representing the nitrogen content in dry ash free fuel sample,
FC_{daf} representing the fixed carbon content in dry ash free fuel sample, and
VM_{daf} representing the volatile matter content in dry ash-free fuel sample.

As supported by other researchers with their experimental work, 0.7 is used as the split ratio between the N_{char} vs the N_{vol} mass fractions [12,15]. Consequently, Table 7.6 summarises the parameters used in the model.

SOx emissions were modelled assuming there is no formation of SO_3, only SO_2 and H_2S. As supported by Manovic and Grubor [16] during their investigations with coal, the split ratio between sulphur in char (S_{char}) vs sulphur in volatiles (S_{vol}) is within the range of 0.65 on average. As such, Equations (7.18) and (7.19) were used to determine the S_{char} and S_{vol} parameters and Table 7.7 summarises these parameters [12].

TABLE 7.6
Nitrogen Content in Fuel Blends

Fuel	Nitrogen in Char	Nitrogen in Volatile
100 HC	0.02092	0.02095
90HC10PS	0.02103	0.01452
80HC20PS	0.01957	0.01222
70HC30PS	0.01739	0.01016
100PS	0.01604	0.00133

TABLE 7.7
Sulphur Content in Fuel Blends

Fuel	Sulphur in Char	Sulphur in Volatile
100 HC	0.00714	0.00899
90HC10PS	0.00769	0.00667
80HC20PS	0.00778	0.00610
70HC30PS	0.00756	0.00555
100PS	0	0

$$S_{char} = \frac{S_{split} * S_{daf}}{FC_{daf}} \tag{7.18}$$

$$S_{vol} = \frac{\left(1\text{-}S_{split}\right) * S_{daf}}{VM_{daf}} \tag{7.19}$$

with

S_{char} representing the sulphur in char mass fraction,
S_{vol} representing the sulphur in volatile mass fraction,
S_{split} representing the split ratio between S_{char} and $S_{vol,}$
S_{daf} representing the sulphur content in dry ash free fuel sample,
FC_{daf} representing the fixed carbon content in dry ash free fuel sample, and
VM_{daf} representing the volatile matter content in dry ash free fuel sample.

A pressure-based solver is employed because the effect of density variation towards fluid trajectories, particle trajectories, and particle temperatures is constrained. This assumption holds because, between the expected inlet and maximum temperatures, the average density of all the gases varies from an average of 0.4395 kg/m^3 to 0.4284 kg/m^3. As a result, a density-based solver is regarded as an unnecessary complication. The second-order upwind linearisation technique for all the discretised equations is used. The Green Gauss cell-based gradient evaluation is preferred since it is based on all the neighbouring cells rather than just a few adjacent cells. A coupled algorithm is used to solve for the equations, which quickens the convergence time and improves the accuracy. This means the linearised equations, particularly Navier-Stokes or modified momentum equations, are solved as simultaneous equations rather than single algebraic equations. These simultaneous equations are all solved at once to give back all the variables at once, which then becomes the input for the next iteration (this is opposite to the segregated approach of using the SIMPLE or SIMPLEC algorithm). Since solving the linearised equations is an iterative process, moving from one iteration to the next needs to be monitored to avoid any instability. This was controlled by using appropriate under-relaxation values for each equation or variable (values closer to 1 led to more instability but decreased the calculation time significantly). Summary of the under-relaxation factors used by the researcher is noted down in Table 7.8. In as much as residual monitoring is not foolproof in

TABLE 7.8
Summary of Solver Setup Inputs

Variable	Under Relaxation Factor
Pressure	0.3
Density	1
Momentum	0.7
Energy	0.7
Swirl velocity	0.9
Species	0.9
Discrete phase sources	0.5

terms of getting accurate results, it is important to follow CFD solver thumb rules. The energy residual limit has to be maintained at 1×10^{-6} with little room for compromise with respect to this variable.

7.4.3 User Defined Functions

The main strength with regard to combustion modelling for ANSYS FLUENT 2021 R1 [12] lies within its ability to handle user defined functions (UDFs). UDFs allow users to include customised submodels other than those built in by default. Due to their versatility and conciseness, C++ codes are written to act as the UDFs for incorporation into ANSYS FLUENT 2021 R1 [12] overall combustion model.

The UDFs for experimental wall furnace temperatures are based on the fourth-order polynomial functions. An example of a UDF used to customise the furnace temperature for a 1,273 K set temperature is presented below with its corresponding flowchart in Figure 7.9.

```
/*****************************************************/
/* User-Defined Functions for 1273 K set temperature profile */

/*****************************************************/
#include "udf.h"
   DEFINE_PROFILE(temperature_profile, thread, position)
   {
   real pos[ND_ND]; /* this will hold the position vector */
   real z;
   face_t f;
   begin_f_loop(f, thread)
   {
   F_CENTROID(pos,f,thread);
   z = pos[2];
```

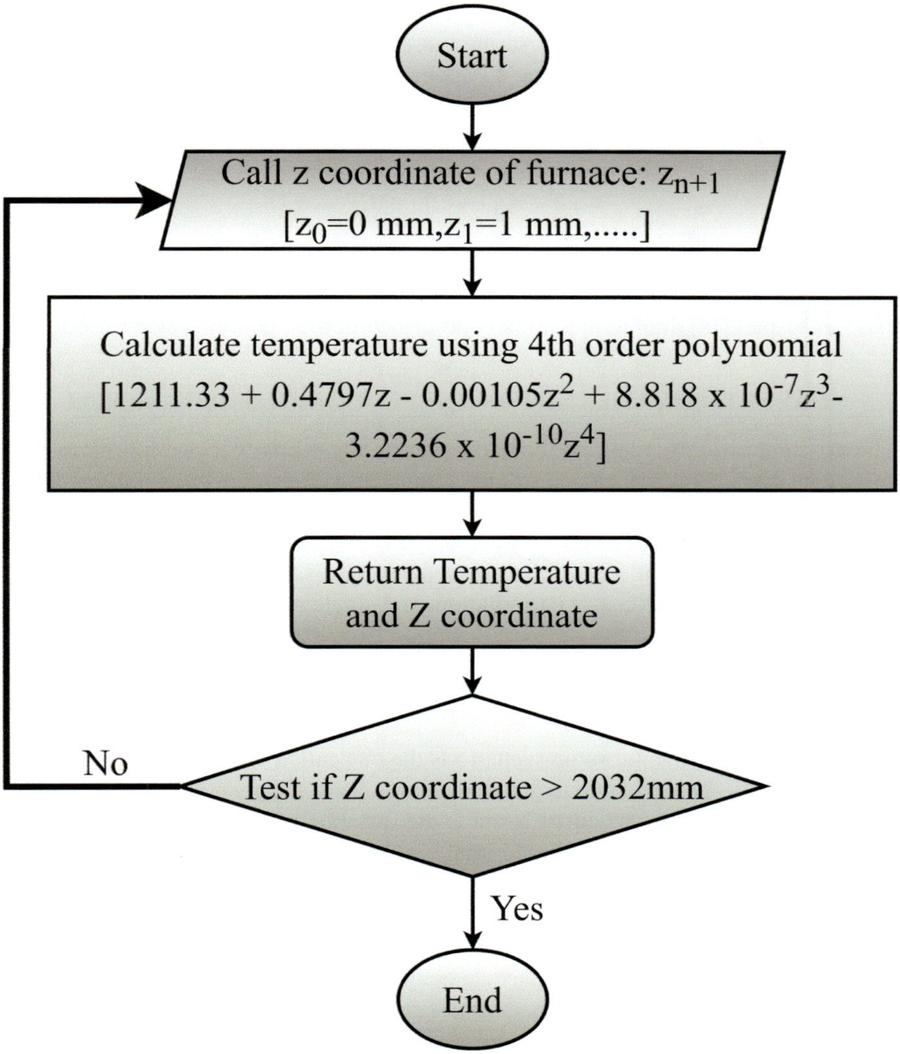

FIGURE 7.9 UDF for a furnace setting of 1273 K.

```
F_PROFILE(f, thread, position) = 1211.33 + 0.4797*z – (0.00105*pow(z,2)) + 8.8184e-7*pow(z,3) – (3.2236e-10*pow(z,4));
}
end_f_loop(f, thread)
}
```

Four UDFs that captured the variation of specific heat capacity (c_p) as a function of temperature for the different species involved in the combustion process were also written. It was deduced that the main species during the combustion process H_2O,

CO_2, N_2, and O_2 had a great implication towards radiation heat transfer through the weighted grey gases model described in the literature review. The variation of specific heat capacity was based on a mathematical piecewise polynomial function represented in Equation (7.20). The coefficients varied depending on the combustion species and temperature range under consideration as demonstrated by Turns [17].

$$c_p = a.T + b.T^2 + c.T^3 + d.T^4 + e.T^5 \tag{7.20}$$

A UDF was also written to capture the particle drag properties of the coal and biomass species as they travelled through the furnace. The UDF was based on the Haider and Levenspiel [18] drag model already described in the literature review section.

7.5 CASE STUDY: VALIDATION OF CO-COMBUSTION MODEL

The co-combustion model is validated concurrently whilst performing mesh sensitivity analysis. This was done through comparison of experimental and modelled particle residence times. The grid convergence index is employed to ascertain the influence of changing the mesh; hence, a 60,208 cell mesh is deduced to be the most appropriate mesh. Turbulence, radiation, and devolatilisation model sensitivity analysis showed that the RNG k-ε, P1 radiation, and single rate kinetic model produced particle residence times with the least deviation from experimental parameters.

7.5.1 Sensitivity Analysis and Validation

Stage 1 DTF experimental setup scenario is the one called upon to perform the sensitivity analysis. The grid convergence index (GCI) as propounded by Roache [19] looks at the effect of mesh density towards the variation of performance parameters. This index is adapted for fluid-based analysis through its ability to analyse boundary layer effects by use of near wall inflation monitoring. Performance parameters are those variables of interest to the user during the modelling practice. As such, mesh independence is achieved when the index approaches unity. In this case, the particle final residence time and particle peak temperature are used to evaluate the GCI. As illustrated in Table 7.3, the shortest edge had a distance of 1.109 mm which was used to guide the initial seed element size for the coarse mesh. Refinement was then carried out depending on the required emphasis as presented. The 399,200 cells mesh (mesh 1), 159,755 cells mesh (mesh 2), and 60,208 cells mesh (mesh 3) evaluate to give a refinement ratio (r) of around 2.576. Equations (7.21) to (7.23) were thus employed to evaluate GCI based on the parameters of interest as presented in Table 7.9.

$$p = \ln\left|\frac{f_3 - f_2}{f_2 - f_1}\right| / \ln(r) \tag{7.21}$$

TABLE 7.9
Mesh Sensitivity Based on Grid Convergence Index

Mesh Cells	Mesh Label	Performance Parameter	p	GCI_1	GCI_2	Asymptotic Range Value
		[Particle Residence Time (s)]				
399,200	1	1.2991	0.4986	0.0932	0.1565	1.0471
159,755	2	1.2407				
60,208	3	1.3343				
		[Particle Peak Temperature (K)]				
399,200	1	1,006.14	1.1840	0.0137	0.0413	0.9778
159,755	2	1,029.03				
60,208	3	958.85				

$$GCI_{12} = \frac{F_s x \left| \frac{f_2 - f_1}{f_1} \right|}{r^p - 1}; GCI_{23} = \frac{F_s x \left| \frac{f_3 - f_2}{f_2} \right|}{r^p - 1} \tag{7.22}$$

$$\frac{GCI_{23}}{r^p x GCI_{12}} \approx 1 \tag{7.23}$$

with

p representing the order of convergence,
f representing the performance parameter,
r representing the refinement ratio = 2.576,
GCI representing the grid convergence index, and
F_s representing the factor of safety (in this case two levels of refinement = 1.25).

As illustrated in Table 7.9, the asymptotic range values are both within the range of unity, highlighting how the solution is mesh independent for the three meshes under consideration. Potgieter et al. [20] made a similar conclusion with regard to the grid convergence index during their heat transfer studies, which further supports the robust nature of this approach.

As a thumb rule, the residuals are monitored for convergence with respect to continuity, energy, and species transport. On average, the energy residual converges by requiring more iterations than the other residuals. Mass conservation for discrete phase modelling is best monitored through analysis of Equation (7.24), which acknowledges the mass balance between the inlets, discrete phase model masses, and the outlets.

$$\text{Mass flow @ inlets} + \text{DPM Mass Source} = \text{Mass flow @ outlets} \tag{7.24}$$

TABLE 7.10
Mesh Sensitivity Analysis by Mass Conservation

	1st Mesh	2nd Mesh	3rd Mesh
Cells	60,208	159,755	399,200
Nth iteration for energy residual < 1×10^{-06}	218	188	217
Net mass flow rate across all boundaries (kg/s)	4.2556×10^{-07}	4.2514×10^{-07}	4.2461×10^{-07}
DPM mass source (kg/s)	4.2517×10^{-07}	4.2512×10^{-07}	4.2516×10^{-07}
Relative error (%)	0.0941	0.0053	0.1294

Results of the sensitivity analysis by use of Equation (7.24) were thus presented in Table 7.10. All the meshes produce an acceptable mass conservation, though the medium density mesh produced the least error of 0.0053%. This highlights how the structured mesh employed in all cases is good enough to ascertain mass conservation. Mesh refinement to 399,200 cells from 159,755 cells does not lead to any visible changes besides prolonging the number of iterations.

Energy conservation principles are also employed in carrying out the mesh sensitivity analysis, as demonstrated by Equation (7.25), which best describes how particle energy balances are obtained during chemical reactions.

$$\text{Net Heat Transfer} = \text{Net Sensible Heat Transfer} + \text{Heat of Reaction Source} \tag{7.25}$$

The net sensible heat transfer represents the heat energy gained or lost by a particle from a reference temperature due to homogeneous reactions; the heat of reaction source represents the heat energy gained or lost by a particle from a reference temperature due to heterogeneous reactions. The overall net heat transfer is thus the sum of the two sources of heat. Table 7.11 summarises the results of the energy sensitivity analysis by use of Equation (7.25). The coarse mesh results in an error of

TABLE 7.11
Mesh Sensitivity Analysis by Energy Conservation

	1st Mesh	2nd Mesh	3rd Mesh
Cells	60,208	159,755	399,200
Total net heat transfer (W)	732.42	821.08	1,183.18
Total net sensible heat transfer (W)	732.56	821.37	1,183.39
Calculated heat of reaction (W)	0.1414	0.2961	0.212
Modelled heat of reaction source (W)	0.137	0.344	0.271
Relative error (%)	3.43	16.16	27.73

3.43% against a comparison of higher errors from the medium and finer mesh. This signifies that the coarse mesh is relatively sufficient to produce a converged solution at minimum time with acceptable particle energy balance. Although this approach is not foolproof, it has been employed by several researchers with success since it is based on conservation principles [21,22].

The increase in relative error as the mesh became finer is mainly related to the structure of the boundary layer. As propounded by Kolmogorov [23], the boundary layer comprises the log layer followed by the viscous layer as we move from the wall. The mesh near the wall must be able to capture the boundary layer growth sufficiently; thus, equations adapted for the log layer cannot be used to solve for the viscous layer. High mesh refinement results in having a smaller log layer and larger viscous layer than necessary, resulting in local flow instability [11]. The conservation of energy equation is sensitive to more parameters such as radiation, enthalpy of reaction, etc.; hence the larger relative errors in comparison to the conservation of mass equation.

Particle residence time as an experimental parameter is also employed in monitoring the mesh sensitivity. This approach also intertwines with the validation of the CFD model since modelled results are compared with experimental results. As such, the particle residence time experimental findings were traced against the modelled findings for the DTF stage 1 experimental setup.

Figure 7.10 shows a trace of the average particle residence time as it passes through the furnace for the South African fuel sample under a 1,273 K furnace temperature with a 520 mm probing position setting. All CFD meshes demonstrated a good initial estimation of particle residence time up to around 250 mm from the injection point. Eventually, the coarse mesh (60,208 cells) produced the least deviation from experimental values by staying closer to the 95% confidence band up to the furnace exit.

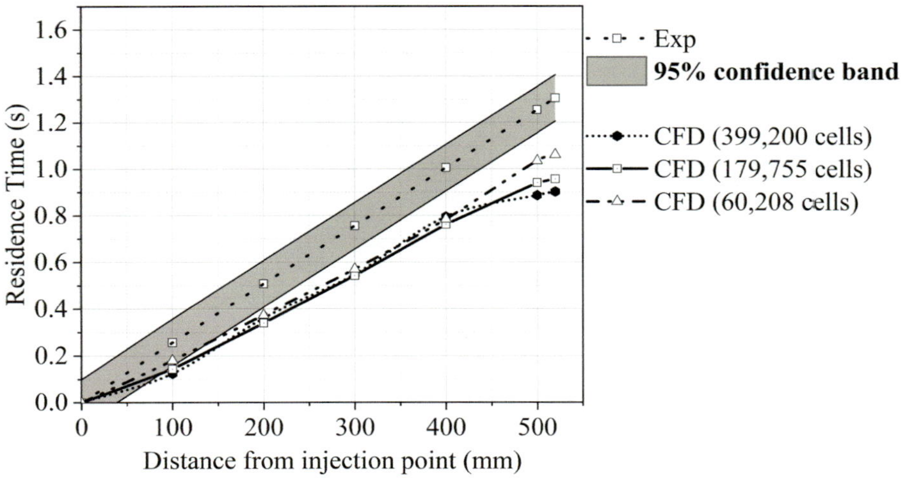

FIGURE 7.10 Mesh sensitivity analysis.

TABLE 7.12
Turbulence Sensitivity Analysis

	RNG	Standard
Nth iteration for energy residual < 1 E-06	218	221
Net mass flow rate across all boundaries (kg/s)	4.2556×10^{-7}	4.2597×10^{-7}
DPM mass source(kg/s)	4.2517×10^{-7}	4.2503×10^{-7}
Relative error (%)	0.09406	0.2209

In conclusion, mass conservation suggested use of a medium density mesh, whilst analysis based on energy conservation suggested use of a coarse density mesh. Correlating with experimental parameters suggested the use of the coarse mesh as well; thus, all subsequent analyses employed a mesh density similar to the coarse mesh.

It was also of interest to perform a sensitivity analysis with regard to turbulence modelling. Theoretically, the k-ε two-equation turbulence model is sufficient for most industrial applications, though the varied RNG k-ε model is known to converge much quicker for moderate swirling flows.

Table 7.12 shows how the RNG k-ε model converged quicker whilst also producing a lower error in terms of mass imbalance. Figure 7.11 further highlights how the RNG k-ε predicted residence times lay closer to the 95% confidence band as compared to the standard k-ε model. As such, all subsequent analyses employed the RNG k-ε model.

Sensitivity of the co-combustion model with regard to radiation models was also of interest to this research since a lot of parameters related to energy conservation

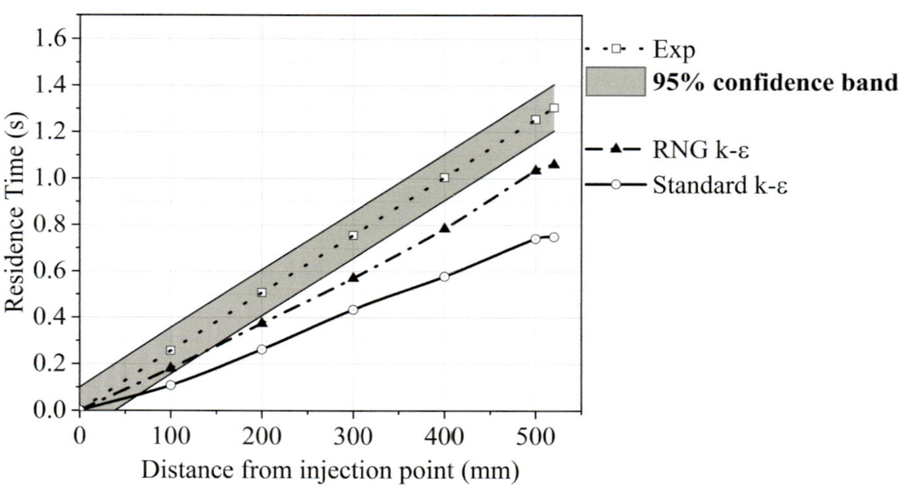

FIGURE 7.11 Turbulence sensitivity analysis.

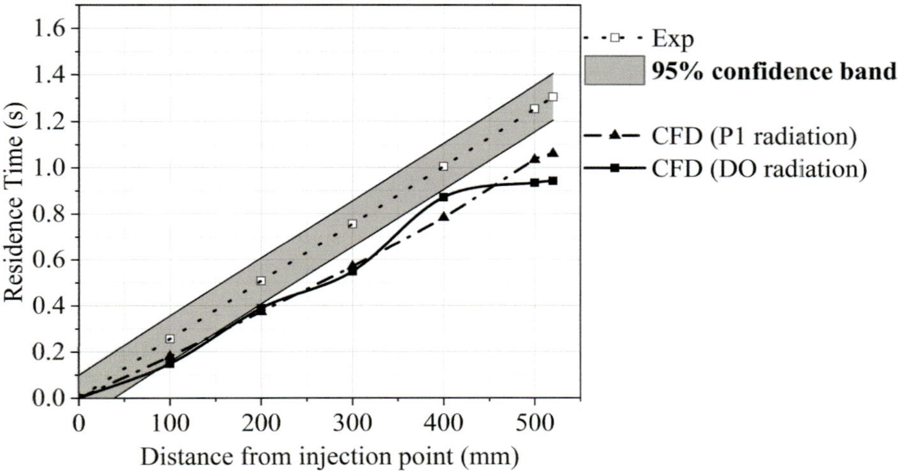

FIGURE 7.12 Radiation sensitivity analysis.

needed analysis. Both radiation models produced parameters that were close to experimental parameters, as shown in Figure 7.12. However, the P1 radiation model produced a predictable variation that was closer to the quasi-linear variation demonstrated by experimental results. Although both models are considered directional, the coupled solver that was employed in this analysis favoured the P1 radiation model since it produces the much-needed simultaneous equations easily.

Devolatilization occupies an important step in the combustion of solid fuels, thus the interest in performing a sensitivity analysis as well. A constant devolatilization model (constant = 50 s^{-1}) was compared to the customised single kinetic rate model for the South African fuel sample, as shown in Figure 7.13. Both models produced

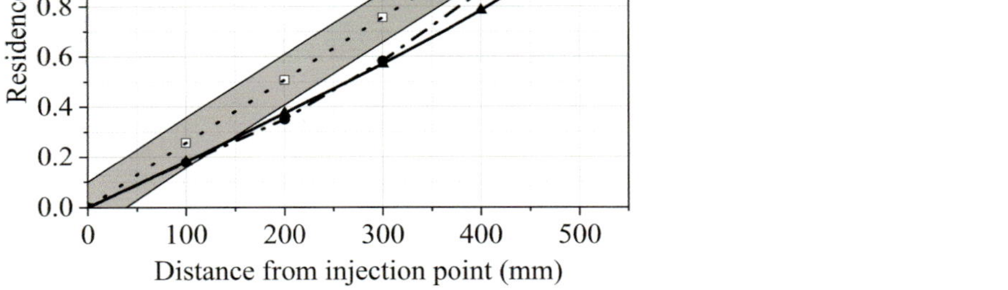

FIGURE 7.13 Devolatilization model sensitivity analysis.

comparable residence times, though the single kinetic rate resulted in a more predictable variation. The single kinetic rate model allowed the researcher to capture the activation energy and pre-exponential factor with more precision. As such, all subsequent analyses employed the single rate kinetic devolatilization model.

7.6 CASE STUDY: EVALUATION OF CO-FIRING COAL AND PINUS SAWDUST

7.6.1 Effect of Blending on Emissions in a Drop Tube Furnace

The emissions from NOx profiles as a function of fuel blending and DTF furnace temperature are analysed using CFD. The emissions from the devolatilisation process are combined with those from heterogeneous char combustion and homogeneous volatile combustion. The NOx emissions produced at the DTF furnace outlet for all the blending cases and various furnace temperatures are presented in Table 7.13 and Figure 7.14. The overall NOx is due to both thermal NOx and fuel NOx emissions. Increasing the furnace temperature from 1,273 K to 1,673 K demonstrates an increase from 146.78 to 187.90 ppm at the furnace outlet for the HC70PS30 case. Increasing pine sawdust proportion generally decreases NOx emissions in the products, especially if the nitrogen in the fuel is minimal. Pine sawdust has less nitrogen in char and nitrogen in volatile than Hwange coal; hence, the fuel NOx shows a decreasing trend as blending increases. As illustrated in Figures 7.15 and 7.16 thermal NOx tends to decrease as blending increases, another deduction which will be later supported when reviewing temperature profiles. The peak values of NOx were obtained near the zone where the heat of reaction was highest.

Adding 20% by mass of sawdust reduces the SOx from 7.36 ppm to 5.71 ppm. Given that pine sawdust is a waste product, this pollutant decrease is impressive. According to Xiong et al. [24], a decrease in SOx emissions generally prolongs the life of the exhaust handling system by lowering maintenance expenses.

7.6.2 Effect of Blending on Trajectory Profiles in a Drop Tube Furnace

DTF stage 1 analysis involves passing the fuel samples of interest (100HC, 90HC10PS, 80HC20PS and 70HC30PS) through the modelled furnace with the

TABLE 7.13
NOx Emissions at the Furnace Outlet

	NOx (ppm)		
	1,673 K	**1,473 K**	**1,273 K**
HC100	257.22	232.09	187.66
HC90PS10	233.78	210.78	178.11
HC80PS20	208.24	193.01	163.84
HC70PS30	187.90	179.999	146.78

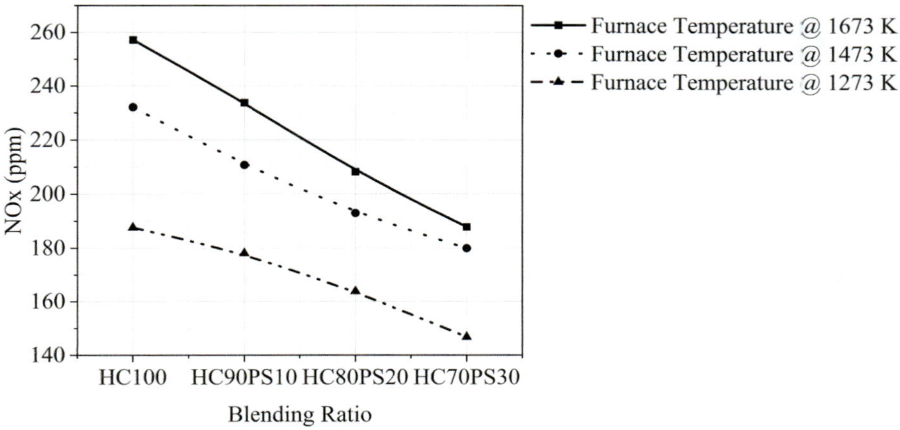

FIGURE 7.14 NOx emissions at the furnace outlet.

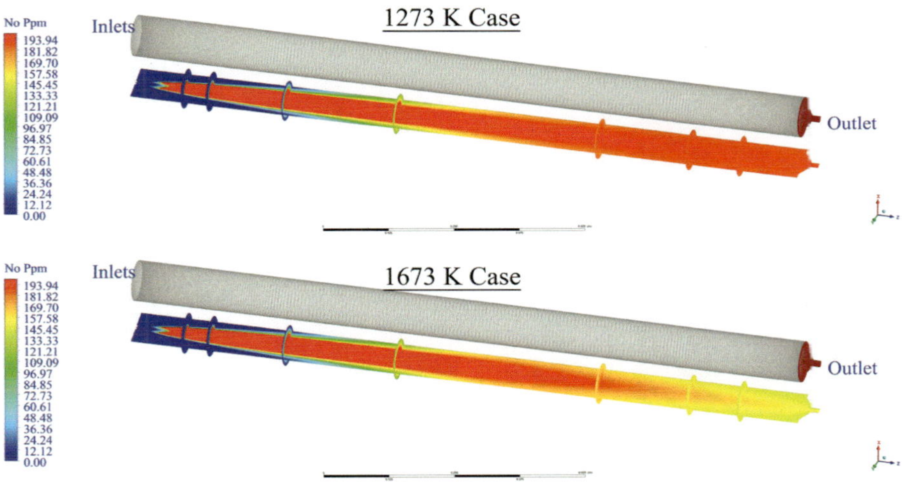

FIGURE 7.15 NOx emissions contours for 70HC30PS fuel sample at 1273 K and 1673 K furnace temperature.

intention of producing char and volatiles only. No oxidation is expected as demonstrated. The secondary carrier gas is modelled as nitrogen at a temperature of 1673 K. The furnace wall temperature is modelled at a temperature of 1,673 K, whilst the exit point is located 520 mm from the furnace injection point. As expected, the devolatilisation process progresses at different rates resulting in different char and volatile yields for the various fuel blends.

Trajectory profiles focus on the analysis of parameters related to the mass conservation equation such as particle position, particle residence time, particle velocity,

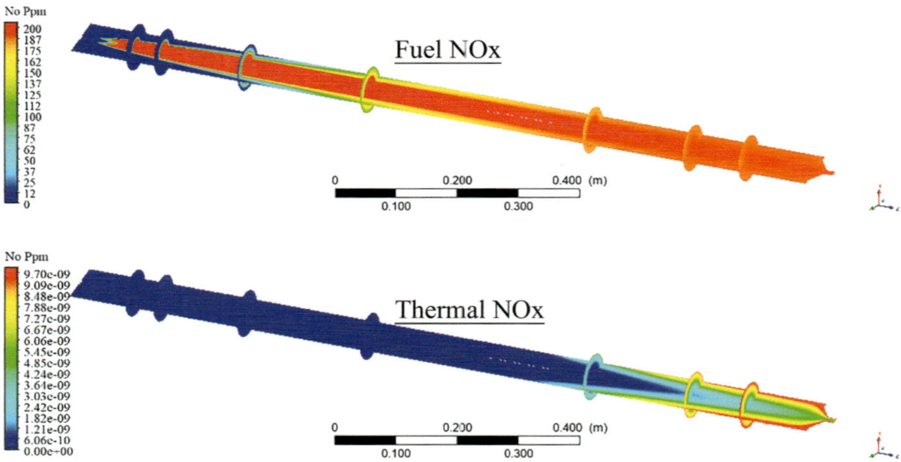

FIGURE 7.16 Fuel and thermal NOx emissions contours for 70HC30PS fuel sample at 1273 K furnace temperature.

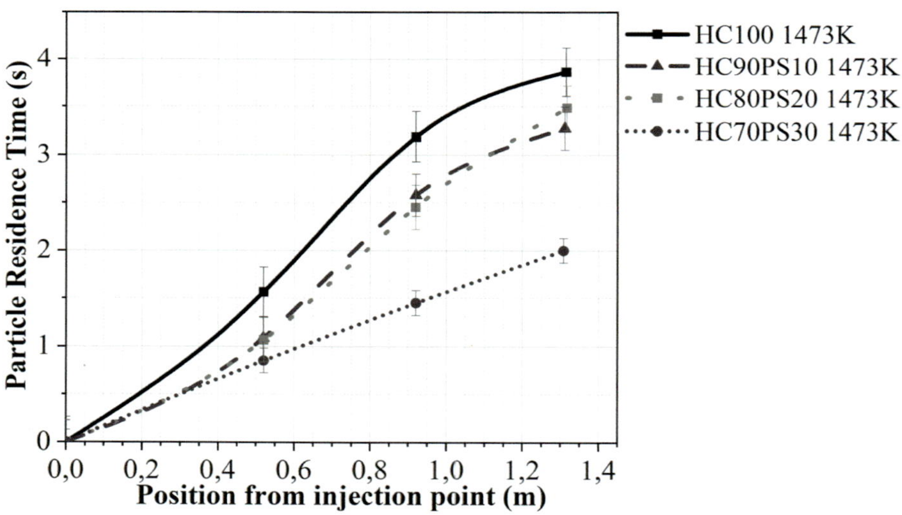

FIGURE 7.17 DTF stage 1 variation of particle residence time with furnace axial distance for various blending ratios.

as well as fluid velocity. The variation in particle residence time with position is represented in Figure 7.17 for the various fuel blends up to 520 mm which corresponds to the experimental probing position. As blending increases from 90HC10PS to 70HC30PS, the final residence time decreases by 24.62%, meaning the 70HC30PS fuel sample travels faster through the furnace. It is of interest to note that the final residence time first increases when passing from an unblended fuel sample of 100HC

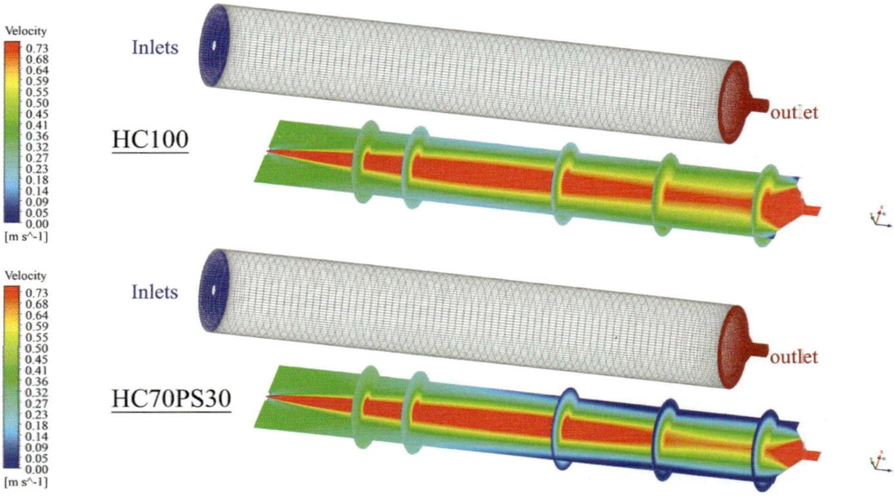

FIGURE 7.18 DTF stage 1 contours of velocity for 100 HC and 70HC30PS fuel samples.

to 90HC10PS before it starts decreasing as blending increases. Since particle velocity is affected by both particle mass and drag, continued blending with biomass beyond 20% by mass tends to decrease the overall particle mass whilst increasing the drag.

Eventually, the overall decrease in final residence time can thus be linked to the fact that lighter particles under high drag, as in the case of 70HC30PS, can be carried with ease through the furnace unlike heavier particles. Katarzyna [25] correlated these findings whilst investigating the combustion of biomass only inside a furnace originally purposed for coal combustion. The difference in particle shape and orientation during coal vs biomass combustion was studied in detail thus giving insights on why residence times for biomass combustion are relatively shorter.

Figure 7.18 shows the fluid velocity contours, whilst Figure 7.19 is a close up of vectors of velocity. Figure 7.18 shows a 3D mesh of the furnace with the corresponding 2D contours along the furnace axis and radial direction. The 2D contours are translated outside the mesh for better visualisation. This approach of translating the 2D contours outside the 3D mesh is extended throughout this study.

Flow develops axially along the furnace axis and radially across the furnace in a gradual manner which supports the experimental procedure suggestions of operating within laminar flow. The high velocities are concentrated along the centreline with an expected minimum value at the furnace walls for all the cases. High velocities along the centreline have a forward direction. Combustion principles prefer early development of flow with recirculation near the burner region which guarantees turbulence [17]. However, the DTF setup needs to function within laminar conditions thus the low tendency to experience recirculation as demonstrated in Figure 7.19.

During stage 2 the co-combustion model is used to monitor the particle residence time and fluid velocity during the combustion of fuel blends of interest (100HC, 90HC10PS, 80HC20PS, and 70HC30PS). Figure 7.20 demonstrates the variation of

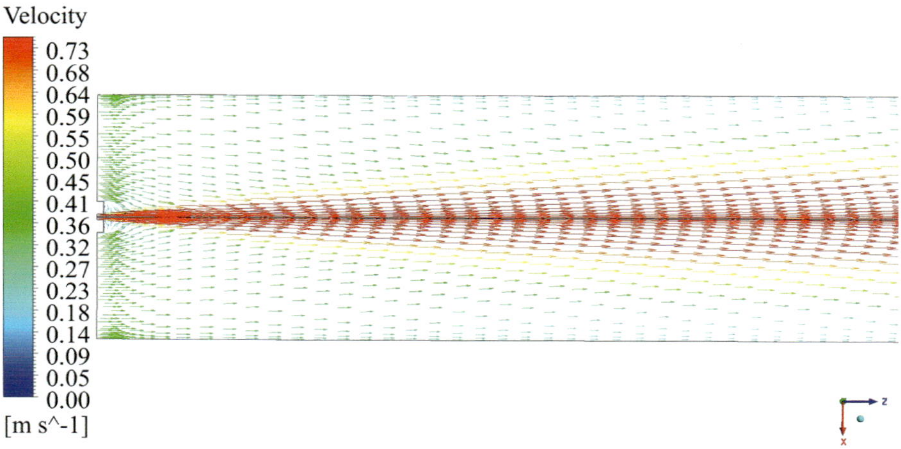

FIGURE 7.19 DTF stage 1 close view of vectors of velocity near injection for 100 HC fuel sample.

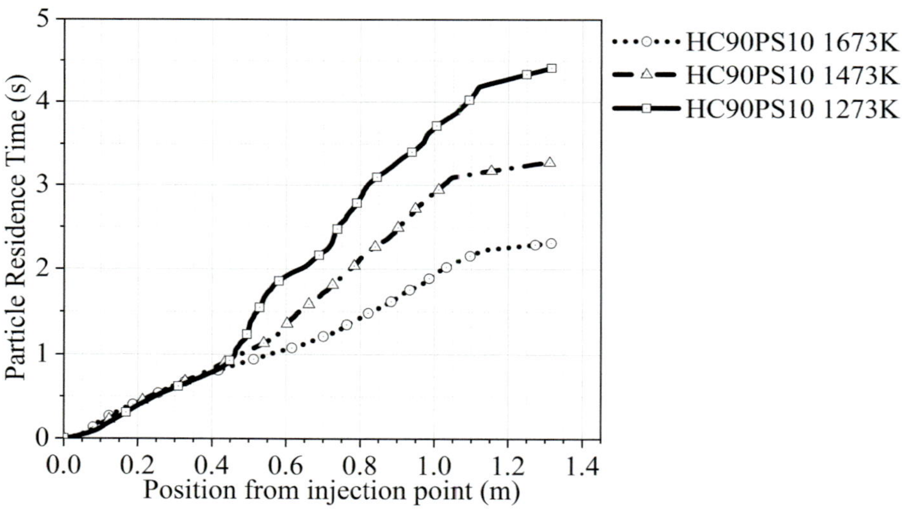

FIGURE 7.20 DTF stage 2 effect of heating rate on predicted residence times of 90HC10PS fuel sample.

residence time with respect to axial position for the 90HC10PS fuel sample as the heating rate increases due to increase of furnace wall temperatures.

An increase in heating rate due to furnace wall temperature increase results in a decrease of particle residence times. The 90HC10PS fuel sample demonstrates a decrease of 47.8% from 4.42 s to 2.31 s as the heating rate due to furnace wall temperature increases from 1,273 K to 1,673 K. Almost similar trends are experienced for the 100HC, 80HC10PS, and 70HC30PS fuel samples of 38.10%, 46.40%, and

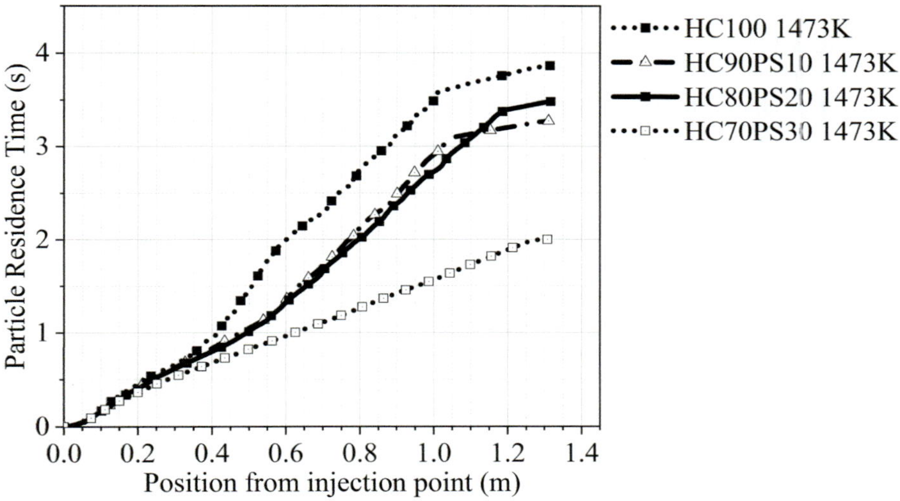

FIGURE 7.21 DTF stage 2 effect of blending ratio on particle residence time at a furnace temperature of 1473K.

38.10% respectively. Unfortunately, the heating rate due to furnace wall temperature increase has not been employed as an optimising tool for combustion particle residence time directly but rather monitoring the consequences of residence time such as burnout percentage.

Figure 7.21 highlights the effect of blending on residence time for a furnace wall temperature of 1,473 K. Figure 7.22 then summarises the percentage variation in

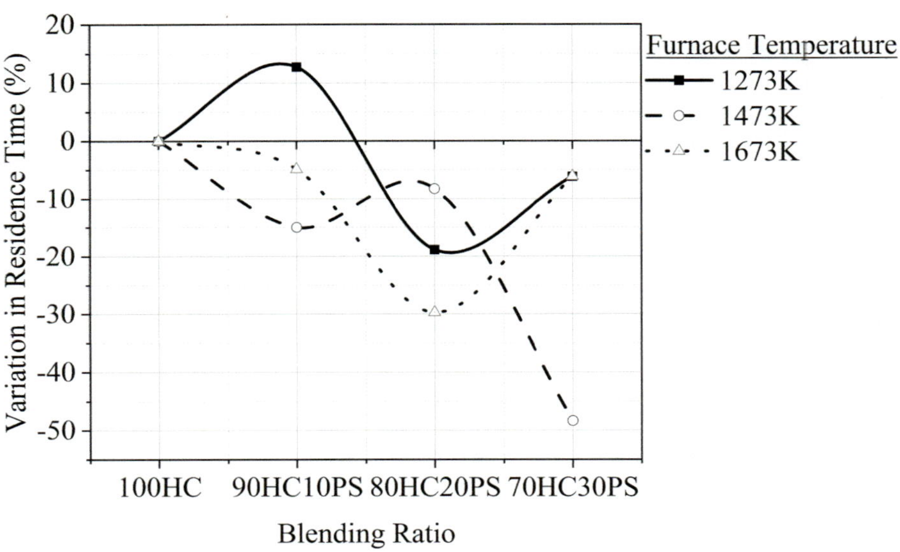

FIGURE 7.22 DTF stage 2 variation of final particle residence time with blending ratio.

residence time with respect to the unblended case for the different blending ratios at various furnace wall temperatures. At a 1,473 K set furnace wall temperature, blending results in an overall decrease of particle residence time from 3.81 s to 1.97 s which is shared for the other furnace wall temperatures though not to the same substantial degree. Further analysis of blending ratio on residence time with respect to furnace temperature demonstrates a localised increase at a furnace temperature of 1,273 K.

Biomass char as employed in the CFD model has a shape factor of 0.83 which assumes a cylindrical shape, a view shared by some researchers [26]. An increase in particle drag means the biomass particle should be entrained easily by the bulk fluid flow thus reducing the residence time. However, since biomass char is modelled as having a lighter weight compared to coal char, the gravitational force acting on a biomass particle is expected to be less which should result in a smaller terminal velocity thus increased residence time.

As blending increases from 100HC to 90HC10PS for the 1273 K furnace temperature case, particle residence time increases from 3.92 s to 4.42 s highlighting the increased influence of particle mass (biomass particles are lighter thus result in reduced terminal velocity due to particle mass). Continued increase in blending results in a reduction of residence time to 3.68 s for the 70HC30PS case which is linked to the increase in terminal velocity caused by particle drag. This contradictory effect becomes less pronounced as the furnace temperature increases to 1,473 K and eventually 1,643 K. Since most of the final particle residence times are relatively below 4.00 s for all cases, a variation of 10% which equates to 0.40 s is relatively difficult to monitor experimentally but easy through CFD.

An example of a particle track is illustrated in Figure 7.23 as a function of residence time and temperature for the 90HC10PS fuel sample at 1,273 K furnace temperature. The illustrated particle track is helpful in explaining the alluded notion that residence time determination requires a three-dimensional domain approach. By using a three-dimensional domain approach, particle-to-particle and particle-to-fluid interactions are captured sufficiently. This is further demonstrated in Figure 7.24 that shows particle velocity magnitudes on axial and radial planes located at 0.1, 0.3, 0.52, 0.92, and 1.32 m from the injection point. The particle flow field tends to diverge from the injection point with the highest velocity magnitudes being concentrated along the centreline.

A fully developed flow is only achieved about 0.52 m from the injection point. It is also of interest to note that flow develops in almost a similar manner for all the cases that are investigated. All cases initially attain similar particle velocities regardless of blending or furnace temperatures, hence similar flow fields.

7.6.3 Effect of Blending on Species Profiles in a Drop Tube Furnace

The species mole and mass fractions are used to analyse parameters related to chemical reactions during the co-combustion process. Nitrogen is supplied in excess using the secondary carrier gas inlet. As the devolatilisation process occurs, the volatile

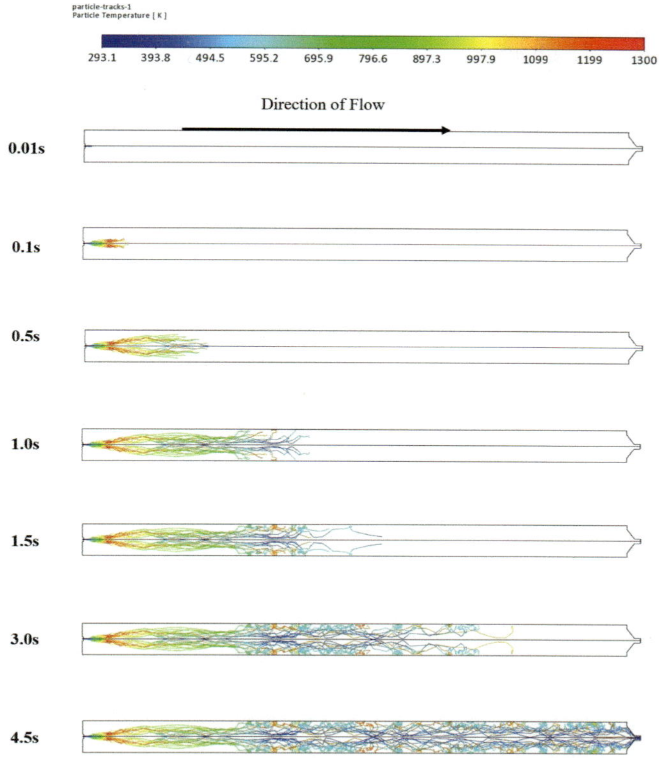

FIGURE 7.23 DTF stage 2 particle track as a function of particle temperature for a 90HC10PS fuel sample at 1273 K furnace temperature.

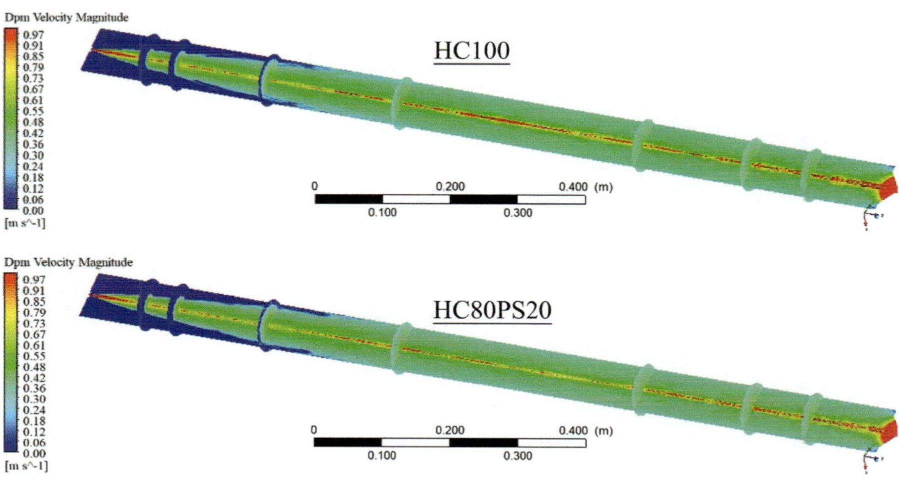

FIGURE 7.24 DTF stage 2 axial and radial contours of particle velocity magnitudes for (a) 100HC and (b) 80HC20PS fuel samples at 1273K furnace temperature.

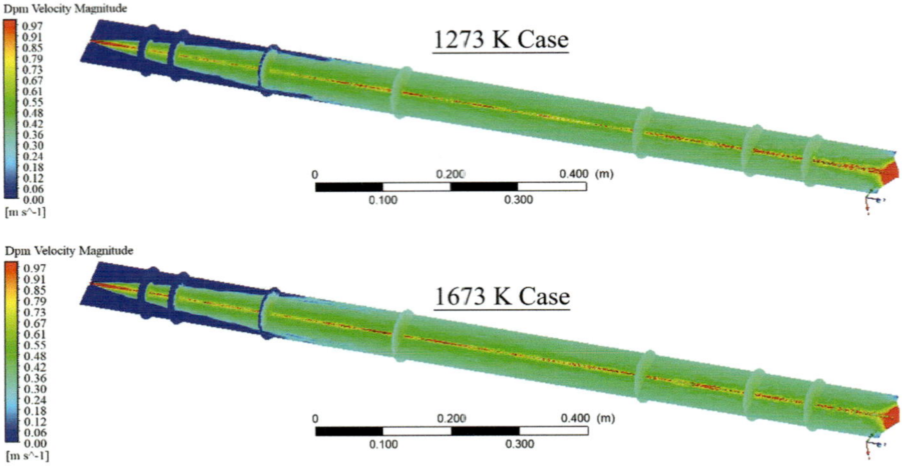

FIGURE 7.25 DTF stage 2 particle velocity magnitudes on axial and radial planes for 80HC20PS at (a) 1273K and (b) 1673K furnace temperature.

molar fraction for each species increases whilst the nitrogen decreases, as illustrated in Figure 7.25.

The final volatile molar fraction increases from 0.01166 for the 100HC case to 0.01338 for the 70HC30PS case, reflecting increased release for highly blended fuels. The high gradients synonymous with the 70HC30PS and 80HC20PS cases closely mirror the plot for the rate of volatile release, which reflects high combustion intensity indirectly [27]. The release of volatiles attains a maximum at 0.043 m from the injection point for the 70HC30PS case compared to 0.103 m for the 100HC case. Biomass flames are formed closer to the burner mainly because volatiles are released earlier compared to coal combustion. This further helps to understand why blended fuel samples tend to release volatiles early on and closer to the burner head [25,28].

Analysing Figure 7.26 that shows volatile molar fraction together with Figure 7.27 that shows contours of Nitrogen on axial and radial planes highlights how areas of low nitrogen correspond to high volatile regions. Figure 7.28 however demonstrates an interesting aspect with regard to the actual rate of devolatilisation. The regions corresponding to high devolatilisation zones are distributed radially outwards rather than axially for the 100HC case. This is due to the flow field development, as already shown by the vectors of velocity in Figure 7.19 which seem to develop more radially for the 100HC case. Hence, it can be expected that the flame becomes narrower as blending increases for laminar combustion regimes, as also demonstrated by Mohanna [29] during the study of biomass flames.

One of the main causes of global warming that has led to numerous extreme weather events is carbon dioxide. Since CO_2 is a byproduct of the oxidation of carbon species, thermal power plants typically produce considerable amounts of it [30]. Figure 7.29 shows the CO species variation along the axial centreline for all fuel blends at 1,273 K in the DTF furnace. At a distance of 0.094 m from the injection

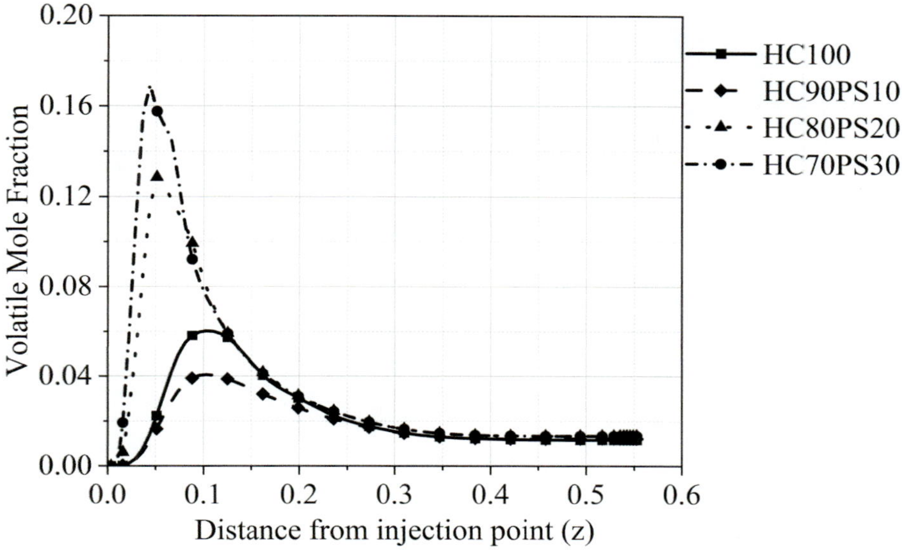

FIGURE 7.26 DTF stage 1 variation of volatile molar fraction with furnace axial distance for various blending ratios.

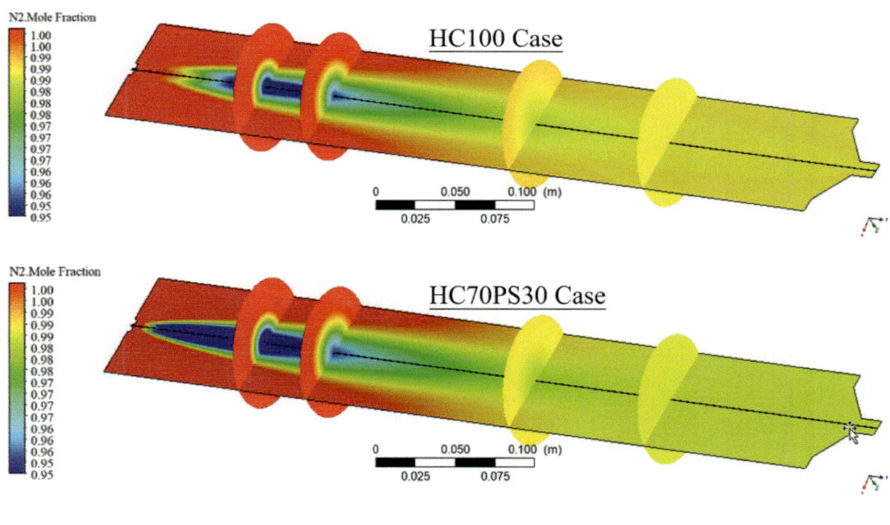

FIGURE 7.27 DTF stage 1 contours of nitrogen molar fraction for 100 HC and 70HC30PS fuel samples.

point, CO molar fraction reaches maximum values, which leaves enough time for CO to oxidise to CO_2. Blending did not cause notable changes in the variation of the CO molar fraction.

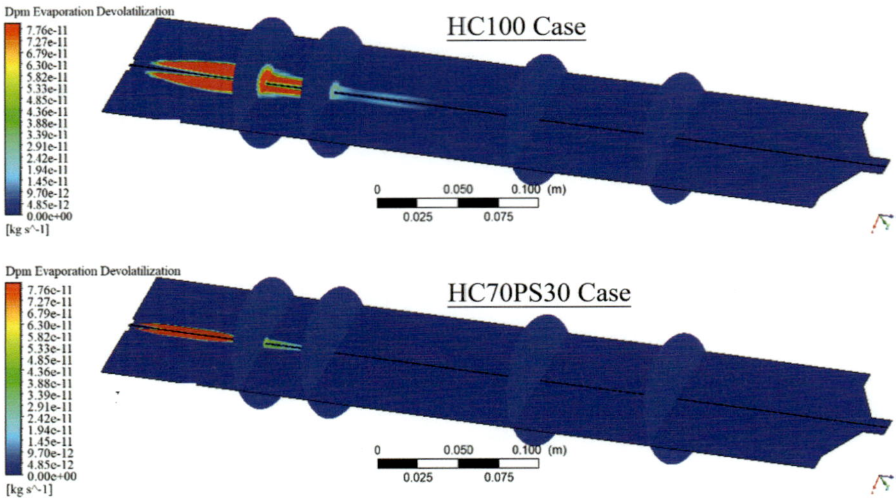

FIGURE 7.28 DTF stage 1 contours of devolatilisation rate for 100 HC and 70HC30PS fuel samples.

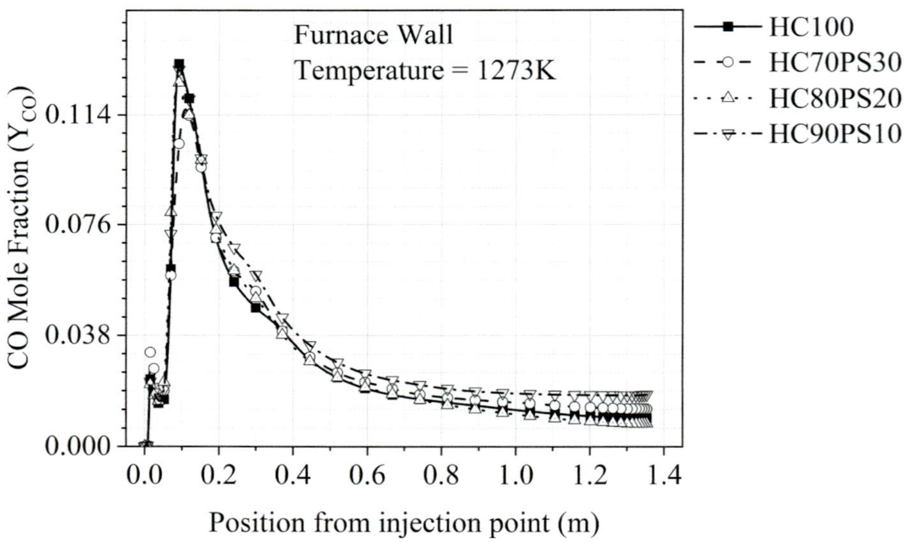

FIGURE 7.29 Axial variation in CO mole fraction for 100HC, 90HC10PS, 80HC20PS, and 70HC30PS at 1273K furnace temperature.

The initial molar fractions of CO_2 and CO in the DTF furnace are 0% because the carrier gas that transports the fuel is 100% N_2. CO starts to form as soon as the fuel from the primary inlet combines with the O_2 from the secondary inlet. It is clear that the initial interaction of fuel with O_2 to create CO predominates in this area. It's

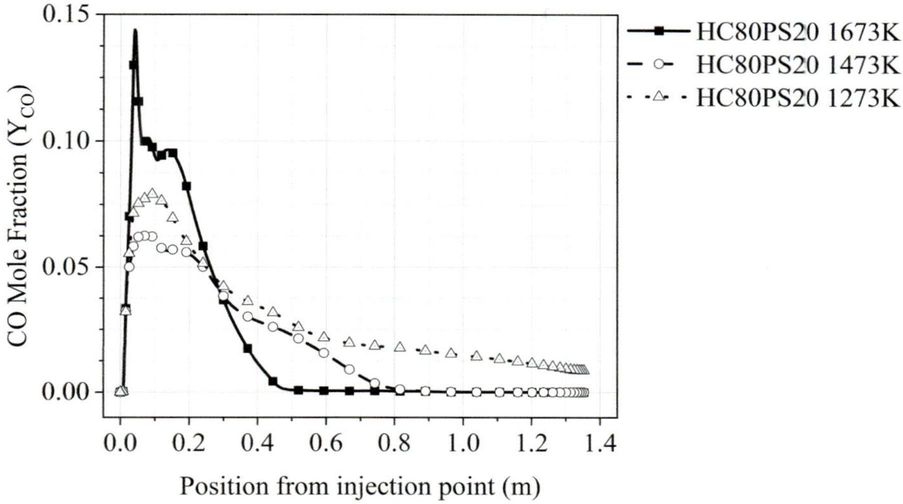

FIGURE 7.30 Effect of furnace temperature on carbon monoxide mole fraction for 80HC20PS fuel.

also critical to understand that as soon as CO is present, it starts to oxidise with O_2 to generate CO_2. Theoretically, the reaction between char and O_2 happens first because it occurs in the boundary layer and has a lower activation energy than the reaction between CO and O_2. Further from the injection point, CO molar fraction begins to steadily decrease as the CO_2 molar fraction increases, which is attributed to the reaction of CO with O_2.

Furnace temperature had a bearing on the species distribution as demonstrated by the plots of 80HC20PS in Figure 7.30. Elfasakhany et al. [31] illustrated that during pulverised wood combustion, the ratio between CO and CO_2 is highly dependent on furnace temperature rather than char composition. Higher furnace temperatures promote the formation of CO early on in the furnace, as demonstrated by the peak increase from 0.06 to 0.15 when furnace temperature increased from 1,273K to 1,473K. High furnace temperatures aid in the oxidation of CO to the point where, at a furnace temperature of 1,673K, all of the CO was consumed about 0.5 metres from the injection point. More residence time is required to consume the CO that had generated at lower furnace temperatures. However, because of the way the flow forms, all examples show a maximum CO molar fraction around the same distance from the injection point.

The effect of furnace temperature on CO_2 axial distribution is closely linked to the CO profiles. As CO molar fractions decrease, the opposite happens to CO_2 as shown in Figure 7.31. A CO_2 peak is even visible in the 1,673 K furnace temperature instance at a distance of about 0.5 m from the injection point, which is the point at which all of the CO is exhausted. If sufficient residence time is provided, the final CO_2 should reach a comparable value for the same fuel combination. Additionally,

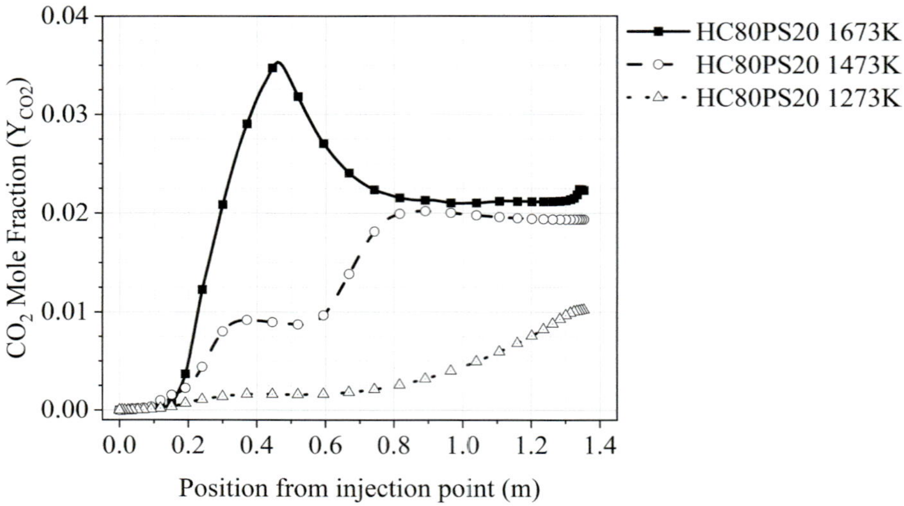

FIGURE 7.31 Effect of furnace temperature on carbon dioxide mole fraction for 80HC20PS fuel.

high temperatures encourage the use of oxygen, which is why the values are low at high temperatures.

Conservation of mass, when applied to chemical reactions, stipulates that the mass of an element must be conserved before and after a reaction. During stage 2, char is predicted to be oxidised to CO_2 in a two-step reaction process with CO acting as the intermediate species. CFD allows monitoring of these reactions against other useful parameters such as furnace temperature and kinetic reaction variables. Since the oxidant is supplied in excess through the secondary inlet, complete combustion is expected for all the cases under consideration.

Figure 7.32 highlights detailed plots on the variation of N_2 species along the axial centreline for all fuel blends at a furnace temperature of 1,273K. Figure 7.33 then shows the contours of the N_2 molar fraction along the axial plane of the DTF for one of the fuel blends cases, 80HC20PS.

The primary inlet (Fuel + N_2) coincides with the centreline, whilst the secondary carrier gas (N_2 + O_2) is introduced concentrically. With the reactions occurring within small residence times, blending seems to not cause any variation of the N_2 molar fraction as the minimum molar fraction occurs around 0.09450 m from the injection point for all cases. However, upon closer examination, it can be seen that the N_2 molar fraction for the 100HC case attains a minimum of 0.866 whilst the 80HC20PS case attains a minimum of 0.883. The 100HC case results in the formation of more CO_2 and CO due to a higher value of char content, thus localised reduction in the N_2 content.

Figure 7.34 highlights detailed plots on the variation of O_2 molar fraction along the axial centreline for all fuel blends at a furnace temperature of 1,273K. Figure

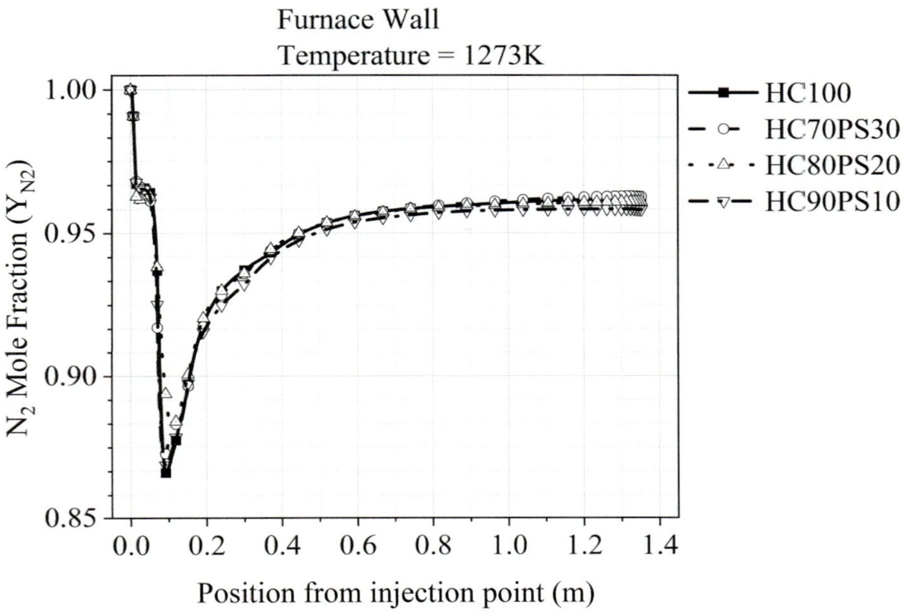

FIGURE 7.32 DTF stage 2 axial variation in nitrogen mole fraction for 100HC, 90HC10PS, 80HC20PS, and 70HC30PS at 1273K furnace temperature.

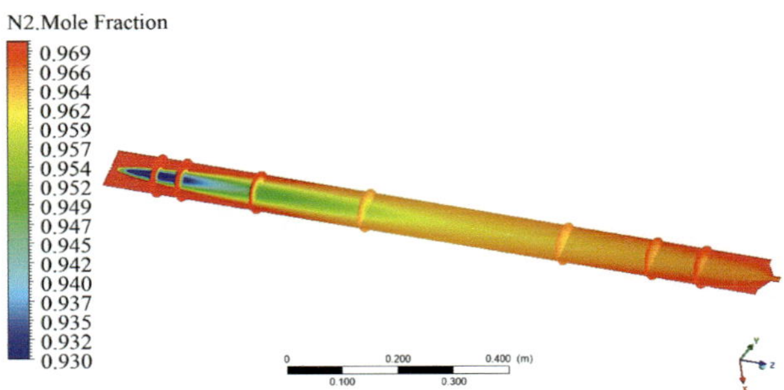

FIGURE 7.33 DTF stage 2 contours of nitrogen mole fractions for 80HC20PS fuel at 1273K furnace temperature.

7.35 then shows contours of O_2 molar fraction on the axial plane for the 80HC20PS case. O_2 is depleted within the initial reaction zone, which corresponds to the dominant formation of CO. Fortunately, since O_2 is supplied in excess, the molar fraction rises steadily along the furnace until reaching steady values at the furnace exit. The final O_2 molar fraction decreases from 0.225 to 0.174 with an increase in blending from 90HC10PS to 70HC30PS.

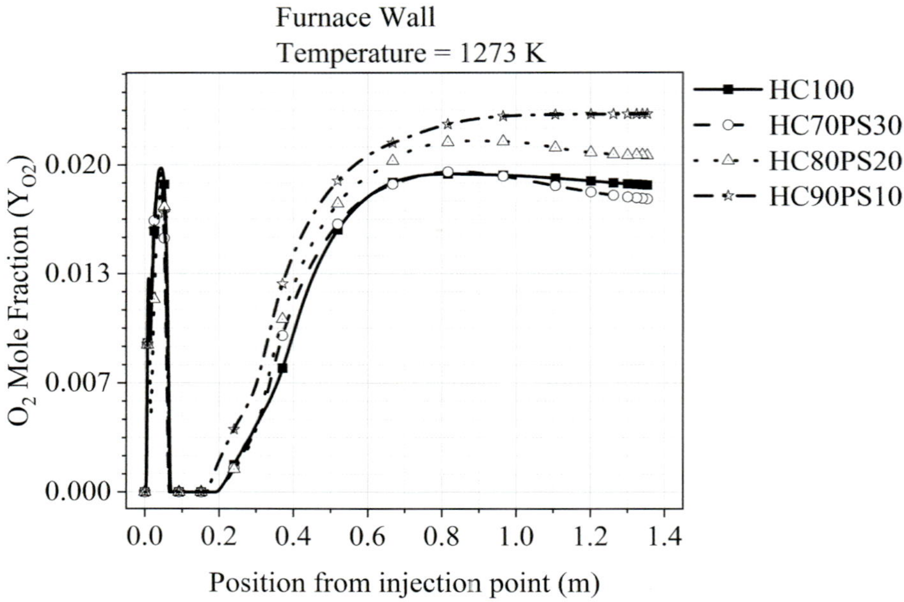

FIGURE 7.34 DTF stage 2 axial variation in oxygen mole fraction for 100HC, 90HC10PS, 80HC20PS, and 70HC30PS at 1,273K furnace temperature.

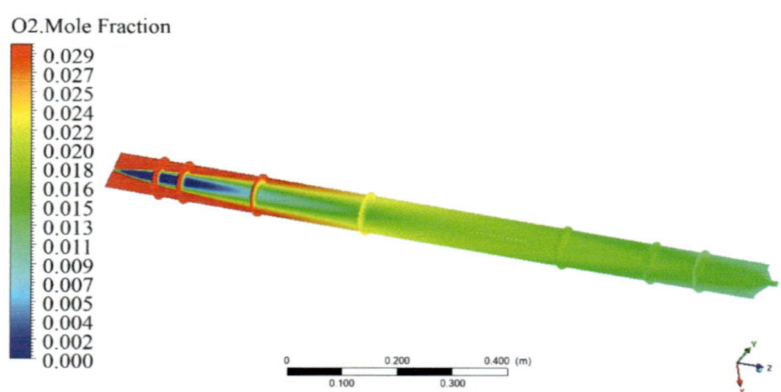

FIGURE 7.35 DTF stage 2 contours of oxygen mole fractions for 80HC20PS fuel at 1273K furnace temperature.

Since the oxidation of char and CO is promoted by high temperatures, the molar fraction of O_2 remains at a minimum for a longer distance when considering the 1,673 K case, as shown in Figure 7.36. In addition, high temperatures promoted the consumption of O_2, hence the lower trend with respect to higher temperatures. Since the carrier gas is supplied in excess, the final molar fraction of N_2 does not vary significantly with temperature, as demonstrated in Figure 7.37.

A radial variation of species molar fractions is also experienced within the DTF furnace as supported by the assertion that flow occurs in a three-dimensional manner. The particles are shown to move radially and axially. Radial information is very

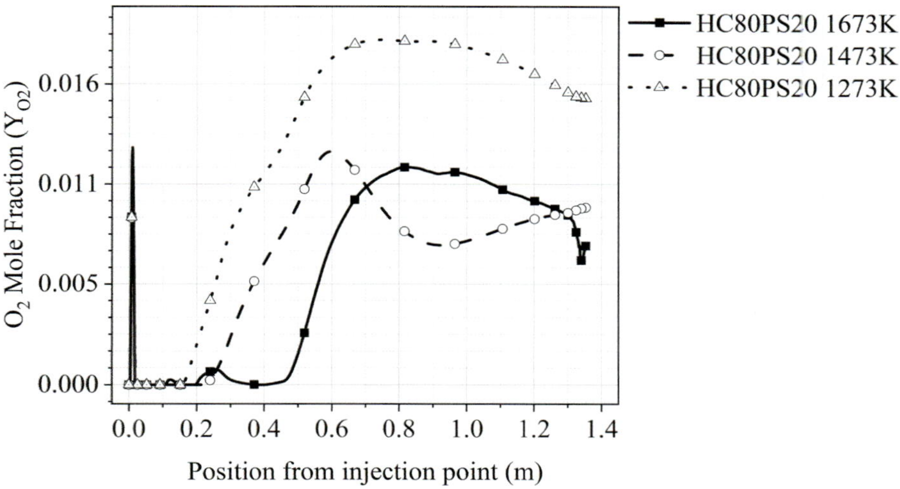

FIGURE 7.36 DTF stage 2 effect of furnace temperature on oxygen mole fraction for 80HC20PS fuel.

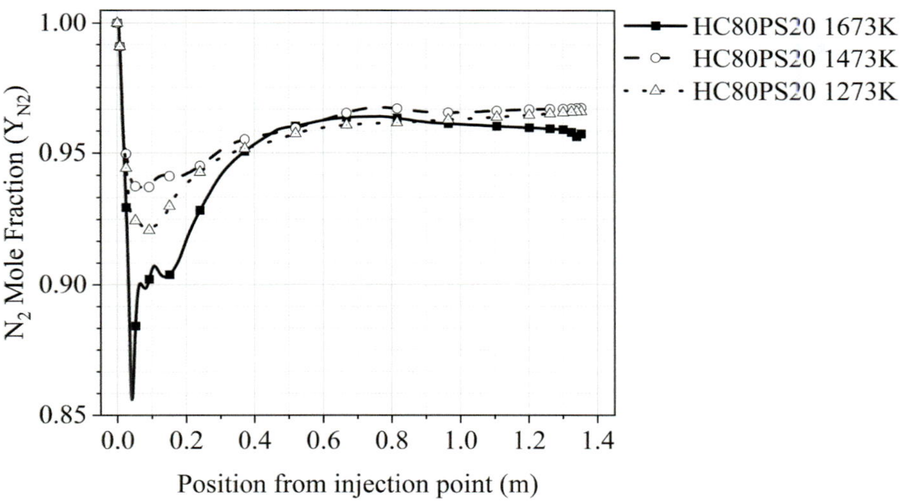

FIGURE 7.37 DTF stage 2 effect of furnace temperature on nitrogen mole fraction for 80HC20PS fuel.

important in monitoring turbulence or swirl, an indispensable parameter as far as combustion science is concerned.

Figures 7.38 to 7.40 highlight plots of the radial variation in species molar fractions for the 100HC case at 0.1, 0.3, 0.52, and 0.92 m distances from the injection point. Near the injection point, at an axial distance of 0.1 m, O_2 molar fraction attains a maximum value of 0.03 towards the walls and a minimum value near the centreline.

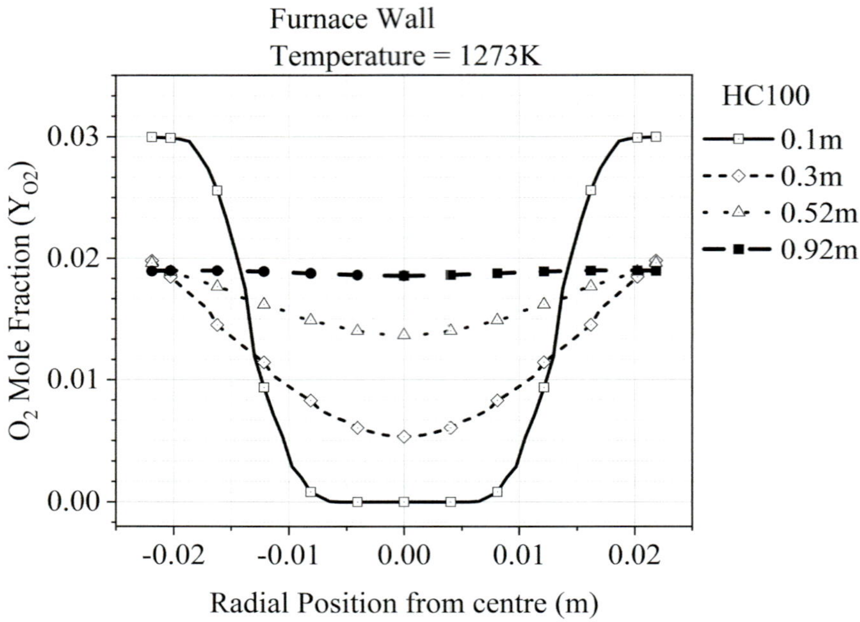

FIGURE 7.38 DTF stage 2 radial plots of oxygen mole fraction of 100HC at different axial positions for 1273K furnace temperature.

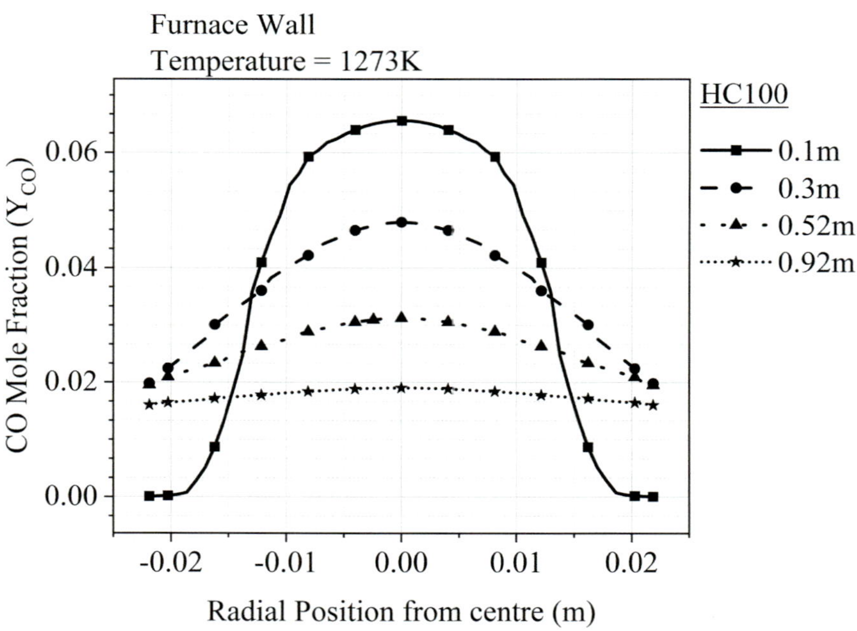

FIGURE 7.39 DTF stage 2 radial plots of carbon monoxide mole fraction of 100HC at different axial positions for 1273K furnace temperature.

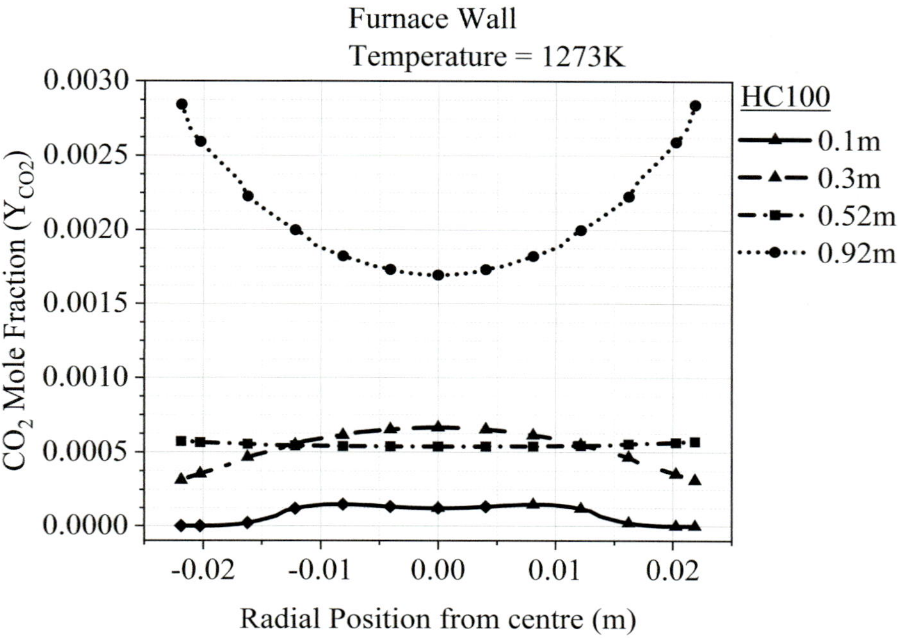

FIGURE 7.40 DTF stage 2 radial plots of carbon dioxide mole fraction of 100HC at different axial positions for 1273K furnace temperature.

These high gradients between the walls and centreline species molar fractions are more pronounced near the injection point. As the axial distance increased from the injection point, the high gradient zones between wall and centreline species molar fractions began to redress until attaining a uniform radial distribution towards the exit.

The effect of blending on radial species profiles is demonstrated in Figures 7.41 and 7.42 at a furnace temperature of 1,273 K for the 100HC and 80HC20PS cases. The chemical reaction mechanism responsible for the formation of CO from char is concentrated near the injection point for all cases, as shown in Figures 7.43 and 7.44. All cases are able to surpass a kinetic rate of 100 mol/m³·s with respect to the formation of CO from char. In comparison, the kinetic rate for the formation of CO_2 from Carbon Monoxide is relatively lower at low furnace temperatures. This means that the rate at which CO is formed from char is quite higher than the rate at which CO_2 is formed from CO. As such, CO_2 formation is deduced to require a larger residence time than CO formation when operating at low temperatures. Increase in furnace temperature is able to increase the average kinetic rate for the formation of CO_2.

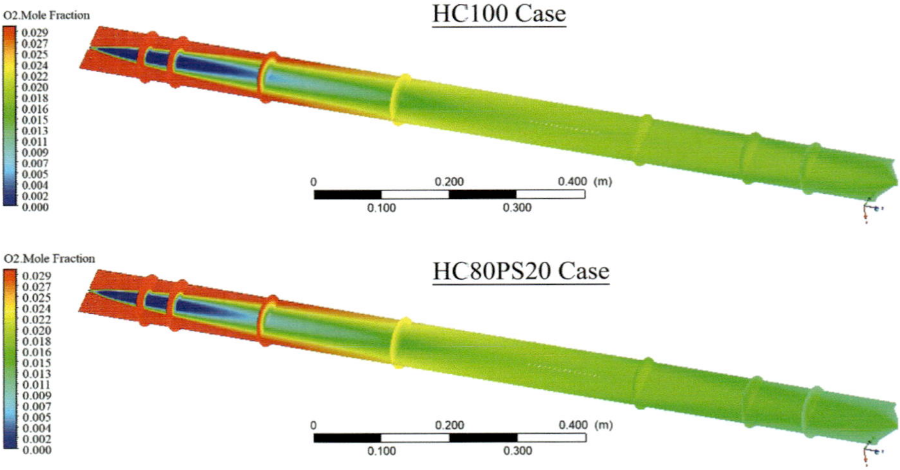

FIGURE 7.41 DTF stage 2 oxygen mole fraction of 100HC and 80HC20PS at 0.1m, 0.15m, 0.3m, 0.52m, 0.92m, and 1.32m axial positions for 1273K furnace temperature.

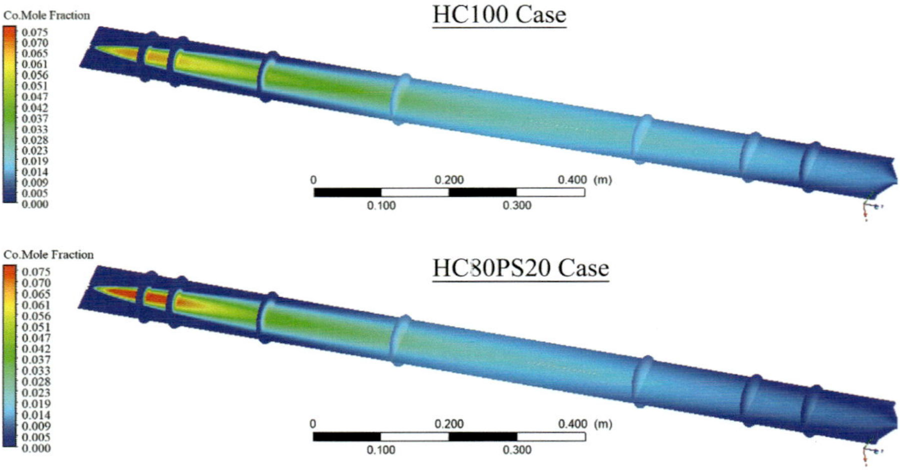

FIGURE 7.42 DTF stage 2 CO mole fraction of 100HC and 80HC20PS at 0.1m, 0.15m, 0.3m, 0.52m, 0.92m, and 1.32m axial positions for 1273K furnace temperature.

7.6.4 Effect of Pinus Sawdust on Temperature Profiles

Due to the release of heat during the combustion of char, parameters related to the conservation of energy needed to be monitored as well. These parameters included particle temperature, sensible heat of reaction, absolute heat of reaction, ignition, as well as burnout properties of the different char samples. The fuels of interest (100HC, 90HC10PS, 80HC20PS, and 70HC30PS) were thus investigated using the

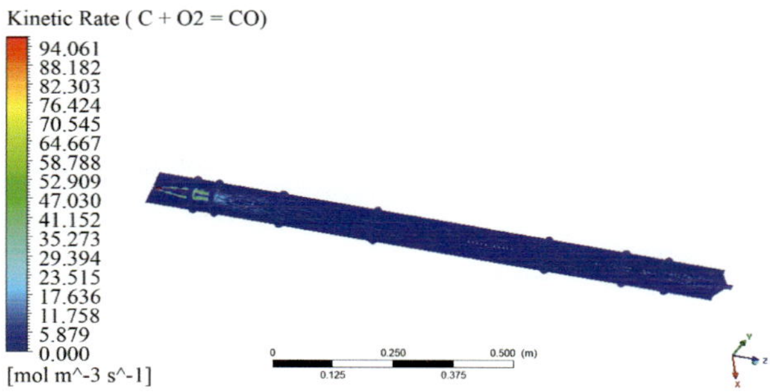

FIGURE 7.43 DTF stage 2 kinetic reactions rates for 80HC20PS fuel sample at 1673K furnace temperature.

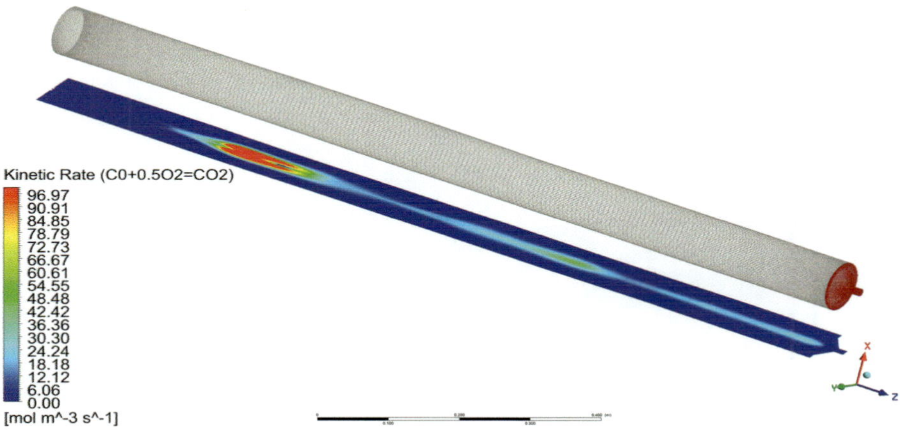

FIGURE 7.44 DTF stage 2 kinetic reactions rates for 80HC20PS fuel sample at 1673K furnace temperature.

co-combustion model that was developed. The heat released during char oxidation was modelled as a two-step reaction by assuming formation of CO within the boundary layer and its subsequent oxidation to CO_2 in the bulk flow. Since coal char had a higher combustible fraction than char from sawdust, more heat was expected from a pure sample as compared to its blends.

Figure 7.45 shows particle temperature of the 100HC case as well as the 70HC30PS case when combusted in a furnace with a temperature of 1,473 K. Blending did not result in any significant variation of particle temperatures along the axis or radially. One of the main characteristics that influence solid fuel combustion is surface area. The particle sizes that were employed were relatively small to warrant influence due to the size difference. Mandø et al. [26] suggested that pulverised samples adopt a quasi-similar spherical shape during combustion unlike large particles, which was demonstrated in this study.

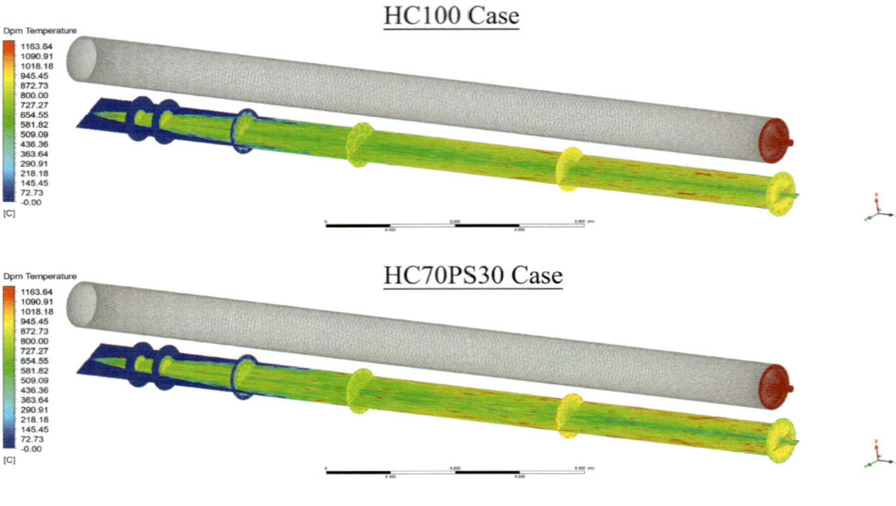

FIGURE 7.45 DTF stage 2 particle temperature on axial and radial planes for 100 HC and 70HC30PS fuel samples at 1473 K furnace temperature.

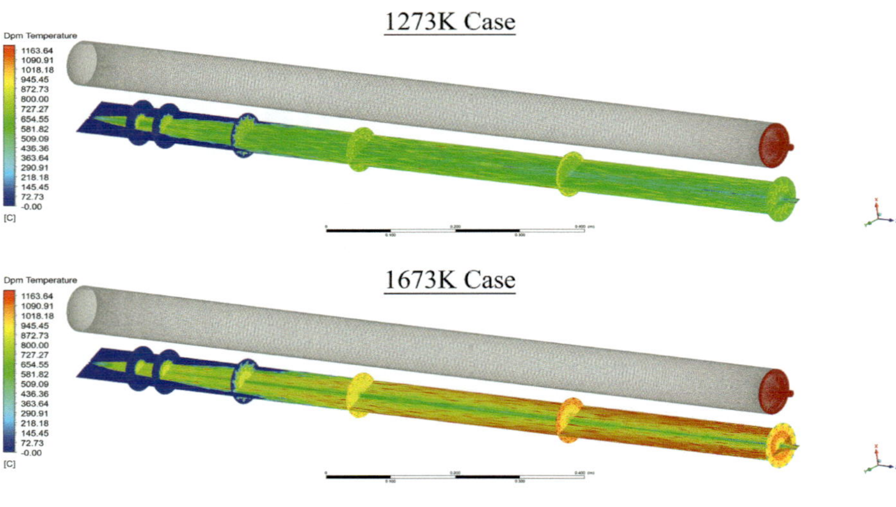

FIGURE 7.46 DTF stage 2 particle temperature against axial particle position for 70HC30PS at 1273K and 1673K.

Blending was done on a mass basis; thus, the fuel flow rates that were employed were constant regardless of fuel blending. With the small flow rates that were employed in this study, furnace wall temperatures rather than combustion reactions were able to influence particle temperature, as demonstrated in Figure 7.46. The figure demonstrates the influence of 1673 K and 1273 K furnace temperatures on char combustion. As the flow developed in an almost similar fashion, the particles under

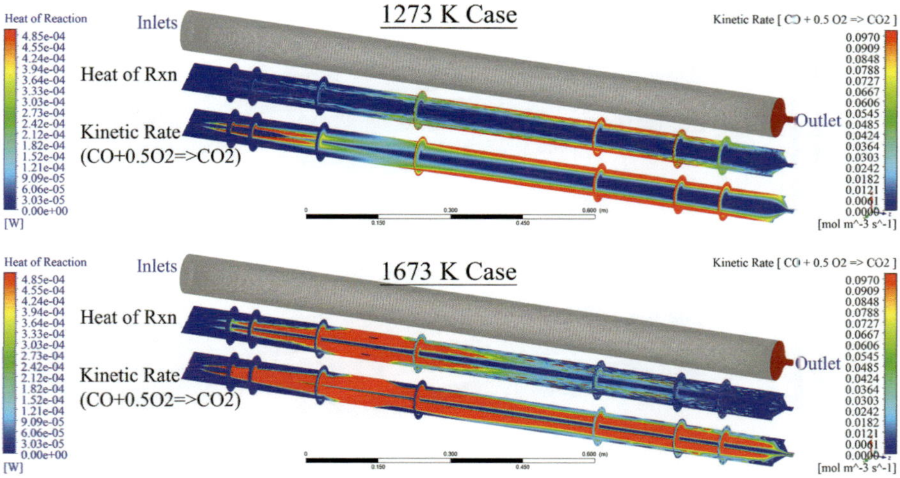

FIGURE 7.47 DTF stage 2 heat of reaction contours and kinetic reaction rate contours for 70HC30PS fuel at 1273K and 1673K.

a high furnace temperature were able to attain a higher temperature along the axial direction. Correlating with species molar fractions of CO_2, it is evident that particle temperature plays an important role in aiding the formation of CO_2 as higher temperatures promote the formation of CO_2.

With respect to the heat released, Figure 7.47 shows heat of reaction contours along the axis and radially for the 70HC30PS case as a function of furnace temperature. Since the oxidation of char is treated as a two-step reaction, the contours of kinetic reaction rates were also superimposed on the figures for better analysis. Since the setting up of the co-combustion model had to resemble the experimental conditions, which required an artificial atmosphere containing 97% N_2 and 3% O_2 by volume, Equation (7.27) was employed instead after assuming a 100% excess O_2 scenario. This was deduced through calculating the average experimental O_2 to fuel ratio for the various fuel blends.

$$C + 0.5\left[O_2 + 32.333N_2\right] \rightarrow CO + 16.167N_2 \tag{7.26}$$

$$CO + 1\left[O_2 + 32.333N_2\right] \rightarrow CO_2 + 32.333N_2 + 0.5O_2 \tag{7.27}$$

The high N_2 content in the products of combustion due to the artificial atmosphere resulted in reduced heat of reaction for all the cases. This can be attributed to the fact that N_2 has a higher specific heat capacity than O_2. This allows N_2 to absorb more heat than O_2 just to raise the temperature by a unit temperature [17]. By correlating with Figure 7.48 which shows particle burnout, it was quite clear that most of the heat was not released during oxidation of char but rather of CO. Initial heat released during char oxidation corresponded to the particle burnout profiles.

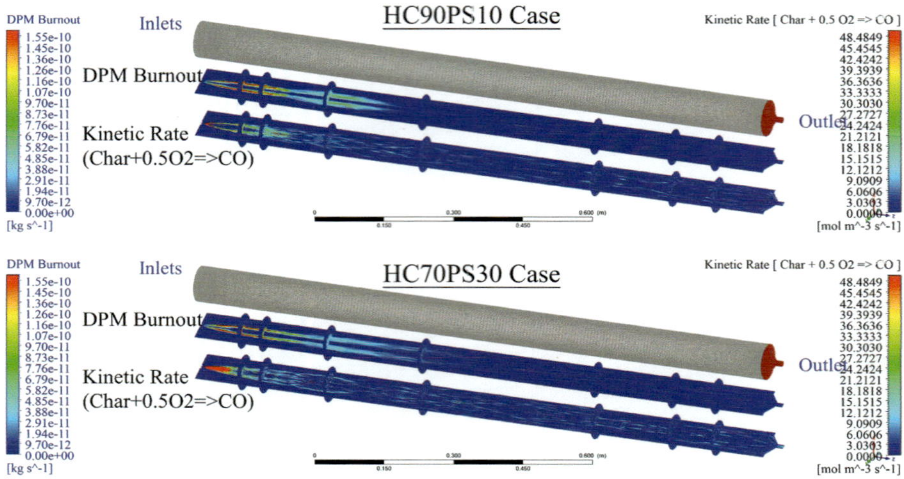

FIGURE 7.48 DTF stage 2 DPM burnout contours and kinetic reaction rate contours for 70HC30PS fuel at 1273K and 1673K.

With respect to the study of heat of reaction as a function of heating rate due to the set furnace wall temperature, a low furnace wall temperature resulted in a reduced heat of reaction, as shown in Figure 7.47. As expected, low temperatures hindered the oxidation of CO to form CO_2 thus delaying the reaction zone to an average location of 0.8 m from the injection point as compared to the initial 0.4 m obtained at high furnace temperatures. The nature of the flow development resulted in a reaction zone that tended to be concentrated towards the centreline for high temperatures and near walls for low temperatures. The sensible heat released during combustion increased with temperature, as shown in Figure 7.49. As already discussed, high furnace temperatures increased the subsequent kinetic reaction rate associated with char oxidation as with heat of reaction.

The effect of blending on heat of reaction is represented in Figure 7.47 with an illustration of axial and radial contours. Blending affected the heat of reaction by delaying the reaction zone as well as increasing the intensity of combustion within the zone.

Thermogravimetric experiments demonstrated an increase with respect to this index for all heating rates as blending increased, e.g., a heating rate of 20 K/min resulted in an increase from 17.0470×10^{-8} to 38.9440×10^{-8} as blending increased from 90HC10PS to 70HC30PS. Figure 7.48 supports these findings by further highlighting where these high combustion intensity zones are located within the furnace with respect to the injection point.

90HC10PS contours show a gradual release of heat directly linked to the gradual burnout of its char particle. In contrast, the 70HC30PS fuel sample showed more intensity within the reaction zone. Even though a higher combustible matter fraction was associated with the 90HC10PS fuel, the rapid release of CO and hence burnout

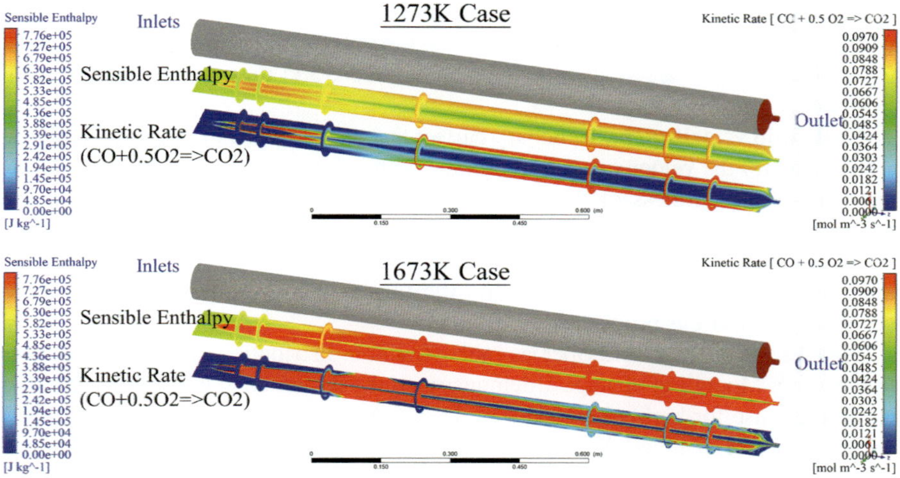

FIGURE 7.49 DTF stage 2 sensible heat contours and kinetic reaction rate contours for 70HC30PS fuel at 1273K and 1673K.

was delayed due to its lower porosity. A low porosity is associated with high-density particles, which reflects on the surface area available for reaction. This assertion is supported by other researchers such as Tang et al. [32] during their studies with demineralised coal, Di Blasi [33] during his pyrolysis of wood studies, as well as Sadhukhan et al. [34] who were particularly interested in large coal particles. The burnout contours showed dependence on the kinetic reaction rate of char with O_2 to form CO, as demonstrated in Figure 7.48. Blending increased DPM burnout maximum values marginally, though the reaction zones were located within the same region for all the cases.

7.7 SUMMARY

The co-combustion model was successfully employed to predict the combustion behaviour of various fuel blends of interest. The modelling process was split into two distinct stages to emulate the DTF experimental setup. The first step involved passing the fuel samples through an inert atmosphere containing N_2 only. This stage resembled the devolatilisation process a solid fuel undergoes just after its drying. Blending with PS resulted in more volatiles being released as mirrored by the proximate composition of the fuel blends. Co-combustion modelling was employed in the monitoring of various parameters that could not be measured experimentally.

Particle residence time: Blending resulted in a decrease in residence time in a down-fired furnace, mainly due to drag effects. Gravitational effects were higher for the coal particles due to their weight in comparison to PS particles. As such, a lower residence time would have been expected for blends with higher coal percentages. Drag effects were higher for PS particles due to their shape factor in comparison to coal particles. As such, a lower residence time would have been expected for blends with higher PS percentages. However, due to the pulverised nature of the fuel blends under study, drag effects became more pronounced, and thus the trends that were exhibited.

Bulk fluid velocity: The secondary carrier that was supplied in excess was mainly responsible for the velocity profiles obtained within the furnace. Flow quickly developed radially regardless of the blending ratio. Since this study tried to keep the flow laminar, no zones of recirculation were experienced near the injection point. Boundary layer effects were distributed uniformly regardless of the blending ratio, meaning the movement of gases within the furnace was mainly controlled by the secondary carrier gas properties.

Species molar fraction: Species molar fraction distribution varied with respect to blending. Modelling of stage 1 demonstrated the existence of a reaction zone that is more intense for highly blended samples since more volatiles are easily released within a shorter distance. The onset of devolatilisation is shown to also commence closer to the injection point for highly blended samples, suggesting combustion flames are sustained closer to the burner. As such, burner modification is also necessary for cases with high blends to avoid any tendency of flame flashback. Modelling of stage 2 demonstrated the existence of a reaction zone that is also dependent upon blending. The highly blended samples had a lower initial char mass fraction that led to lower mole fractions of CO and CO_2 at the exit of the furnace. High furnace temperatures promoted high gradients on the plots of species mole fractions against position from the injection point. Radial variation of species mole fractions at different furnace positions was closely linked to the manner in which the bulk fluid velocity developed as well.

Particle temperature variation: With regard to stage 2 modelling, blending did not result in any significant change of particle temperature as it progressed along the furnace due to the low fuel flow rates. Increase in furnace wall temperatures actually caused an increase in the particle temperatures, which was expected.

Reaction kinetics: The first stage demonstrated higher values for devolatilisation rates as blending increased. Highly blended fuel samples lost their volatiles quickly in a short period of time, unlike low blended fuel samples that lost their volatiles gradually over a longer period. This reflected on the increased combustion intensity that is experienced for highly blended fuels, thus there is the need to modify the burner appropriately to avoid any flashback as noted previously. The second stage demonstrated increased peak values of heat of reaction as blending increased. This was correlated to the reaction between carbon monoxide and oxygen which increased in intensity as blending increased. The other reaction of char with oxygen demonstrated moderate variation when blending was increased.

NOx profiles: Increased blending decreased the amount of NOx emissions that exited the furnace. Further, it was demonstrated that fuel NOx produced during the combustion process forms an integral part of the final NOx as compared to thermal NOx when blending is being considered.

REFERENCES

1. Ji, L.; Wu, X.; Gao, Y.; Zhang, Y.; Yan, Z.; Pu, L.; Liu, X. Coal Blending with Peanut Shells: Thermal Behavior, Kinetics, Carbon Reduction and Pollution Reduction Analyses. *Water, Air, & Soil Pollution* **2024**, *235*, 634, doi:10.1007/s11270-024-07425-6.
2. Kumar, P.; Nandi, B.K. Impact of Wheat Straw and Petroleum Coke Blending on Combustion Behavior of High Ash Coal. *Energy Sources, Part A: Recovery, Utilization, and Environmental Effects* **2023**, *45*, 3125–3137, doi:10.1080/15567036.2023.2193152.
3. Gao, H.; Runstedtler, A.; Majeski, A.; Boisvert, P.; Campbell, D. Optimizing a Woodchip and Coal Co-Firing Retrofit for a Power Utility Boiler Using CFD. *Biomass Bioenergy* **2016**, *88*, 35–42, doi:10.1016/j.biombioe.2016.03.006.
4. Bian, L.; Li, Z.; Lu, Y.; Chen, Z.; Li, Z.; Sun, D. Numerical Simulation on Combustion and Slagging Characteristics of a Down-Fired Boiler Under Different Blending Ratios of Anthracite/Bituminous Coal. *Combustion Science and Technology* **2025**, 1–29, doi: 10.1080/00102202.2025.2541851.
5. Wang, H.; Cao, Y.; Huang, Z.; Liu, Y.; Hu, P.; Luo, X.; Song, Z.; Zhao, W.; Liu, J.; Sun, J.; et al. Recent Advances on Machine Learning for Computational Fluid Dynamics: A Survey. *ArXiv* 2024, *1*, 1–22, doi:10.48550/arXiv.2408.12171.
6. .Marangwanda, G.T.; Madyira, D.M. Evaluating Combustion Ignition, Burnout, Stability, and Intensity of Coal–Biomass Blends Within a Drop Tube Furnace Through Modelling. *Energies* **2025**, *18*, 1322, doi:10.3390/en18061322.
7. White, F.M. *Viscous Fluid Flow*; 3rd ed.; McGraw-Hill Higher Education: Boston, 2006.
8. Bergman, T.; Lavine, A.; Incropera, F.; DeWitt, D. *Fundamentals of Heat and Mass Transfer*; Wiley, Ed.; 7th ed.; Wiley, 2011.
9. Cengel, Y.; Ghajar, A. *Heat and Mass Transfer: Fundamentals and Applications*; 6th ed.; McGraw-Hill Education: New York, 2020.
10. ANSYS. *SpaceClaim SpaceClaim User's Guide*; SpaceClaim: Pennsylvania, 2021.
11. ANSYS. *ICEM User Guide*; ANSYS: Pennsylvania, 2021, 1–1294.
12. ANSYS. *FLUENT User's Guide*; ANSYS: Pennsylvania, 2021, 1–170.
13. Marangwanda, G.T.; Madyira, D.M.; Babarinde, T.O. Coal Combustion Models: A Review. *Journal of Physics: Conference Series* **2019**, *1378*, 032070, doi:10.1088/1742-6596/1378/3/032070.
14. de Souza-Santos, M.L. *Solid Fuels Combustion and Gasification*; CRC Press: Boca Raton, 2010.
15. Kim, G.-M.; Jeong, J.-W.; Jeong, J.; Kim, D.-Y.; Kim, S.; Jeon, C. Empirical Formula to Predict the NOx Emissions from Coal Power Plant Using Lab-Scale and Real-Scale Operating Data. *Applied Sciences* **2019**, *9*, 2914, doi:10.3390/app9142914.
16. Manovic, V.; Grubor, B. Correlation for the Total Sulfur Content in Char after Devolatilization. *Energy and Fuels* **2006**, *20*, 133–137, doi:10.1021/ef050244a.
17. Turns, S. *An Introduction to Combustion: Concepts and Applications*; McGraw-Hill series in mechanical engineering; 2nd ed.; McGraw-Hill: New York, 1996.
18. Haider, A.; Levenspiel, O. Drag Coefficient and Terminal Velocity of Spherical and Nonspherical Particles. *Powder Technology* **1989**, *58*, 63–70, doi:10.1016/0032-5910(89)80008-7.
19. Roache, P.J. Perspective: A Method for Uniform Reporting of Grid Refinement Studies. *Journal of Fluid Engineering* **1994**, *116*, 405–413, doi:10.1115/1.2910291.
20. Potgieter, M.S.W.; Bester, C.R.; Bhamjee, M. Experimental and CFD Investigation of a Hybrid Solar Air Heater. *Solar Energy* **2020**, *195*, 413–428, doi:10.1016/j.solener.2019.11.058.

21. Li, X.; Knudsen Kær, S.; Condra, T.; Yin, C. A Detailed Computational Fluid Dynamics Model on Biomass Pellet Smoldering Combustion and Its Parametric Study. *Chemical Engineering Science* **2021**, *231*, doi:10.1016/j.ces.2020.116247.
22. Yin, C.; Kaer, S.K.; Rosendahl, L.A.; Hvid, S.L. Modelling of Pulverized Coal and Biomass Co-Firing in a 150 KW Swirling-Stabilized Burner and Experimental Validation. *The Proceedings of the International Conference on Power Engineering (ICOPE)* **2009**, *2009.2*, 305–310, doi:10.1299/jsmeicope.2009.2._2-305_.
23. Versteeg, H.K.; Malalasekera, W. *An Introduction to Computational Fluid Dynamics: The Finite Volume Method Approach*; Longman Scientific & Technical: Essex, 1995.
24. Xiong, J.; Zhao, H.; Zheng, C. Techno-Economic Evaluation of Oxy-Combustion Coal-Fired Power Plants. *Chinese Science Bulletin* **2011**, *56*, 3333, doi:10.1007/s11434-011-4707-5.
25. Katarzyna, T.; Stęchły, S. CFD Modelling of Pulverised Coal and Biomass Combustion, Sheffield, 2019.
26. Mandø, M.; Rosendahl, L.A.; Yin, C.; Sørensen, H. Pulverized Straw Combustion in a Low-NOx Multifuel Burner: Modeling the Transition from Coal to Straw. *Fuel* **2010**, *89*, 3051–3062, doi:10.1016/j.fuel.2010.05.016.
27. Marangwanda, G.T.; Madyira, D.M.; Ndungu, P.G.; Chihobo, C.H. Combustion Characterisation of Bituminous Coal and Pinus Sawdust Blends by Use of Thermo-Gravimetric Analysis. *Energies (Basel)* **2021**, *14*, 1–20, doi:10.3390/en14227547.
28. Yin, C.; Rosendahl, L.A.; Condra, T.J.; Kær, S.K. Use of Numerical Modeling in Design for Co-Firing Biomass in Wall-Fired Burners. *Chemical Engineering Science* **2004**, *59*, 3281–3292, doi:10.1016/j.ces.2004.04.036.
29. Mohanna, H. Combustion of Pulverized Biomass: Impact of Fuel Preparation and Flow Conditions, Normandie Université, 2020.
30. Newell, R.G.; Raimi, D.; Aldana, G. *Global Energy Outlook 2019: The Next Generation of Energy*; Washington, 2019.
31. Elfasakhany, A.; Klason, T.; Bai, X.S. Modelling of Pulverised Wood Combustion Using a Functional Group Model. *Combustion Theory and Modelling* **2008**, *12*, 883–904, doi:10.1080/13647830802094344.
32. Tang, L.; Xiao, J.; Mao, Q.; Zhang, Z.; Yao, Z.; Zhu, X.; Ye, S.; Zhong, Q. Thermogravimetric Analysis of the Combustion Characteristics and Combustion Kinetics of Coals Subjected to Different Chemical Demineralization Processes. *ACS Omega* **2022**, *7*, 13998–14008, doi:10.1021/acsomega.2c00522.
33. Di Blasi, C. Modeling Chemical and Physical Processes of Wood and Biomass Pyrolysis. *Progress in Energy and Combustion Science* **2008**, *34*, 47–90, doi:10.1016/j.pecs.2006.12.001.
34. Sadhukhan, A.K.; Gupta, P.; Saha, R.K. Modeling and Experimental Investigations on the Pyrolysis of Large Coal Particles. *Energy and Fuels* **2011**, *25*, 5573–5583, doi:10.1021/ef201162c.

8 Conclusion and Future Trends

8.1 CONCLUSION

This book looked into deepening the technical understanding of solid fuel combustion by the use of computational fluid dynamics (CFD). CFD has become an indispensable tool with regard to optimisation of thermal systems. Due to the high temperatures and low residence times exhibited by combustion, modelling the process allows for intricate analysis without the need for costly and dangerous experimental setups. Combustion will remain the backbone of the heating industry for the foreseeable future; hence, there is a need for continued clean coal technologies. Use of solid combustion waste as a raw material for the heating and processing industry is a viable pathway towards low carbon transition, diversification of energy portfolio, and circular economy adoption. The actual application of this technology needs review, which is the reason for presenting this book. This book presented a practical way towards the process of modelling biomass and coal co-combustion. The detailed approach allows the reader to understand firstly the theory that governs combustion modelling, followed by the minimum experimental parameters needed for setting up and validating such a co-combustion model.

8.2 EXPERIMENTAL RESULTS

The theoretical and practical foundations needed to understand coal and biomass co-combustion are established first. Experimental characterisation needs to be integrated to set up a successful combustion model. Proximate and ultimate analyses showed that blending bituminous coal with sawdust (biomass) decreases fixed carbon from 53.97% to 15.62% as biomass blending increases to 30%. The volatile matter content increases from 23.10% to 33.74%, whilst the ash content decreases from 22.93% to 20.94%. The oxygen content increases by 29.11%, carbon content decreases by 3.04%, and hydrogen content increases by 38.93%. The thermal characterisation shows a decrease in heating value from 22.51 MJ/kg to 18.07 MJ/kg. Thermo-gravimetric analysis showed that at a heating rate of 12.5°C/min, the combustion intensity index increases by 301.50%, combustion stability decreases by 60.85%, ignition index increases by 206.87%, and burnout index increases as well. This highlights how blending enhances ignition at the expense of combustion stability with an optimum blending ratio of 20% Pinus sawdust blending.

Further analysis of TGA results using the Distributed Activation Energy Model gave insights into important Arrhenius equation parameters such as activation energy and pre-exponential factor. These parameters were determined for specific

DOI: 10.1201/9781003647942-8

stages of combustion, particularly devolatilisation, homogeneous volatile combustion reaction, and heterogeneous char combustion reactions. As blending increased from 100HC to 70HC30PS, the homogeneous volatile combustion activation energy changed from 92.98 to 106.05 kJ/mol, whilst the pre-exponential factor changed from 5.84×10^5 to 2.72×10^9 s^{-1}. Similarly, the heterogeneous char combustion activation energy changed from 52.90 to 103.85 kJ/mol, whilst the pre-exponential factor changed from 1.16×10^3 to 6.12×10^7 s^{-1}. An important phenomenon was made apparent, which is related to the volatile cloud that forms around a particle. Inasmuch as biomass exhibits higher porosity and volatile content, these same volatiles create a cloud around the char particle, hindering its oxidation. This contradictory effect was demonstrated in detail within this study.

The combustion experiments within the drop tube furnace were analysed under a downfired furnace scenario. The experiments were carried out at furnace temperatures of 1,273 K, 1,473 K, and 1,673 K, which allowed for intricate monitoring of combustion parameters. Primary measurements that were obtained from the experimental measurements were the velocity of the carrier gas and the mass flow of particles, which were used to deduce the particle residence time and temperature as they travelled through the furnace. At a furnace temperature of 1,473 K, as the blending ratio increased from 100HC to 80HC20PS, the particle residence time at furnace locations of 0.52 m, 0.92 m, and 1.32 m decreased by 29.40%, 24.86%, and 8.27%, respectively. This was attributed to the increase in drag, which meant that highly blended samples could be carried with less effort along the furnace. Fuel blends with higher percentages of Pinus sawdust exhibited shorter residence times which meant the particles moved through the furnace more quickly; hence, combustion stability is decreased. As the blending ratio increased from 100HC to 80HC20PS, the particle temperatures at furnace locations of 0.52 m, 0.92 m, and 1.32 m only showed a marginal increase of around 1.56% at basically all locations. This can be attributed to the size of the particles participating in the combustion process, which are relatively small regardless of the blending ratio.

8.3 MODELLING RESULTS

A co-combustion model was successfully set up after the pertinent experimental parameters that function as essential boundary conditions were determined. The book began with a thorough analysis of the modelling techniques that were required, with a particular emphasis on the solid fuel combustion sequence as it relates to coal and biomass co-combustion. To replicate the DTF experimental setup, the modelling procedure was divided into two separate phases. The fuel samples were initially passed through an inert environment that contained solely N_2. This phase was similar to the devolatilisation that a solid fuel goes through shortly after drying. Several parameters that could not be measured physically were monitored using the co-combustion model.

As experienced in the experimental setup, blending resulted in a decrease in residence time for the down-fired furnace mainly due to drag effects. Through modelling, the particle tracking analysis was used to monitor the variation of particle diameter, velocity, and position, as it combusted within the furnace. Because coal particles are heavier than sawdust particles, their gravitational impacts were greater. Therefore, blends with larger coal percentages would have been expected to have a shorter residence duration. However, because of the shape factor, sawdust particles had greater drag effects than coal particles. Therefore, blends with higher sawdust percentages showed a predicted shorter residence time. Fortunately, because the fuel blends under investigation were pulverised, drag effects became more noticeable, hence the observed trends. Through modelling, the contours of velocity and velocity vectors were used to analyse gas velocity profiles within the furnace, important parameters related to fuel mixing within the combustion furnace. The velocity profiles found inside the furnace were mostly caused by the secondary carrier, which was provided in excess. Regardless of the blending ratio, the flow rapidly developed radially. There were no zones of recirculation close to the injection point because this study attempted to maintain a laminar flow. Regardless of the blending ratio, boundary layer effects were consistently distributed, indicating that the secondary carrier gas properties mostly regulated the flow of gases within the furnace.

Modelling allows for species molar fraction analysis within the furnace. Because more volatiles are readily released over a shorter distance, stage one modelling revealed the existence of a more intense reaction zone for highly blended samples. For strongly blended samples, devolatilisation is also observed to start closer to the injection point, indicating that combustion flames are more intense and closer to the burner. Therefore, in cases with high blends, burner adjustment is also required to prevent any potential flashback flames. Stage two modelling revealed the presence of a reaction zone that depends on blending as well. At the furnace's exit, the heavily blended samples showed lower mole fractions of CO and CO_2 due to their lower beginning char mass fraction. Plots of species mole fractions against injection point position showed strong gradients due to high furnace temperatures. The way the bulk fluid velocity grew was closely related to the radial fluctuation of species mole fractions at various furnace sites. High furnace temperatures promoted high gradients of species mole fractions along the furnace. Radial variation of species mole fractions at different furnace positions was closely linked to the manner in which the bulk fluid velocity developed. Increased blending decreased the amount of NOx emissions that exited the furnace. Further, it was demonstrated that Fuel NOx produced during the combustion process forms an integral part of the final NOx as compared to Thermal NOx when blending is being considered.

As far as particle temperature is concerned, modelling results showed that blending does not result in any significant change in particle temperature. An increase in furnace wall temperatures actually causes an increase in the particle temperatures, which is an expected variation. Combustion reactions can be modelled with more ease by the use of CFD. The first stage demonstrated higher values for devolatilisation rates as blending increased. Highly blended fuel samples lost their volatiles quickly in a short period of time, unlike low blended fuel samples, which lost their volatiles gradually over a longer period. This reflected on the increased combustion intensity that is experienced for highly blended fuels; thus, there is a need to modify the burner appropriately to avoid any flashback, as noted previously. The second stage demonstrated increased peak values of the heat of reaction as blending increased. This was correlated to the reaction between CO and O_2, which increased in intensity as blending increased. The other reaction of char with O_2 demonstrated moderate variation when blending was increased.

8.4 BEST PRACTICES TO ACHIEVE CONVERGENCE

It is quite difficult for a reacting model to converge due to the interaction between chemistry, turbulence, radiation, and multiphase transport. The interaction between the continuous phase (gaseous flow) and the discrete phase (solid particles) in solid co-combustion modelling creates numerical hurdles that, if improperly handled, can lead to divergence. It is easier to first perform a run under non-reacting (cold flow) simulation so as to achieve mass flow continuity. This is generally checked by comparing the mass flux entering with that leaving the system. To reduce oscillations of residuals, the timestep needs to be refined and used to calculate the Courant number, which should be maintained below 0.5. When the meshing is being performed, grid refinement around flame zones and near-wall regions needs to be prioritised to capture the high reaction zones. Verification needs to be performed through systematic testing of grid sensitivity, time-step independence, and convergence stability. The solver is deemed converged when all residuals dropped below 1×10^{-6} though continuity-related residuals can be accepted at 1×10^{-3}.

The simplest way to compare experimental and modelled data is through the use of root mean square error (RMSE). The higher the coefficients, the higher the model accuracy. It is also important to acknowledge that experimental measurements related to combustion are highly prone to errors due to the low residence time of the combustion process. As such, RMSE values that are acceptable for combustion models tend to be a bit relaxed when compared to other applications such as cold flow.

8.5 APPLICATIONS

Having successfully developed and applied the co-combustion model, the insights derived from this study provide measurable evidence to support technical, operating guidelines, standards, and regulatory policies that aid in the adoption of co-combustion within industry. Examples of standards that stand to benefit from this study are thermal power plant combustion emission limits (SANS 1535), greenhouse-gas inventory quantification (ISO 14064-1), and energy-management systems (ISO 50001). Integrating CFD validated results, process parameters, and emission factors enables regulators to set evidence-based standards, for example, achievable NO_x and SO_2 caps, when biomass blending is employed. Also, as the technology is expanded towards other biomass waste, more benefits can be derived from the CFD-based results. Validated CFD models such as the one developed here can serve as a training and decision support tool for engineers and researchers. In retrospect, co-firing technologies require comparatively low capital investment because existing boiler and burner infrastructure can be retrofitted with minimal modification.

8.6 FUTURE TRENDS

The validated co-combustion model presented in this study forms a foundation for integrating artificial intelligence (AI), particularly machine learning (ML), to accelerate design, control, and optimisation of co-combustion systems. CFD and machine learning, when coupled together, can reduce computational cost and improve the fidelity of results obtained from the co-combustion model. This approach, commonly known as physics-informed machine learning, allows for instantaneous estimation of key parameters without the need to rerun the full CFD model. ML algorithms can be trained on CFD-generated datasets to predict optimal co-combustion parameters.

An extension of this approach within industrial combustion control systems is an AI-assisted boiler management system, which can automatically tune burner air-fuel ratios based on feedback from continuous emission monitoring sensors. This digital twin allows for the creation of a live, continuously updating digital replica of a physical plant. CFD-based digital twins of furnaces can model transient scenarios such as fuel quality variation or partial load operation, assisting with predictive maintenance. As cloud-based computing and IoT infrastructures expand, digital twins will enable virtual commissioning of power plant upgrades, testing, and retrofits in the digital domain before onsite implementation. In terms of scale, high-performance computing (HPC) is also playing a big role with regard to turbulence modelling, particularly direct numerical simulation (DNS).

Appendix A Nomenclature

CHAPTER 1

CFD:	computational fluid dynamics
SDG:	sustainable development goals
CHP:	combined heat and power
WtE:	waste to energy
MSW:	municipal solid waste
GIS:	geographical information system
GDP:	gross domestic product
HPC:	high-performance computing
GPU:	graphic processing unit
ROM:	reduced order model
DPM:	discrete phase model
PPM:	particle phase model
DTF:	drop tube furnace
EFR:	entrained flow reactor
MSW_{AE}:	MSW Annual Energy
Pop:	population
η_{eff}:	collection efficiency
MSW_{gpc}:	MSW generation per capita
MSW_{CV}:	MSW calorific value
ϕ:	variable (e.g. temperature (K), pressure (Pa), species mass fraction)
c_i:	velocity vector (u: x-component, v: y-component, w: z-component) (m/s)
x_i:	direction (x_1: x-direction, x_2: y-direction, x_3: z-direction for Cartesian coordinates)
ρ:	density (kg/m^3)
Γ:	diffusion coefficient
S_ϕ:	difference between variable source and sink
P:	pressure (Pa)
μ:	dynamic viscosity (Pa.s)
c_v:	specific heat capacity at constant volume (J/kg)
T:	temperature (K)
Ψ:	dissipation function
T_p:	particle temperature (K)

DOI: 10.1201/9781003647942-9

T_{vap}:	local vapourisation temperature of the volatiles (K)
$T\infty$:	bulk fluid temperature (K)
T_w:	furnace wall temperature (K)
m_p:	particle mass (kg)
$c_{p,p}$:	specific heat capacity of particle (J/kg.K)
A_p:	surface area of particle (m^2)
ε_p:	emissivity of particle
h_{fg}:	specific latent heat of vapourisation of the particle (J/kg)
h_{conv}:	convective heat transfer coefficient ($W/m^2.K$)
σ:	Stefan Boltzmann constant

CHAPTER 2

CFD:	computational fluid dynamics
HHV:	higher heating value
HCV:	higher calorific value
LHV:	lower heating value
LCV:	lower calorific value
SAXS:	small angle X-ray scattering
NMR:	nuclear magnetic resonance
SEM:	scanning electron microscopy
MIP:	mercury intrusion porosimetry
100HC:	100% coal by mass
90HC10PS:	90% coal and 10% biomass by mass
80HC20PS:	80% coal and 20% biomass by mass
70HC30PS:	70% coal and 30% biomass by mass
100PS:	100% biomass by mass
MC:	moisture content
VM:	volatile matter
Ash:	ash content
FC:	fixed carbon
d_p:	particle diameter
Y_d:	mass fraction of particles with a diameter greater than d_p
$d_{p,mean}$:	mean particle diameter
s:	spread parameter
$(A/F)_S$:	stoichiometric air to fuel ratio
$(A/F)_A$:	actual air to fuel ratio
ϕ:	equivalence ratio
h:	specific enthalpy (J/kg)
z:	altitude (m)
g:	gravitational acceleration (m/s^2)
c:	velocity (m/s)
q:	specific heat transfer (J/kg)
w:	specific work done (J/kg)
h_f:	specific enthalpy formation (J/kg)

c_p:	specific heat capacity at constant pressure for the species at a temperature T
T_r:	reference temperature (K)
h_i:	absolute specific enthalpy of species i
n_i:	number of moles of species i
$H_{f,vol}$:	enthalpy of formation for volatile
A:	pre-exponential factor
E:	activation energy (J)
$\Re$ **:**	universal gas constant (J/mol·K)
m_{evap}:	amount of moisture in sample
m_{dry}:	final sample mass after drying
h_{fg}:	latent heat of vaporisation
k_c:	thermal conductivity
h_{conv}:	heat transfer coefficient
ϵ**:**	emissivity
σ**:**	absorptivity
T_f:	gas temperature
T_p:	particle temperature
Q_{vap}:	heat of vaporisation
Q_{vol}:	heat of volatilisation
Q_c:	heat of combustion
A_p:	surface area of the fuel particle
T_{sur}:	temperature at the particle surface
T_c:	temperature at the centre of the particle
r_p:	radius of the particle
ar:	as received (subscripts)
daf:	dry ash free (subscripts)
init:	Initial (subscripts)

CHAPTER 3

CFD:	computational fluid dynamics
HPC:	high performance computer
GPU:	graphic processing unit
ML:	machine learning
ρ**:**	density
φ**:**	variable (mass, specific enthalpy, temperature, etc.)
x_i:	direction (x_1: x direction, x_2: y direction, x_3: z direction for Cartesian coordinates)
c_i:	velocity (u: x component, v: y component, w: z component)
Γ**:**	diffusion coefficient of variable
S_φ:	source or sink parameter of variable
δ**:**	normal stress (pressure)
τ**:**	shear stress (viscous)

$\frac{D\varphi}{Dt}$: material derivative of a variable
P: pressure
μ: dynamic viscosity
c_v: specific heat capacity at constant volume (J/kg)
T: temperature (K)
k_c: thermal conductivity
Ψ: dissipation function
DPM: discrete phase model
PPM: particle phase model
J_i: diffusion flux of a species
X_i: local mass fraction of a species
R_i: rate of production of a species
S_i: source term of a species
F: convective mass flux per unit area (F=ρu)
D: diffusion conductance at cell faces $D = \frac{\Gamma}{\delta x}$
$\overline{c_i}$: average velocity $(\bar{u}, \bar{v}, \bar{w})$
x_i: coordinates (e.g. x, y, z)
$\overline{c_i c_j}$: Reynolds stresses($\overline{u'^2}$, $\overline{u'v'}$, $\overline{u'w'}$, $\overline{v'^2}$, and $\overline{w'^2}$)
h_{tot}: total enthalpy
k: turbulent kinetic energy
p': modified pressure
μ_{eff}: effective viscosity ($\mu_{eff} = \mu + \mu t$)
S_M: momentum source
P_k: shear production of turbulence
k: turbulent kinetic energy
ω: turbulent frequency
k-ε: k-epsilon
k-ω: k-omega
RNG: renormalisation group analysis
LES: large eddy simulations
DNS: direct numerical simulation
τ_{ij}: sub-grid stresses in LES

CHAPTER 4

m_p: particle mass (kg)
$c_{p,p}$: particle specific heat capacity at constant pressure (J/kg.K)
T_p: particle temperature (K)
h_{conv}: convective heat transfer coefficient (W/m².K)
A_p: particle surface area (m²)
T_∞: continuous phase local temperature (K)
ε_p: particle surface emissivity

σ: Stefan Boltzmann constant
T_w: wall temperature (K)
Q_v: heat transfer due to vapourisation (W)
N_w: flux of water vapor
P: pressure of the bulk gas
P_{sat}: saturated water vapour pressure
X_w: molar fraction of the water vapour in the bulk gas
R: gas constant
k_w: mass transfer coefficient
D_w: diffusion coefficient of water vapour in bulk gas
Re_p: particle Reynolds number
Sc: Schmidt number
C_{sw}: swelling coefficient
$f_{w,0}$: initial mass fraction of boiling/evaporating material (if wet combustion is modelled)
$m_{p,0}$: initial particle mass
$f_{v,0}$: initial mass fraction of volatiles in the bulk gas
d_p: particle diameter
$d_{p,0}$: initial particle diameter
A_0: rate constant
$f_{v,0}$: initial mass fraction of volatiles in the bulk gas
k: chemical reaction kinetic rate
A: pre-exponential factor
E_a: activation energy
$m_v(t)$: volatile yield at a certain time
α_1 and α_2: mass stoichiometric factor of volatiles evolved at low temperatures and high temperatures
$m_{p,daf}$: mass of dry ash free particle
DAEM: distributed activation energy model
UDPM: universal devolatilisation process model
CPD: chemical percolation devolatilisation
FGD:: functional group devolatilisation model
FG-DVC: functional group depolymerisation, vapourisation and crosslinking devolatilisation model
D_0: diffusion rate coefficient of oxygen in the carrier gas
X_{ox}: mass fraction of oxygen in the carrier gas
$T\infty$: temperature of carrier gas
S_b: stoichiometric coefficient of oxygen during char oxidation ($S_b = 1$ for complete combustion and $S_b = 0.5$ for incomplete combustion)
P_{ox}: partial pressure of oxygen in the carrier gas
D_0: diffusion rate coefficient of oxygen in the carrier gas
Φ_{eh}: enhancement factor dependent on particle shape
Y_{ox}: mole fraction of oxygen in the carrier gas
P_{atm}: atmospheric pressure

η:	ratio of actual combustion rate to rate attainable if no pore diffusion resistance existed
ρ_p:	apparent density of char
A_g:	specific internal surface area of the char particle
k_i:	intrinsic reactivity
k_t:	turbulent kinetic energy
ε_t:	turbulent dissipation energy

CHAPTER 5

RANS:	Reynolds-Averaged Navier-Stokes
φ:	variable (e.g. temperature (K), pressure (Pa), species mass fraction)
$\bar{\varphi}$:	average quantity (e.g. $\bar{u}; \bar{v}; \bar{w}$ are average values for velocity $\overline{c_i}$ in the x, y, and z velocity)
φ”:	fluctuating quantity (e.g. $u''; v''; w''$ are fluctuating values for x, y, and z velocity)
x_i:	direction (x_1: x direction, x_2: y direction, x_3: z direction for Cartesian coordinates)
ρ:	density (kg/m^3)
ν:	kinematic viscosity (m^2.s)
k_t:	turbulence kinetic energy
ε_t:	rate of turbulence dissipation
G_k:	generation of turbulence kinetic energy due to the mean velocity gradients
G_b:	generation of turbulence kinetic energy due to buoyancy
Y_m:	contribution of the ratio between fluctuating dilatation in compressible turbulence to the overall dissipation rate
α_k:	inverse effective Prandtl number for turbulence kinetic energy
α_ε:	inverse effective Prandtl number for rate of turbulence dissipation
μ_{eff}:	effective viscosity
ε_f:	void fraction
ρ_f:	fluid density
u_f:	average velocity of the fluid unit
P:	pressure
τ:	viscous stress tensor
S_p:	source term generated by the interaction between particles and volumetric fluid
u_p:	particle velocity
ρ_p:	particle density
$\bar{\bar{\tau}}_s$:	solid stress tensor caused by particle to particle interaction
F_D:	drag force
another:	acceleration caused by an external force
C_D:	particle drag coefficient

Re_p: particle Reynolds number
ϕ_s: particle shape factor
$\phi_{s\parallel}$: elongated particle shape factor
$\phi_{s\perp}$: elongated particle shape factor
I: radiation intensity which depends on position ($\vec{r}$) and direction ($\vec{s}$)
s: path length
α: absorption coefficient
σ_s: scattering coefficient
n: refractive index
σ: Stefan Boltzmann constant (5.67×10^{-8} W/m^{2}K^4)
T: local temperature
$\vec{s}'$: scattering direction vector
Φ: phase function
Ω: solid angle

CHAPTER 6

TGA: thermogravimetric analysis
100HC: 100% coal by mass
90HC10PS: 90% coal and 10% biomass by mass
80HC20PS: 80% coal and 20% biomass by mass
70HC30PS: 70% coal and 30% biomass by mass
100PS: 100% biomass by mass
T_{init}: temperature at the beginning of TG-DTG analysis
T_{max}: the temperature at which maximum mass loss rate occurs
t_{max}: time at which maximum mass loss rate occurs
DTG_{max}: maximum mass loss rate
Φ_{ig}: ignition index
t_{ig}: time at ignition
Ψ_1: stability index or first combustion index
Ψ_2: intensity index or second combustion index
T_{ig}: temperature at ignition
T_b: temperature at the final burnout
$\Delta T_{1/2}$: temperature range when $|DTG| > |DTG_{max}|/2$
DTG_{mean}: mean combustion rate
ω: synergistic effect coefficient
$U_{experimental}$: experimental combustion parameter of the fuel blend
k_i: mass fraction of individual fuel within the blend
U_i: combustion parameter of individual fuel within the blend
CCS: carbon capture and storage
CLC: chemical looping combustion
LFPI: low fuel plasma ignition
ML: machine learning
KAS:: Kissinger Akahira Sunose
OFW: Ozawa Flynn Wall
ANN: artificial neural network

SVM: support vector machine
CNN: convolutional neural network

CHAPTER 7

CFD: computational fluid dynamics
DTF: drop tube furnace
100HC: 100% coal by mass
90HC10PS: 90% coal and 10% biomass by mass
80HC20PS: 80% coal and 20% biomass by mass
70HC30PS: 70% coal and 30% biomass by mass
100PS: 100% biomass by mass
TGA: thermogravimetric analysis
t: particle residence time (s)
L_i: length of slice (m) (whole length of DTF was sliced into sections)
v_g: velocity of carrier gas (m/s)
v_t: terminal velocity of particle (m/s)
g: gravitational acceleration (m/s^2)
d_p: particle diameter (m)
ρ_p: particle density (kg/m^3)
ρ_g: carrier gas density (kg/m^3)
μ_g: carrier gas viscosity (Pa.s)
μ_0: reference gas dynamic viscosity (Pa.s) [N_2 gas:1.663 x 10^{-05} Pa.s]
T_g: gas temperature (K)
T_0: reference gas temperature (K) [N_2 gas: 273 K]
$C\mu_s$: Sutherland constant for viscosity [N_2 gas:107]
Q_v: corrected gas volume flow rate (m^3/s)
$Q_{v,0}$: inlet gas volume flow rate (m^3/s at STP)
$V_{reactor}$: volume of reactor (m^3)
$L_{reactor}$: length of reactor (m)
P_0: inlet carrier gas pressure at STP
T_0: inlet carrier gas temperature at STP
P_g: combustion zone gas pressure
T_g: combustion zone gas temperature
VM_{DTF}: volatile matter from DTF analysis on a dry basis
Ash_0: mass of ash in initial particle on a dry basis
Ash_1: mass of ash in sampled residue particle on a dry basis (collected char)
m_p: particle mass (kg)
$C_{p,p}$: particle specific heat capacity (J/kg·K)
T_p: particle temperature (K)
H_{rad}: radiation heat transfer (W)
H_{conv}: convection heat transfer (W)
Q_{vap}: heat of vapourisation (W)
Δh: specific enthalpy of reaction (J/kg)
ε_p: particle surface emissivity

σ: Stefan Boltzmann constant
A_p: particle surface area (m^2)
T_w: DTF wall temperature (K)
h_{conv}: convective heat transfer coefficient ($W/m^2.K$)
T_g: combustion zone gas temperature (K)
CE: combustion extent on a dry basis
$Ash_{0,char}$: mass of ash in initial char particle on a dry basis
Ash1, char: mass of ash in sampled residue particle on a dry basis (collected ash)
h_{reac}: enthalpy of reaction (J/kg) (Char + O_2 –> CO_2)
$[m_{p,initial} - m_{p,final}]$: change in mass of combustible matter in char particle (kg)
Re: carrier gas Reynolds number
Pr: carrier gas Prandtl number
μ_g: carrier gas dynamic viscosity (Pa.s)
μ_p: carrier gas dynamic viscosity at particle surface (Pa.s)
λ_g: thermal conductivity of carrier gas (W/m·K)
λ_0: reference thermal conductivity (W/m.K) [N_2 gas:0.0242 W/m.K]
$C_{\lambda s}$: Sutherland constant for thermal conductivity [N_2 gas:150]
Nu: Nusselt number
h_{conv}: convective heat transfer coefficient ($W/m^2.K$)
T_g: combustion zone gas temperature (K)
CAD: computer-aided design
DPM: discrete phase model
E_a: activation energy
A: pre-exponential factor
N_{char}: nitrogen in char mass fraction
N_{vol}: nitrogen in volatile mass fraction
N_{split}: split ratio between N_{char} and N_{vol}
N_{daf}: nitrogen content in dry ash free fuel sample
FC_{daf}: fixed carbon content in dry ash free fuel sample
VM_{daf}: volatile matter content in dry ash free fuel sample
S_{char}: sulphur in char mass fraction
S_{vol}: sulphur in volatile mass fraction
S_{split}: split ratio between S_{char} and S_{vol}
S_{daf}: sulphur content in dry ash free fuel sample
FC_{daf}: fixed carbon content in dry ash free fuel sample
VM_{daf}: volatile matter content in dry ash free fuel sample
UDF: user defined functions
GCI: grid convergence index
p: order of convergence
f: performance parameter
r: refinement ratio = 2.576
F_s: representing the factor of safety (in this case two levels of refinement = 1.25)
ppm: parts per million

Appendix B Thermogravimetric Experimental Data

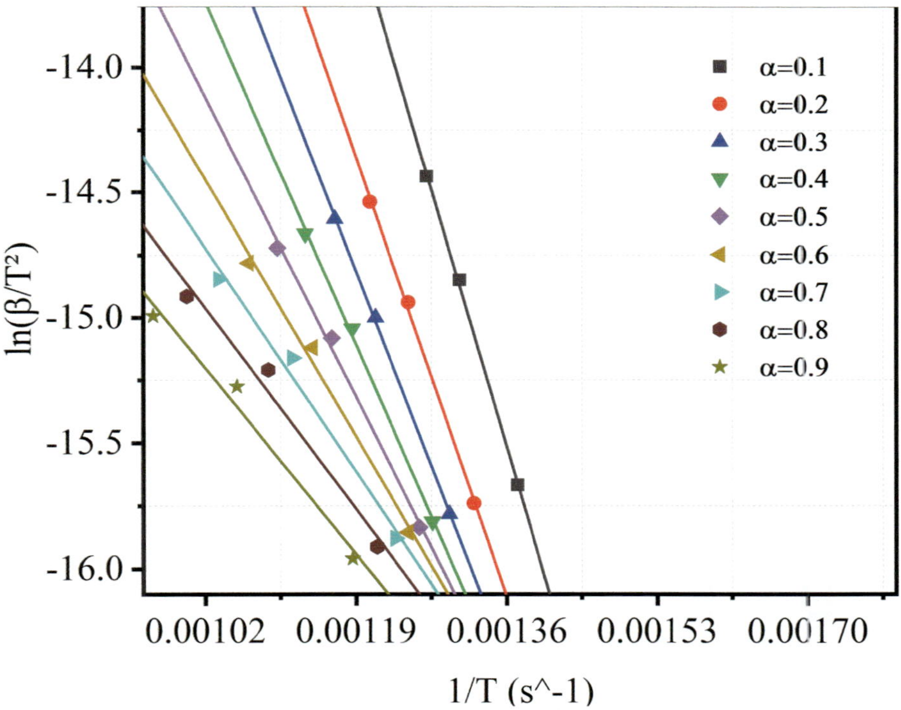

FIGURE AB.1 100HC DAEM Stage 3.

DOI: 10.1201/9781003647942-10

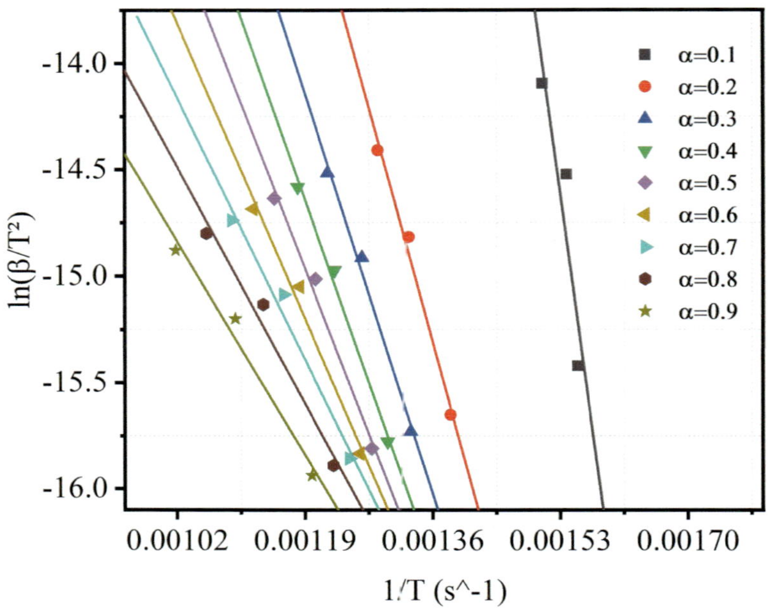

FIGURE AB.2 90HC10PS DAEM Stage 3.

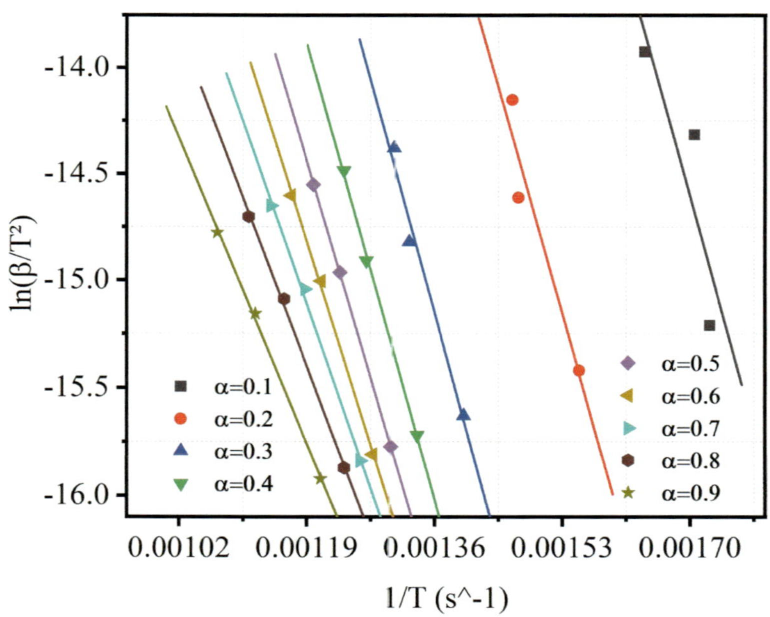

FIGURE AB.3 80HC20PS DAEM Stage 3.

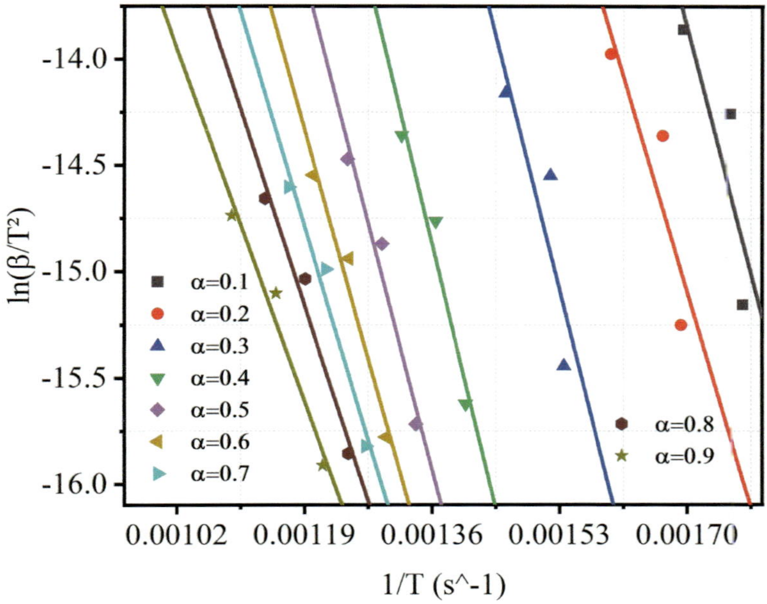

FIGURE AB.4 70HC30PS DAEM Stage 3.

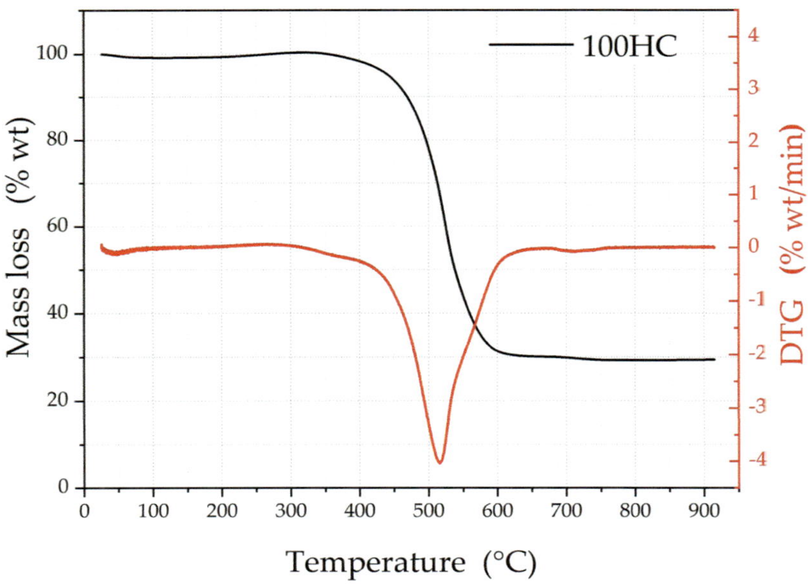

FIGURE AB.5 HC100 at 5 deg per min.

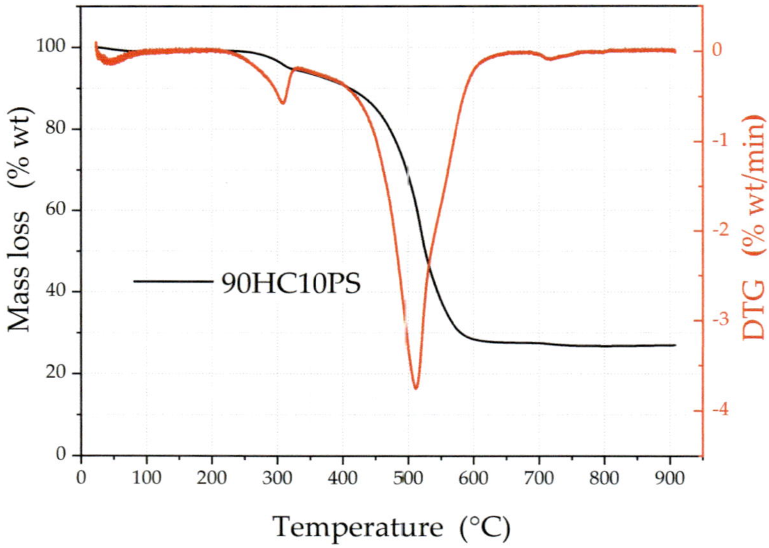

FIGURE AB.6 HC90PS10 at 5 deg per min.

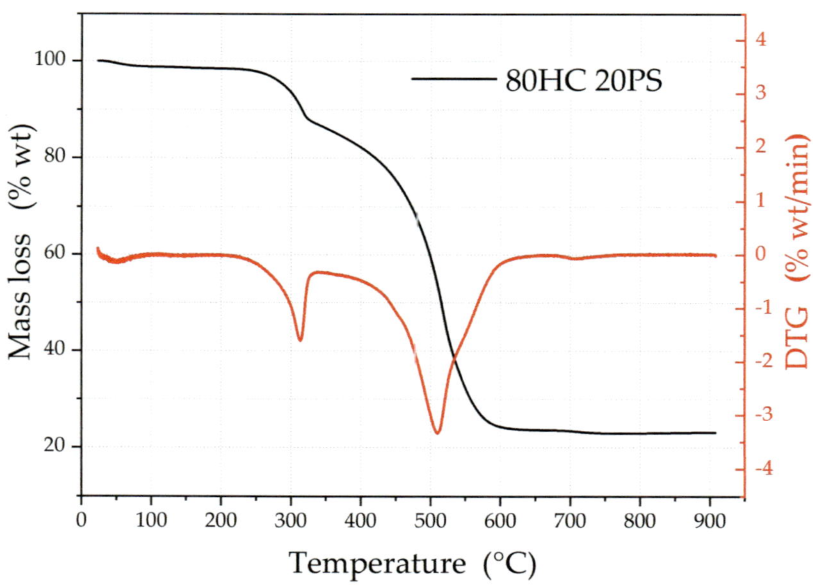

FIGURE AB.7 HC80PS20 at 5 deg per min.

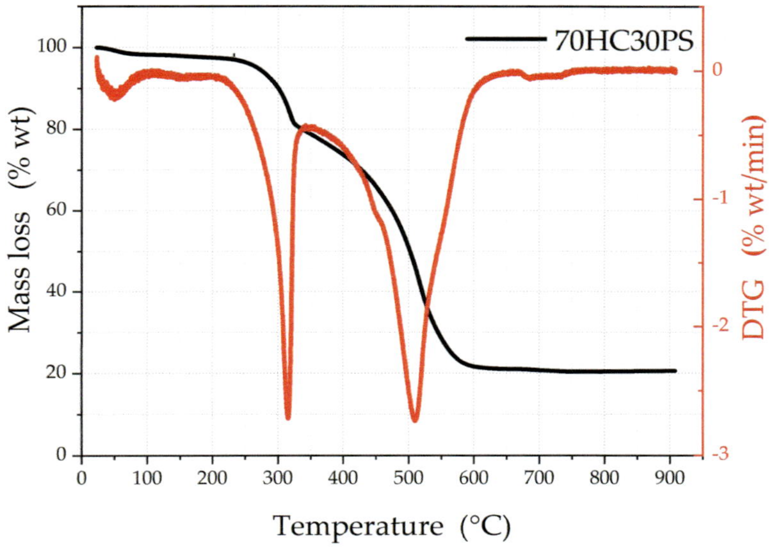

FIGURE AB.8 HC70PS30 at 5 deg per min.

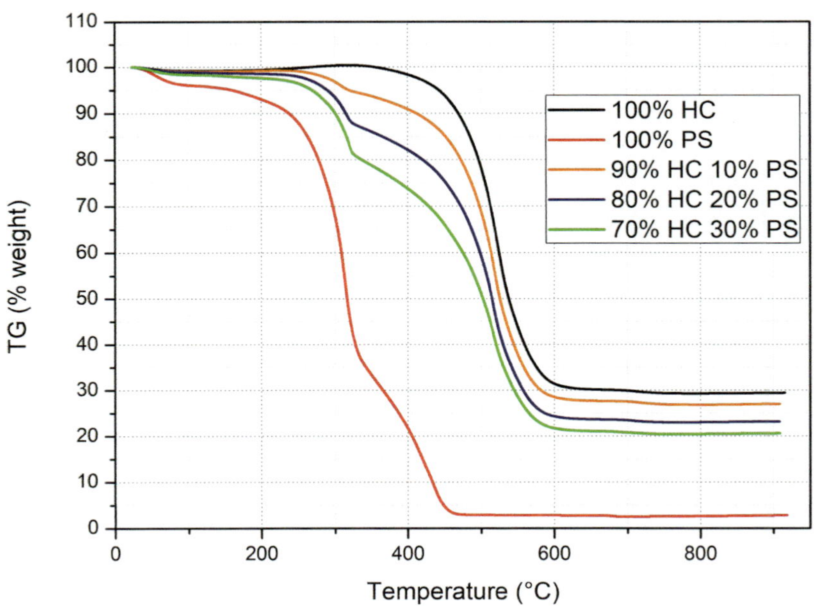

FIGURE AB.9 All samples TG 5 deg per min.

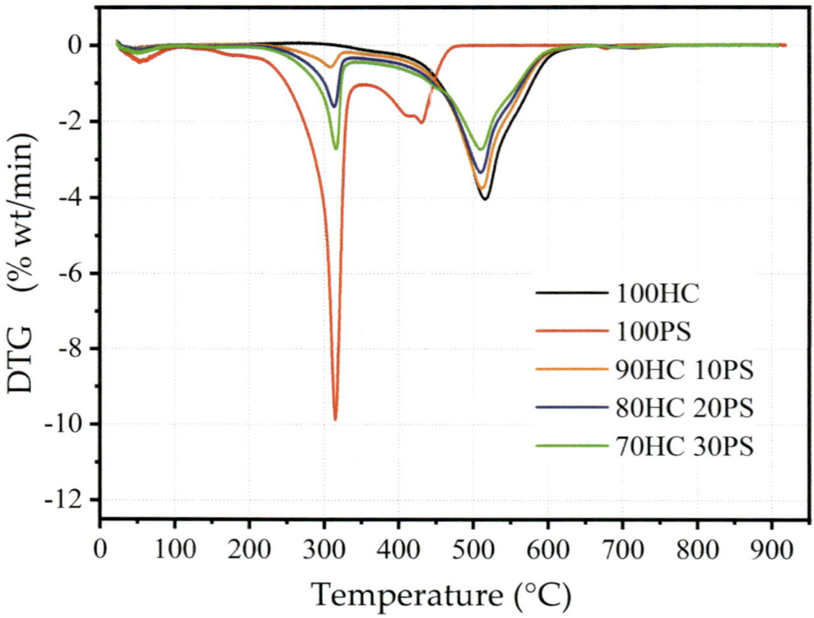

FIGURE AB.10 All samples DTG 5 deg per min.

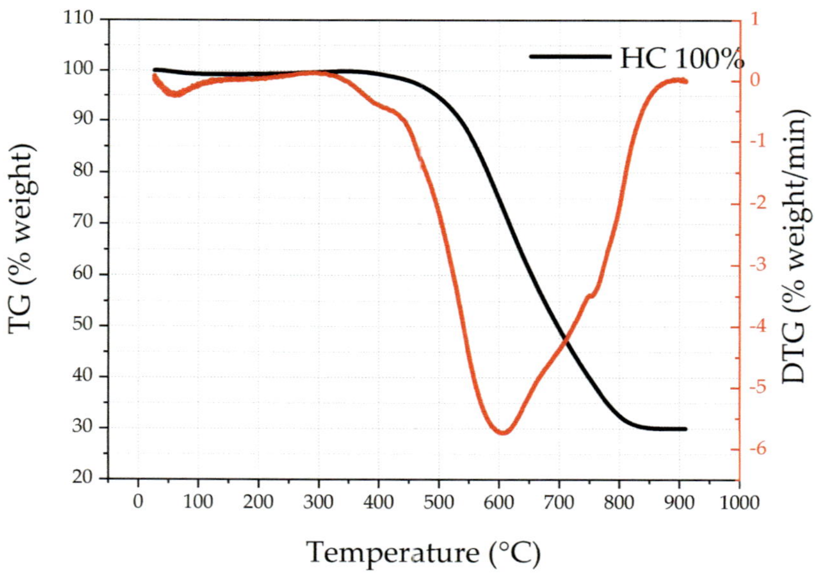

FIGURE AB.11 HC100 at 20deg per min.

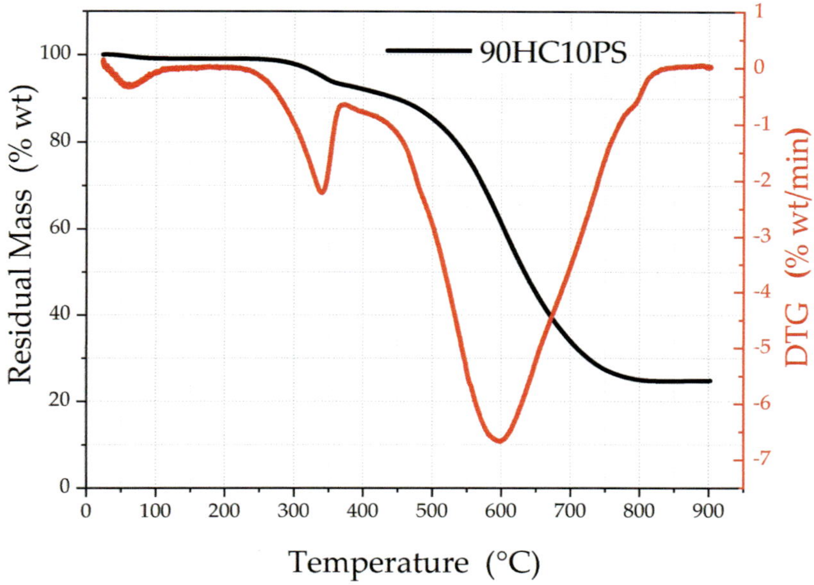

FIGURE AB.12 HC90PS10 at 20 deg per min.

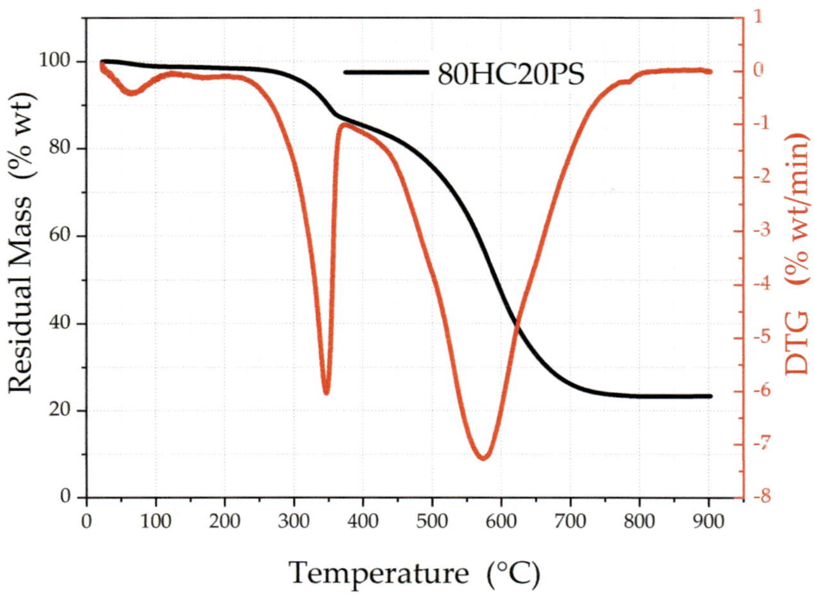

FIGURE AB.13 HC80PS20 at 20 deg per min.

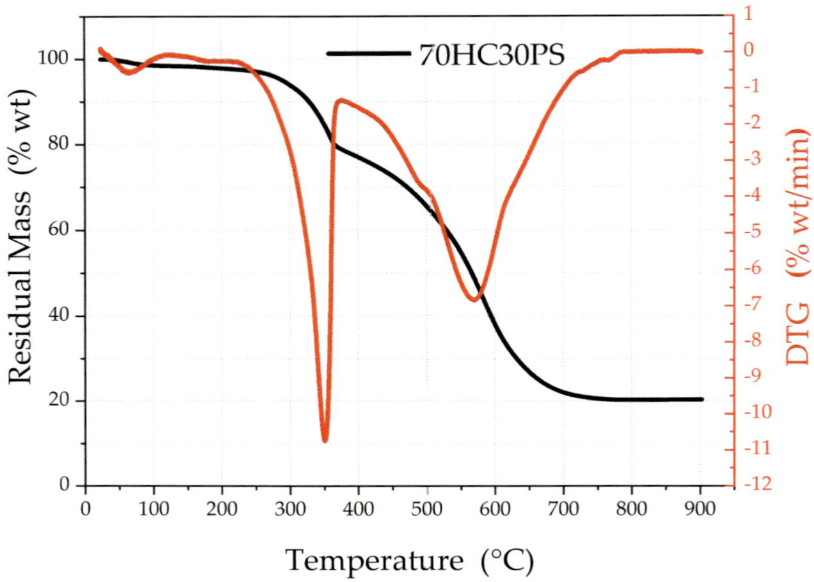

FIGURE AB.14 HC70PS30 at 20 deg per min.

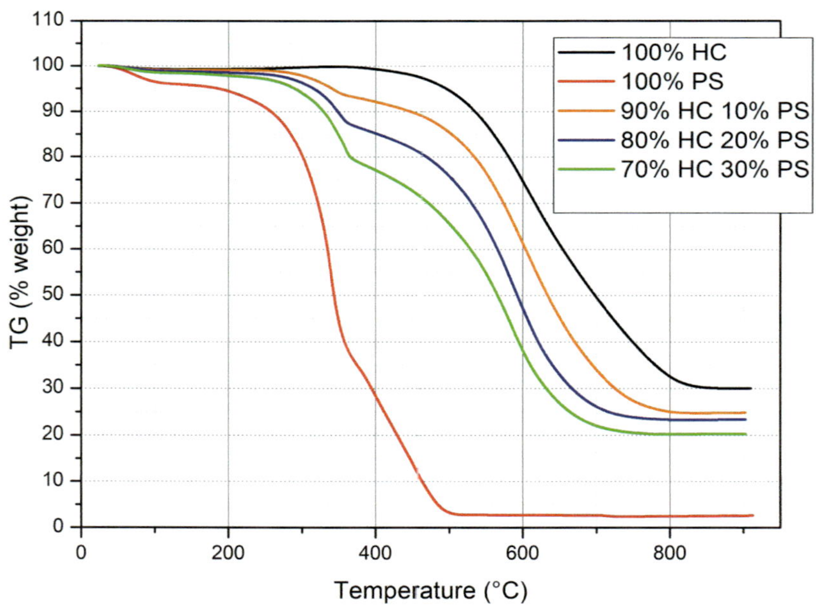

FIGURE AB.15 All samples TG 20 deg per min.

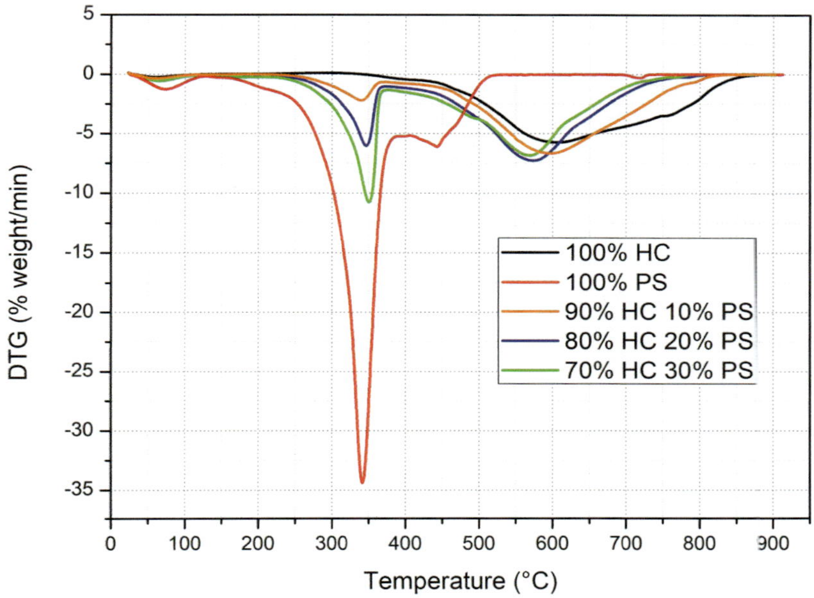

FIGURE AB.16 All samples DTG 20 deg per min.

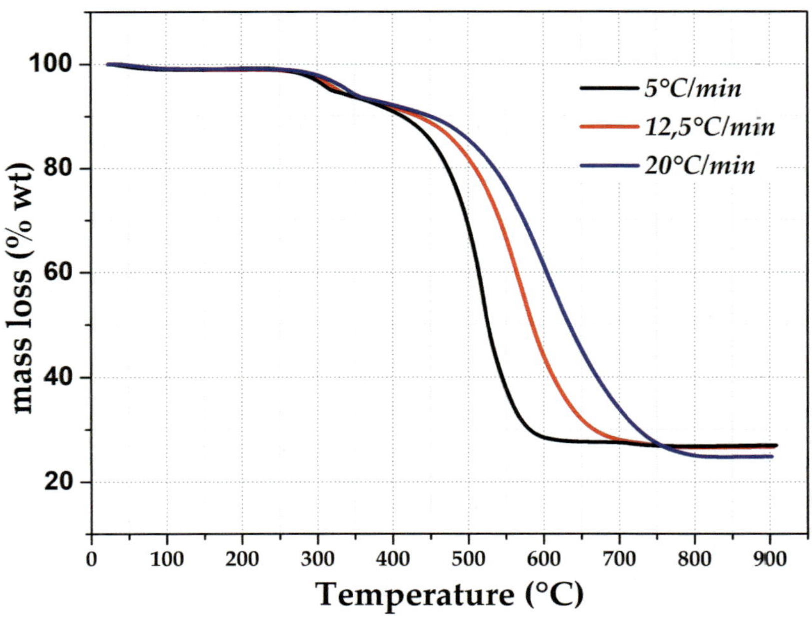

FIGURE AB.17 90HC10PS by heating rate TG.

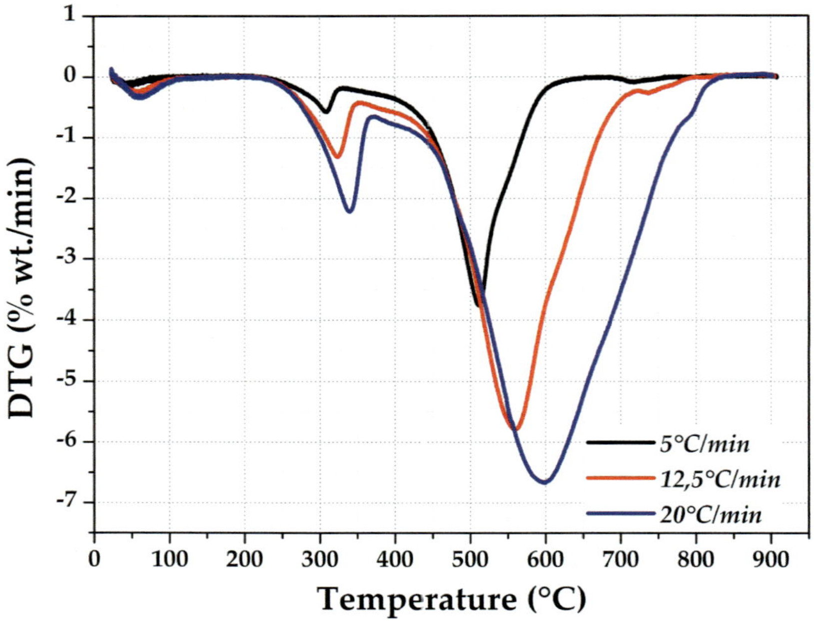

FIG AB.18 90HC10PS by heating rate DTG.

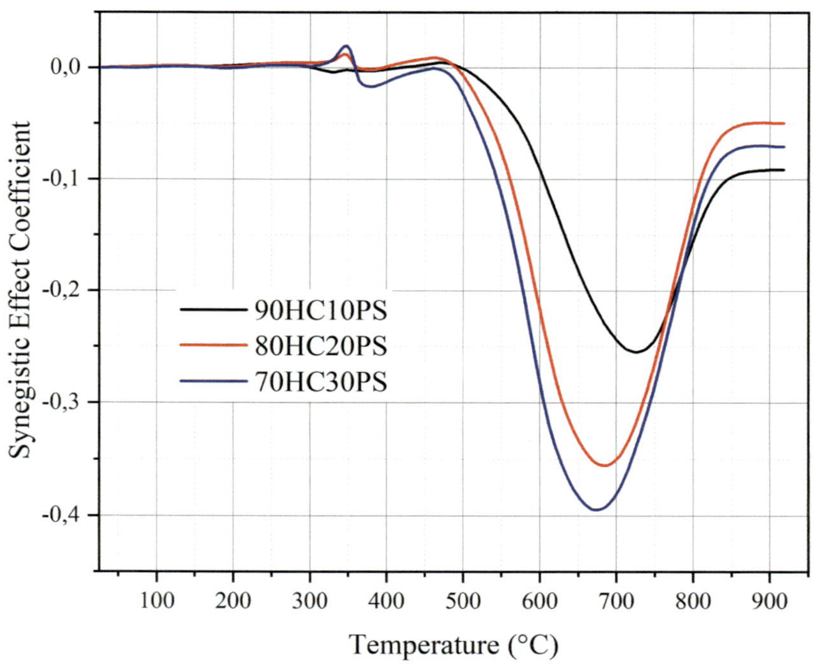

FIGURE AB.19 Synergy TG at 20 deg per min.

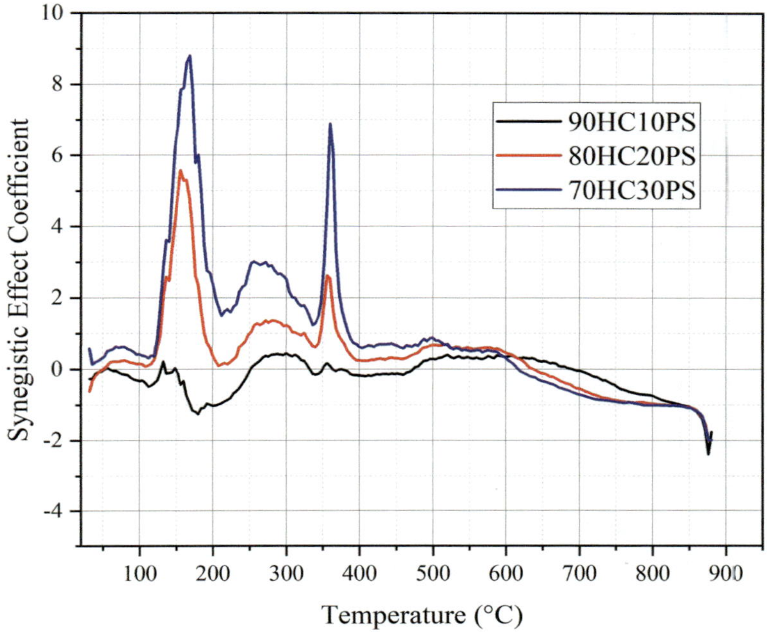

FIGURE AB.20 Synergy DTG at 20 deg per min.

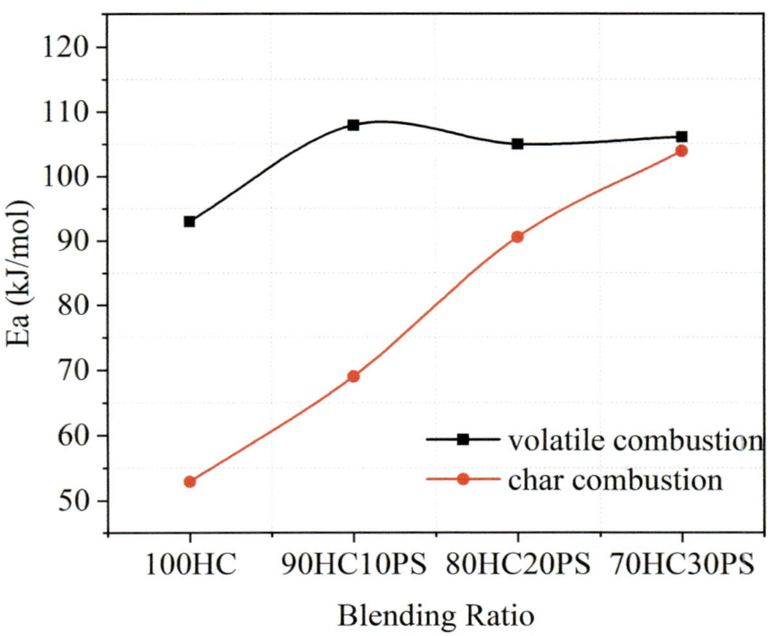

FIGURE AB.21 Activation energy vs blending ratio.

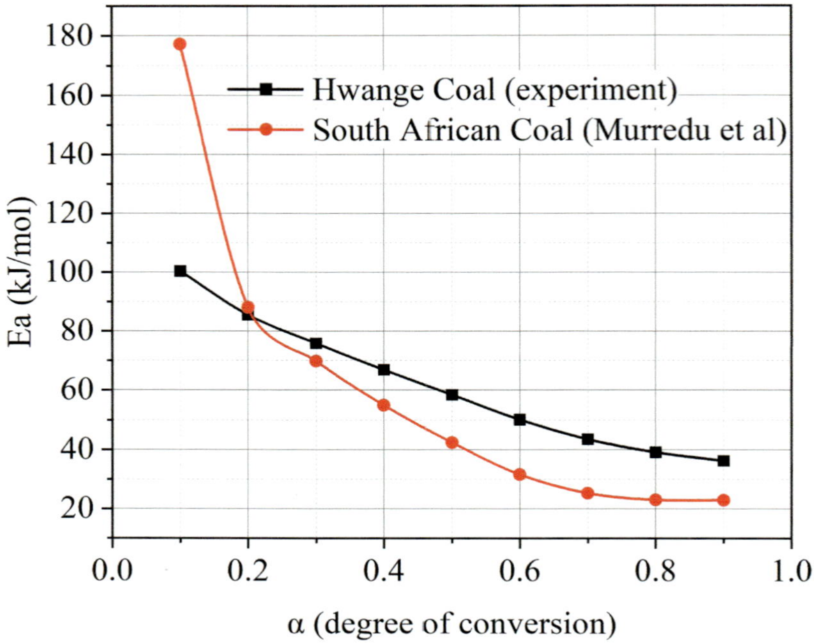

FIGURE AB.22 Activation energy vs conversion comparison with other coals.

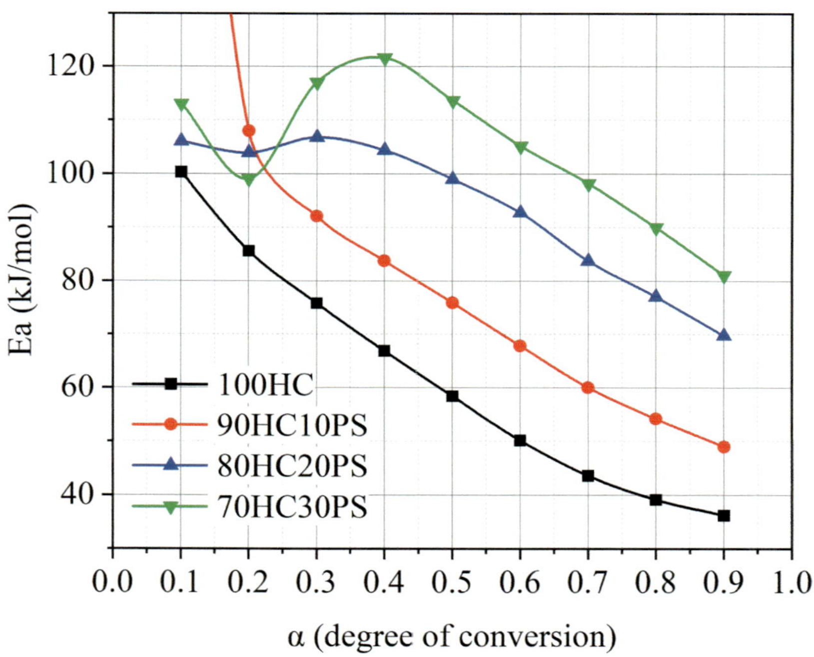

FIGURE AB.23 Activation energy vs conversion.

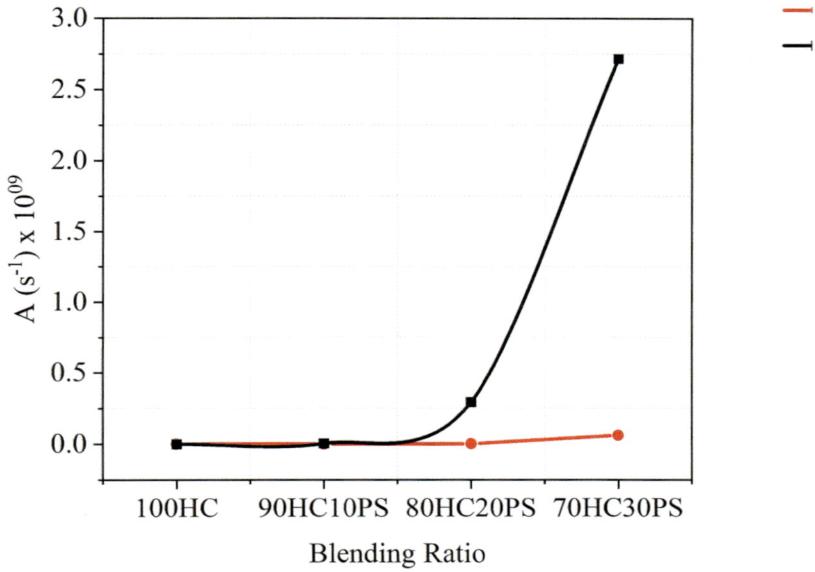

FIGURE AB.24 Pre-exponential vs blending ratio.

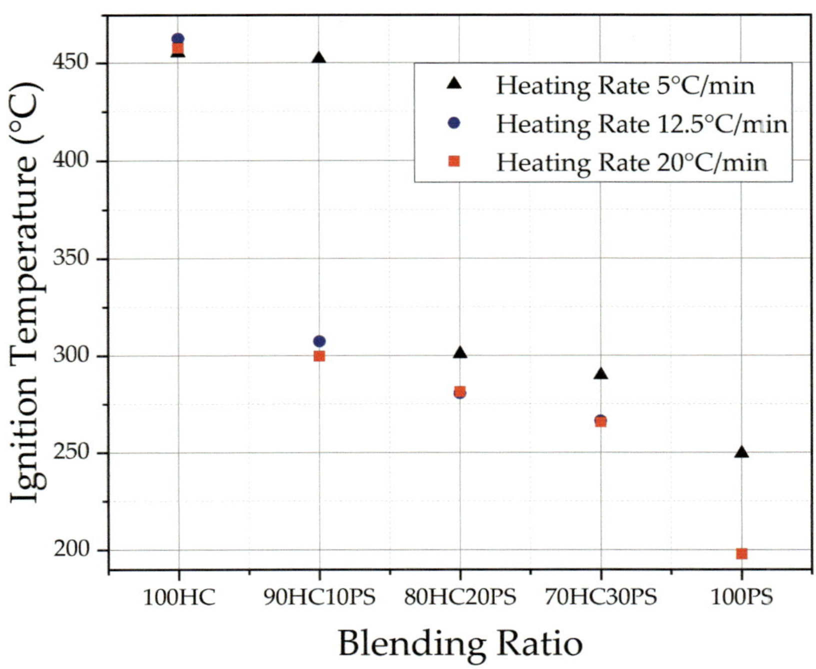

FIGURE AB.25 Ignition index with heating.

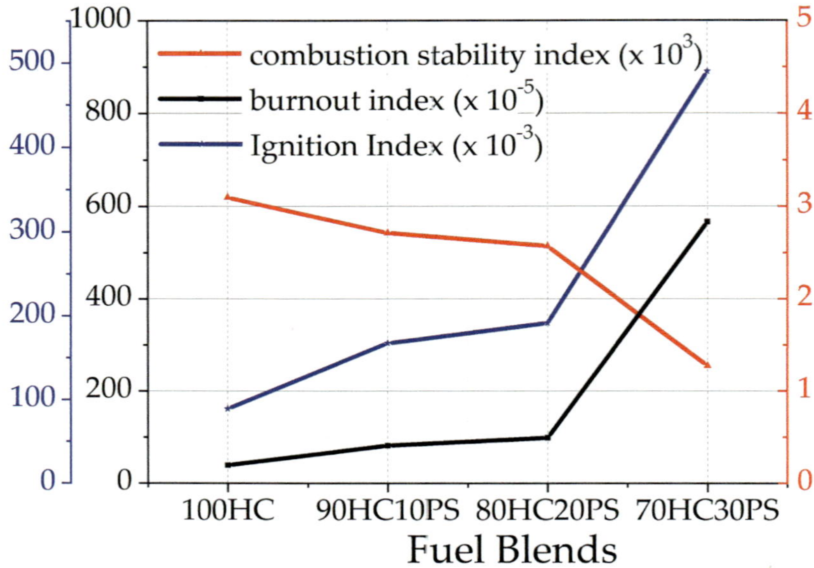

FIGURE AB.26 Index vs blending ratio.

Index

D

E

F

G

H

I

K

L

M

N

O

P

Q

R

S

T

U

V

W

Z

For Product Safety Concerns and Information please contact our EU representative GPSR@taylorandfrancis.com Taylor & Francis Verlag GmbH, Kaufingerstraße 24, 80331 München, Germany